OCCASIONAL PUBLICATIONS OF THE CAMBRIDGE UNIVERSITY
MUSEUM OF ARCHAEOLOGY AND ETHNOLOGY

IV

PREHISTORY AND
PLEISTOCENE GEOLOGY IN
CYRENAICAN LIBYA

PREHISTORY AND PLEISTOCENE GEOLOGY IN CYRENAICAN LIBYA

A RECORD OF TWO SEASONS' GEOLOGICAL AND
ARCHAEOLOGICAL FIELDWORK IN THE GEBEL AKHDAR
HILLS, WITH A SUMMARY OF PREHISTORIC FINDS
FROM NEIGHBOURING TERRITORIES

BY

C. B. M. McBURNEY
M.A., PH.D., F.S.A.

&

R. W. HEY
M.A., PH.D., F.G.S.

CAMBRIDGE
AT THE UNIVERSITY PRESS
1955

CAMBRIDGE UNIVERSITY PRESS
Cambridge, New York, Melbourne, Madrid, Cape Town, Singapore, São Paulo, Delhi

Cambridge University Press
The Edinburgh Building, Cambridge CB2 8RU, UK

Published in the United States of America by Cambridge University Press, New York

www.cambridge.org
Information on this title: www.cambridge.org/9780521109567

First published 1955
This digitally printed version 2009

A catalogue record for this publication is available from the British Library

ISBN 978-0-521-05624-3 hardback
ISBN 978-0-521-10956-7 paperback

CONTENTS

PART I: PLEISTOCENE GEOLOGY (R.W.H.)

A. SHORELINES AND MARINE DEPOSITS

B. CONTINENTAL DEPOSITS

PART II: PREHISTORY (C.B.M.McB.)

CONTENTS

LIST OF PLATES

The plates are bound together, between pages 308–9

LIST OF TEXT-FIGURES

PREFACE

During the late war, one of the joint authors of this memoir (C. B. M. McB.) had an opportunity of travelling through the Gebel Akhdar range of Northern Cyrenaica, and of forming some idea of the potentialities of this region as a field for Quaternary research. Both the observations made at this time and subsequent consideration of data from other sources suggested that conditions here might be favourable for the investigation of a number of general problems affecting the prehistory and Pleistocene chronology of North Africa as a whole. A combined archaeological and geological reconnaissance was accordingly organized, during the summer vacation of 1947, under the auspices of the Department of Archaeology and Anthropology of Cambridge University. The party consisted of C. B. M. McBurney, from the Department of Archaeology and Anthropology, W. Watson, from the Department of British Antiquities of the British Museum, and R. W. Hey.

The results of this first season were sufficiently encouraging to justify the organization of a second expedition on a somewhat larger scale during the summer of 1948, comprising C. B. M. McBurney, R. W. Hey, and three undergraduate members, C. H. Houlder, G. de G. Sieveking, and T. B. Bagenal. Further geological observations were also made by R. W. Hey in 1952.

The main financial support for the first two expeditions was provided by grants from the Percy Sladen Memorial Fund, while further grants were received from the Society of Antiquaries of London, and the Worts Fund of the University of Cambridge. Help towards the expenses of individual members was received from King's College, Cambridge, the Trustees of the British Museum, the Lake Fund of Cambridge University and various other sources. During much of the time in which this work was in preparation, one of us (R. W. H.) was in receipt of a maintenance grant from the Department of Scientific and Industrial Research. Grants towards the cost of publication were received from the Museum of Archaeology and Ethnology, King's College, and Trinity College. We should like to express our gratitude for all these forms of financial assistance.

Surveying instruments were kindly lent by the Royal Geographical Society, and help and services of the most varied and essential kinds were constantly provided by the authorities of the British Military Administration in Cyrenaica.

We are most grateful to Miss K. M. Kenyon, Dr C. R. Metcalfe, Professor F. Zeuner and the late Miss D. M. A. Bate, not only for having contributed reports on various specialized subjects, but also for having given much useful information and advice. Among many others who have helped and advised us, we would especially like to thank the following: Dr P. Allen, Dr W. J. Arkell, Dr M. Black, Mr M. C. Burkitt, Dr C. Chiesa, Miss E. W. Gardner, Miss D. A. E. Garrod, Dr E. Gobert, Dr H. Godwin, Monsieur J. Haller, Dr H. Hamshaw Thomas, Professor W. B. R. King, Dr K. S. Sandford and Dr H. Whitehouse.

To Major P. Sandison, Mr K. S. R. Robinson and Dr C. Willett-Cunnington we are greatly indebted for permission to publish notes and material on the prehistory of neighbouring territories which have helped to put our own finds in their correct context; also, to Signor C. Petrocchi, for kindly discussing his own work in the cave of Hagfet et Tera and allowing us to examine his material in Tripoli.

Finally, we cannot close this list of acknowledgments without recalling with appreciation the assistance and faithful service of many Cyrenaican friends and acquaintances, who contributed greatly to our comfort and security while in the field.

C. B. M. McB.
R. W. H.

CAMBRIDGE
1954

DISPOSAL OF SPECIMENS

The archaeological specimens collected during the two expeditions were distributed as follows. All surface material and all closed finds from beach conglomerates at Ras Aamer were given to the British Museum. A representative series from all the more important closed finds was sent to the Antiquities Officer of the United Kingdom of Libya, and is housed in the Museums of Tripoli and Cyrene. All the remainder was donated to the Museum of Archaeology and Ethnology, Cambridge, where are also the collections of Mr K. S. R. Robinson, Major P. Sandison, and Dr C. Willett-Cunnington. The small series of objects from the Wadi Bei el Kebir site are in the Ashmolean Museum, Oxford.

Of the geological specimens, the invertebrate fossils and most of the Wadi Derna plant-remains have been deposited in the Sedgwick Museum, Cambridge; the registration numbers of these are given in the text. A representative selection of the plant-remains has been presented to the Museo di Storia Naturale, Tripoli. Of the vertebrate remains those of greatest palaeontological interest have been presented to the British Museum (Natural History) and a typical series has also been sent to the Museo di Storia Naturale, Tripoli; the bulk of the bone fragments, which are largely of archaeological interest, have been given to the Museum of Archaeology and Ethnology, Cambridge.

GENERAL INTRODUCTION

I. AIM AND SCOPE OF THE INVESTIGATION

The observations presented in this report were made for the most part during two seasons' field-work on the northern Cyrenaican coast, in 1947 and 1948. Little work had previously been attempted either on the Palaeolithic archaeology or the Pleistocene geology of this region; and although the archaeological material obtained can be supplemented to some extent from published and unpublished sources in territories to the east and west, in Marmarica and Tripolitania; the results are in the main of a pioneering character.

The primary object of our work was to examine the Cyrenaican coast for signs of ancient shorelines, which might serve to co-ordinate any traces of archaeological and natural events found within the area, and conceivably might also provide a basis for correlations with other parts of the Mediterranean. In addition, it was hoped to contribute to two problems of wider significance: the correlation of the North African cultural sequence as a whole with that of Western Europe, and the question of alleged worldwide fluctuations of sea-level during the Pleistocene Age.

This last problem is of fundamental importance to many different aspects of Quaternary research, since any traces left by such fluctuations should provide a means of direct correlation between widely separated coastal areas. It has been suggested, moreover, that the supposed changes of sea-level may be directly connected with the advances and recessions of the Pleistocene ice-sheets. If such a connexion could be proved, it should then be possible to extend the correlations to glaciated areas lying far inland. Unfortunately, however, the reliability of much of the earlier work in this field is now widely regarded as by no means above question. It therefore seemed to us that a careful and objective examination of an entirely new stretch of coastline might provide a valuable check of earlier conclusions. Both the geology and the topography of the Cyrenaican coast encouraged us to believe that a clear record of ancient shorelines might be found there, if the area were to be explored with this specific end in view.

From the archaeological aspect, also, the region appeared to have definite advantages. The widespread occurrence of limestone suggested that even the thinnest of superficial deposits might have been so consolidated, by the action of lime-bearing waters, that their contents would have been protected from later accidental intrusions. Again, the limestone could be expected to contain caves, in which archaeological and organic remains might be preserved. Finally, the geographical and ecological characteristics of the Gebel Akhdar made it seem likely that traces of human activity might occur in sufficient abundance to show a significant relationship to those in adjacent areas of settlement in the Nile Valley and the Atlas Mountains.

These, then, were the main considerations at the start of the investigation. As the work proceeded, however, it became clear that certain aspects of the inquiry would have only a limited interest, while others began to assume greater importance.

The traces of high sea-levels were on the whole satisfactorily preserved, and appeared to represent a period of time much greater than that of the human occupation. Moreover, these traces, as had been hoped, were of a kind which enabled conclusions to be drawn whose interest extended beyond the immediate problem of correlation within the Mediterranean region.

The Pleistocene continental deposits, on the other hand, were almost wholly confined to the final phases of the period. At the same time, despite their short chronological span, these deposits also provided a surprising amount of miscellaneous information, especially with regard to changes of climate and to a movement of the sea-level to a position far *below* that of the present day.

A corresponding deficiency was noted in the archaeological evidence from all sources, and no remains were discovered which could be assigned to a period earlier than the Middle Palaeolithic. In compensation, a certain number of finds from this and later periods shed unexpectedly clear light on coastwise cultural relationships. Finally, the evidence from the Siwa depression and Tripolitania provided interesting evidence regarding the post-Pleistocene 'Neolithic' stage—hitherto largely lacking from Cyrenaica itself—as well as some indications concerning the earlier periods in these regions also. It is accordingly possible to sketch in broad outline the geographical pattern of cultural activities in northern Libya, from the Middle Palaeolithic stage onwards.

In short, both branches of the research developed into something more than the essays in correlation originally intended. The problem of correlation remains none the less the most important single question dealt with; and although results were only fully realized and applied in the Upper Pleistocene, it is hoped that the descriptions and discussions of the methods used may themselves prove of practical assistance to archaeologists and others besides specialists in Quaternary geology.

For the information of future workers in the region, it is probably true to say that the geological and geomorphological evidence has by now been fairly thoroughly examined over about two-thirds of the stretch of coast between Benghazi and Derna.

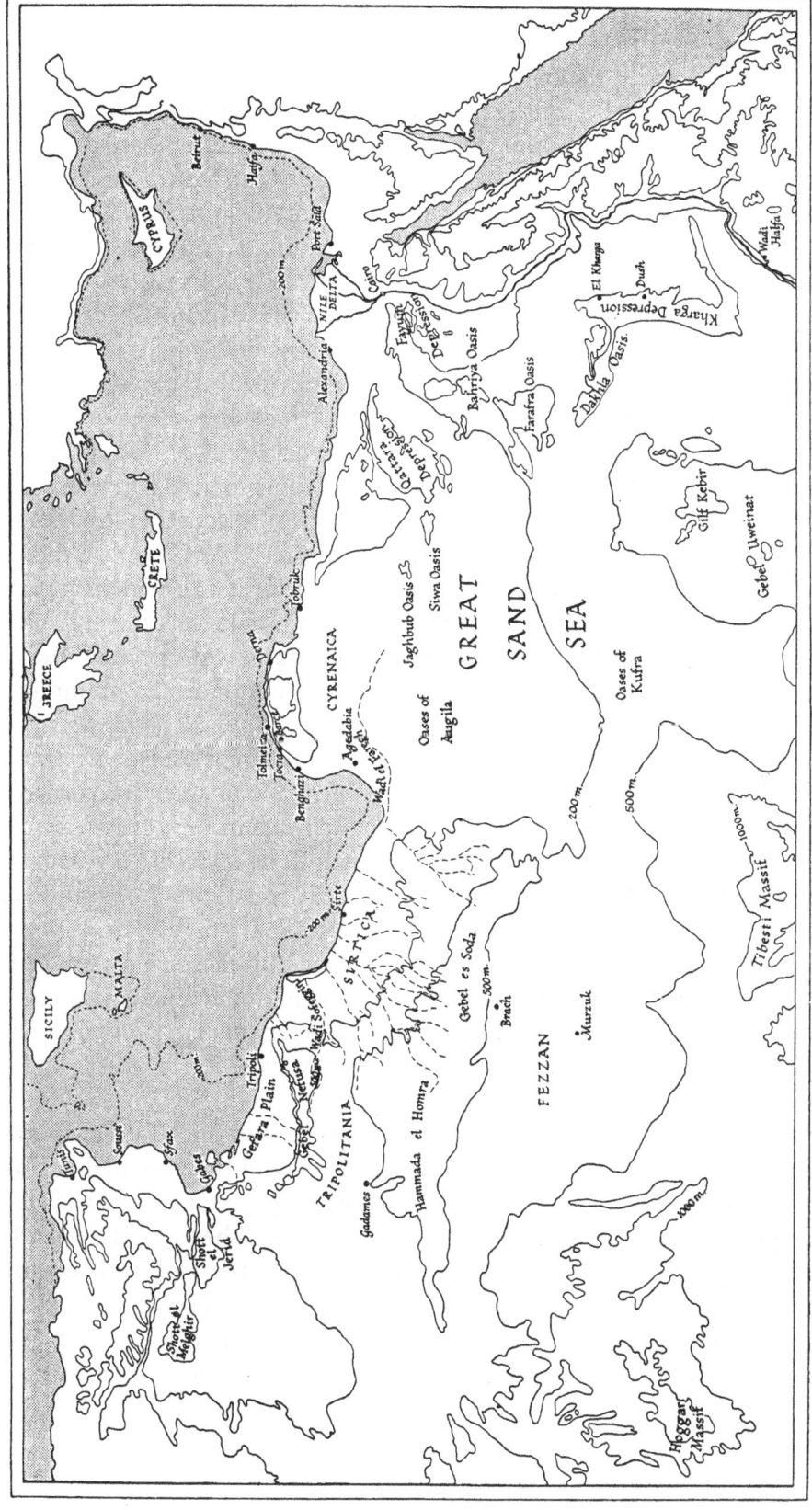

FIG. 1. General map of eastern North Africa

I-2

The same can hardly be claimed for the archaeological and palaeontological data. In the nature of things the traces of prehistoric cultures are appreciably more sporadic and less predictable in their occurrence than the traces of geological phenomena. As a result the cultural sequence contains a number of gaps and tantalizing uncertainties, and there can be little doubt that the region reserves many unsuspected features for future investigators. Nevertheless, it seems likely that the character of some of the main episodes during the second half of the Stone Age in northern Libya are now sufficiently well established to be of use to prehistorians working in other areas of North Africa, and to provide a reasonably reliable starting point for further researches in Cyrenaica itself.

In preparing the material for publication our main purpose has been to provide a clear presentation of the facts, and confine discussion to the more immediate issues. A departure from this policy is offered by Chapter xv, dealing with archaeological material of 'Neolithic' type. Here it seemed that some attempt at a critical analysis of current facts and theories in neighbouring territories was desirable for an intelligible presentation of the new material. Elsewhere, however, the observations are described with the minimum of comment.

At the same time, circumstances of discovery and methods of investigation have often been described at some length. The inclusion of such descriptions has been considered justifiable on several grounds. Apart from their practical interest to others working under similar conditions, such details are often essential for the precise appreciation of field results, particularly in an initial investigation of new territory such as ours.

Finally, a word may be said concerning investigations still in progress in Cyrenaica and neighbouring areas. These are mainly concerned with the new method of radiocarbon dating. An attempt is being made to obtain reliable absolute dates for the later phases of the Stone Age in the Atlas massif, and concurrently from our area also.

It is clear that some time must elapse before the full implications of this work can be assessed, both as regards the results obtained and the potentialities of the materials offered by the region. Since many of the problems dealt with in the present report lie outside the range of the new researches, it was considered inadvisable to withhold publication of the former until they are available, or for that matter until the full examination of certain newly-discovered sites. The latter may contribute important new elements to the archaeological record, but little is known as yet of the time required for their investigation—even supposing practical circumstances will permit it—or the precise nature of the results they may be expected to yield. Carbon readings and other evidence available and relevant to the present work at the time of publication has been included as far as possible in appendices and footnotes.

2. THE GEOGRAPHY OF CYRENAICA IN RELATION TO NORTHERN AFRICA IN GENERAL

Before describing more particularly the main area under examination, it will be convenient to recall briefly the wider geographical pattern of which it forms part (Fig. 1). For so large an area, the topographical and climatic distributions of northern Africa offer a comparatively simple design.

The transition from tropical forest to desert along the southern margins of the Sahara at the present time forms a zone of moderately watered savannah some 300–400 km. wide, falling roughly between the 13th and 18th parallels. The Sahara proper, to the north of this zone, comprises an area some 5000 km. from east to west and 2000 km. from north to south, of which about one-third is true waterless desert, and the remainder desert steppe of extremely arid character. Topographically the greater part is a plateau between 200 and 300 m. in altitude, with little relief apart from a single discontinuous line of hills, starting, with the Hoggar massif, a little north and west of the centre, and extending via the Tummo, Tibesti and Ennedi ranges to Darfur in the south-eastern part of the desert.

North and east of the desert zone the two most important habitable regions are the Atlas massif—or Maghreb—and the Nile Valley. The former comprises an area of mixed prairie and scrub some 2000 × 400 km., while the latter provides the main line of access between the Levant and Central Africa. Along the greater part of the 2500 km. of coast separating the Nile Delta from the Gulf of Gabes and forming the littoral of the ancient province of Libya, the desert is divided from the sea by only a narrow margin of extremely arid steppe. Surface water is here available at most for a few weeks in the year.

There is indeed only one important break in the desolate character of the Libyan littoral—that provided by the Gebel Akhdar or 'Green Mountain', on the coast of Cyrenaica. Here the hilly topography gives rise to an area of greater rainfall and fertility over an area of some 300 × 100 km. Within this region the rainfall ranges between 200 and 550 mm. per annum. Eastwards and westwards along the coast the isohyets of mean annual rainfall at 150 and even 100 mm. lie within a few km. of the coast, and the zone of sparse steppe vegetation rarely extends further than 75 km. inland. The only other area at all comparable to the Gebel Akhdar along the Libyan coast is that of the Gebel Nefusa, of Northern

Tripolitania. The fertile area is here both less marked and very much smaller in extent than that of the Gebel Akhdar, and forms in effect a minor outlier of the Maghreb.

Inland of the Gebel Akhdar, to the south and south-east, lies the most arid portion of the Sahara, sometimes termed the Libyan Sand Sea—a region some 1700 × 700 km., virtually waterless, and occupied in large part by immense dune-fields. A few oases occur in a widely scattered group in the angle between the lower Nile and the coast east of the Gebel Akhdar, and west of the Sand Sea at Augila and Kufra, respectively 200 and 800 km. inland from the Gulf of Sirte. Apart from the last-mentioned oases, the country to the west is but little less inhospitable than the Sand Sea itself, until the more numerous oases near the edge of the Hammada el Homra plateau and the Fezzan depression are reached. These western oases form a chain southwards from Tripolitania to the Tummo and Tibesti Hills mentioned above. Between the Hoggar, at the north-western extremity of the range, and the Maghreb proper, are again a few widely scattered oases, but west again between the Hoggar and the Atlantic littoral lies the second most important expanse of consistently desert conditions.

Before any conclusions regarding the effect of this geographical pattern on the development of human settlement can be drawn, it is of course necessary to form some idea of the extent to which it may have been altered by different climatic and ecological factors in the past. In the earlier days of Saharan exploration, considerable attention was attracted by a variety of indications of former greater abundance of water. It was moreover natural that these should be to some extent connected with historical evidence of former greater fertility along the Mediterranean littoral.

Many of these indications have been to a large extent discounted by more recent investigations. It has for instance been shown that the arrangements for dealing with the run-off in Roman times do not greatly differ from those required at the present day,[1] while a much greater share in causing the spread of the desert is now assigned to soil erosion consequent on over-grazing, deforestation, and the break-down (for social reasons) of former elaborate irrigation installations. Again, dry watercourses that may be found in many parts of the desert can be explained in part by the torrential character of the downpours that still occur at widely separated intervals, even far inland.

There remain, none the less, a number of striking indications that cannot be explained in this fashion, and are generally admitted to demonstrate an appre-

---

[1] 'La plupart des sources qui alimentarient les centres romains, existent encore....Leur débit a-t-il diminué depuis une quinzaine de siècles?...de rares constations permettent de croire qu'en divers lieux ce débit ne s'est pas modifié.' S. Gsell, quoted in E. F. Gautier, 1946.

ciable difference in the pattern of precipitation at more than one period in the geologically recent past. The latest of these seems to have been of minor significance and to have occurred no later than the archaeological stage associated with the earliest spread of pastoral and agricultural activities. Of more importance are the indications associated with the Middle and Lower Palaeolithic Culture stages, while signs have been noted in several areas of a still greater rainfall at some period in the Lower Pleistocene, perhaps before the appearance of tool-making man.

Although nothing approaching a reliable climatic sequence for any considerable part of the Sahara is yet in sight, it is generally agreed that the wetter episodes just mentioned were interspersed with others in which desiccation was as great as or greater than at present. It is now usually held that even during the more favourable periods the larger part of the Sahara scarcely rose above the level of fertility of a dry steppe or prairie.

The conclusion seems to be that, although the relative extent of scrub, prairie and desert zones may have altered on several occasions, the changes were not such as to modify the broad ecological pattern as we see it today. The main areas of human settlement and biological activity in general would be substantially the same, and the main routes of natural intercourse also.

Conclusions regarding these latter are clearly of great importance for the correct interpretation of events in any one area. From what has just been said there can be little doubt that the main east-to-west line of communication lay along the Mediterranean littoral, close to the shore. A subsidiary route further to the south, starting in Tripolitania and running eastwards to Augila and thence along the northern fringe of the sand sea to Siwa and the Egyptian oases, is still used by camels, and probably played some part under favourable conditions in the past.[1]

The north-to-south route along the Atlantic littoral has been studied little as yet, but it is known that an important inland route from north to south in Roman times started in Tripolitania and followed the oases to the Fezzan, continuing thence via the Hoggar or the Tummo Hills to Central Africa. There is also some evidence, at a late prehistoric stage, for the use of a route from southern Tunisia and Algeria along the Tazilli Plateau south-eastwards to the Hoggar. Thence in Neolithic times movement could certainly take place due south through Aïr to the Niger, and south-east along the hills to Darfur.[2]

As regards the Gebel Akhdar it can be said with some confidence that there is at present no sign of an ancient north-to-south route there or indeed anywhere

[1] See Chapters XVI and XVII.
[2] This is proved *inter alia* by the Neolithic site at Tamaya Mellet (Kelley, 1934).

between the Nile and the Hammada el Homra plateau. The existing Augila—Kufra—Tibesti route is of very recent origin.[1] The interest of the Gebel Akhdar is thus enhanced by its character of principal staging post between the Nile and the Maghreb—whatever movements took place between the two it is here that they are likely to have left their clearest traces. Moreover, the distances between Cyrenaica and the two areas in question are such that most major events in either are likely to have found some reflection there also.

3. TOPOGRAPHY AND GEOLOGY OF CYRENAICA

Emphasis has been laid in the foregoing section on the importance of rainfall as the principal controlling factor in the density and character of human settlement. A cursory comparison between topography and rainfall distributions along the whole North African littoral reveals at once the close correspondence between the two; indeed it is clear both here and elsewhere in the Sahara that topographical relief plays an important part as an ultimate factor in human affairs.

The relevance of solid geology to human affairs is of a less direct kind in North Africa. It plays nevertheless a leading role in the present discussion owing to its close connexion with the manner in which the traces of cultural and natural events are preserved, and the background which it supplies to the geological events of the Pleistocene.

It will accordingly be convenient at this point to give a brief introductory account of the topographical and geological features of the area under discussion. Further details relating to particular districts will be given later as required.

The main topographical features of northern Cyrenaica are shown in Fig. 2.[2] As can be seen from the inset, this is a region of generally simple relief. The only area of high ground is that which occupies the most northerly part of the territory, and which is known as the Gebel Akhdar; this is, indeed, the only large area of high ground in the whole of the 2500 km. of flat coastline between Homs, in Tripolitania, and Mount Carmel, in Palestine. In plan, the Gebel is roughly elliptical, and its length from west to east is about 250 km.

The highest points of the Gebel, some of them over 800 m. above sea-level, all lie to the north, within 50 km. of the coast. Southwards, the ground falls away very gradually towards the line of depressions which contain the oases of

---

[1] This route seems to have been first opened by a native of Jalo named Shehaymah, acting for Sabun Sultan of Wadai, about 1810. (See Evans-Pritchard, 1949.)

[2] The best available maps of northern Cyrenaica are in an Italian series to the scale of 1:100,000. For regions south of Benghazi, it has been necessary to use the series G.S.G.S. 2465, to the scale of 1:1,000,000.

Marada, Augila, Jaghbub and Siwa. To the east, the fall is still gradual, towards the much lower plateau of Marmarica (the hinterland of the coast east of the Gulf of Bomba). It is only on its northern and western sides that the limits of the Gebel are sharply defined by steep slopes, and these slopes take the form of two successive escarpments.

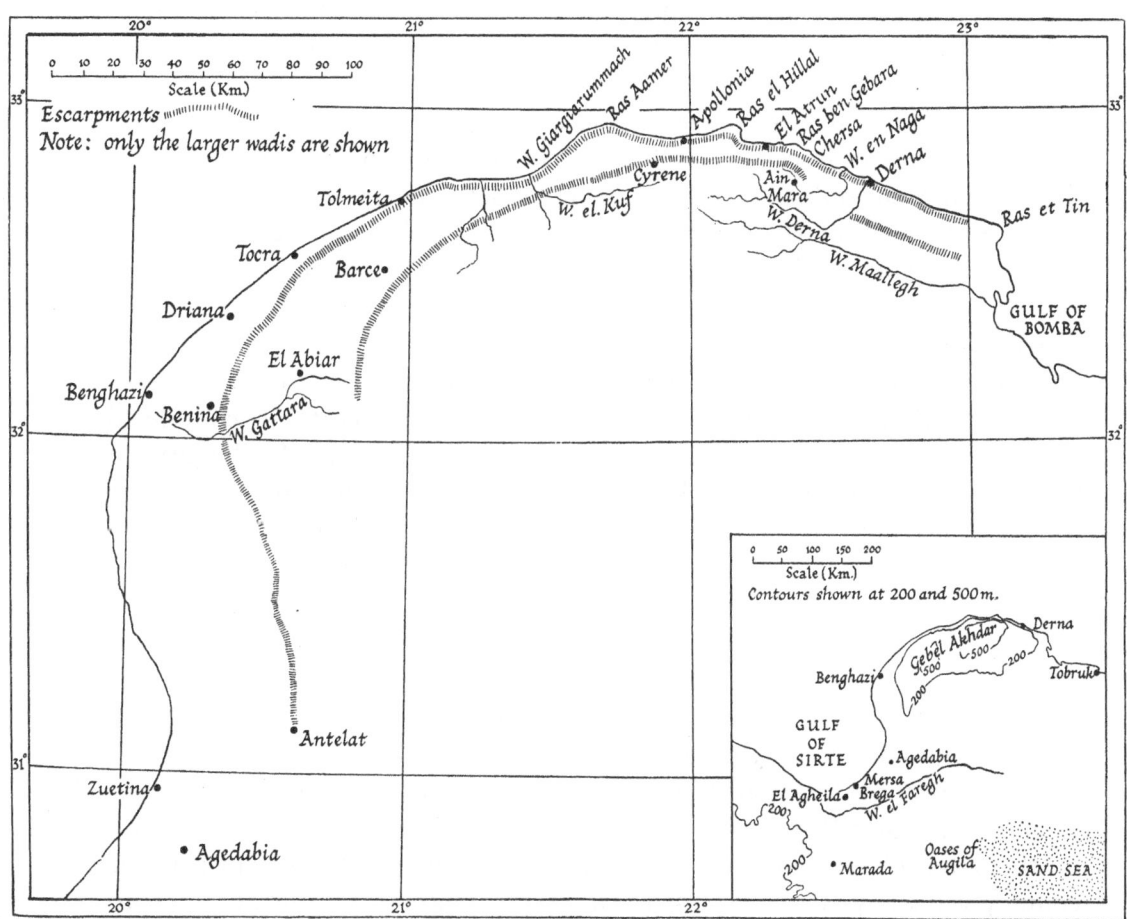

FIG. 2. Map of the area investigated

Inset. Map of Cyrenaica, showing position of the Gebel Akhdar.

The very great extent of these escarpments can be seen from Fig. 2. The lower and outermost of the two runs for over 400 km., between Ras et Tin and Antelat; in the whole of its course, its continuity is unbroken except by the gorges of innumerable wadis. From Ras et Tin to Tolmeita, it runs very close to the sea. South of Tolmeita, however, it gradually retreats from the coast, from which it is here separated by a gently sloping plain. The maximum width of the plain is

9

about 50 km.; at its southern end it merges into the low-lying country at the head of the Gulf of Sirte. The upper escarpment is only about 300 km. long, but this also is continuous for nearly the whole of its course.

The geology of Cyrenaica to the north of the oasis depressions is reasonably well known. The latest general accounts are those of Desio[1] and of Marchetti;[2] the best available geological map is that which is included in Desio's book.

Of the rocks exposed on the surface, all are of sedimentary origin, and almost all are marine limestones. The oldest beds known are of Upper Cretaceous age; these are confined to two inliers on the crest of the Gebel, and a few localities on the north coast. Eocene and Oligocene beds occupy much larger areas, though these again occur only in and around the Gebel. The greater part of the region is covered by rocks of Miocene age, those of the Middle Miocene being especially widespread. Younger beds, on the other hand, apart from those of the Pleistocene, are only doubtfully present, and must in any case be of very limited distribution.

Most of the region appears to have a very simple geological structure, the beds showing nothing more than a gentle southerly dip. Tectonic complications are known only from Marmarica and from the Gebel itself. In the latter area, with which this report is chiefly concerned, these complications are mainly in the form of faults and monoclines, sometimes of considerable downthrow and extent.

4. HUMAN ECOLOGY OF CYRENAICA IN MODERN AND HISTORIC TIMES

Documentary evidence regarding the distribution, size and character of the human communities inhabiting Cyrenaica is available at various periods over the past four thousand years. The classical and earlier sources were exhaustively examined by Oric Bates over thirty years ago in what is still the standard work on the subject.[3] Italian and other statistical reports on the recent state of the country are analysed by Professor Evans-Pritchard in the introduction to his recent study of the Sanusiya religious order.[4] Other summaries have been provided in the last few years by the publications of the British Military Administration. From these and other sources it is possible to form an idea of the relationship of culture and environment in the territory, which is also not without relevance to the more remote periods with which this report is concerned.

The most striking single feature of this picture is the remarkable constancy of the dominant mode of existence. De Agostini, in his statistical summary of

[1] Desio (1935).   [2] Marchetti (1938).
[3] Bates (1914).   [4] Evans-Pritchard (1949).

1922, divides the population—which he estimates as about 200,000 in all—into the following classes:

54 *per cent 'stable'*. That is to say, supporting themselves by mixed grazing and agriculture within an area delimited by permanent boundaries and subject to individual and tribal ownership. Habitations, however, are normally impermanent—tents or other makeshift arrangements—moved in accordance with the requirements of grazing and seasonal tillage or harvesting.

19 *per cent 'semi-nomads'*. Groups practising a greater relative degree of pastoralism, and liable to more extensive migrations outside their own territory.

9 *per cent 'true nomads'*. Living entirely or almost entirely by herding, and who, while they may possess a recognized centre, nevertheless spend most of their time in migratory movements to distant and not always the same outside territories.

4 *per cent live permanently in the oases*, and 14 *per cent in the towns*.

From these figures it is clear that the great bulk of the population live a basically similar type of existence in which agriculture and commerce play only a subsidiary role and the mainstay of the economy is supplied by animal husbandry.

In Cyrenaica people living in this fashion are distinguished by the name *bwadi*, while the townsfolk are called *hadur*. Profound differences of culture, outlook, and ethnic origin divide the two. The latter are in the main descended from Tripolitanian immigrants arrived during the last 150 years, while the former are descended very largely from the Arab invaders of the eleventh century.

During the past century the economic function of the towns seems to have been to act as clearing houses for a small export trade in foodstuffs, and import of minor luxuries, and as centres for the rather meagre manufacturing trades. Only two towns were of any importance, Benghazi and Derna.

The only other period at which towns played an appreciable role in the life of the region, was during the Græco-Roman occupation, when they seem to have owed their existence largely to specialized commerce in the mysterious *Sylphium*, and perhaps to some extent in horses. Both before and after this period the country, to judge by the figures given in historical documents cited below, seems to have got on very well without them. Even in the classical heyday of town life it is clear that the hinterland was occupied by a large section of the population living an existence very closely comparable to that of the modern Beduin. Neither Egyptians nor Phoenicians succeeded in establishing permanent settlements, and the effects of the nomad invaders of the eleventh century seem to have been even more completely destructive of the later Byzantine centres here than elsewhere.

In view of the apparent fertility of the Gebel Akhdar plateau and the coastal plain at its foot, this relative absence of village and town communities, which form so characteristic a feature of most Mediterranean countries, seems to require some special explanation. Evans-Pritchard suggests that one factor is to be sought in the periodic recurrence at short intervals of years of severe drought, which would make any economy largely or wholly based on agriculture very insecure. No doubt a second factor is provided simply by the greater geographical and political isolation of the fertile territory, with its wide surrounding zone of purely steppe environment, suitable only for nomads.

In his study of the Beduin community, Evans-Pritchard draws attention to a further interesting feature, in the distinction between the limited nomadism of the plateau and western littoral—*al jasha* and *al barqa al hamra*—and the now complete nomadism of the regions further to the south and west—*al sirwal* and *al barqa al beda*. 'In Cyrenaica always stand contrasted the mountains and the plain, the forest and the steppe, the red soils and the white, the region of goats and cows, and the regions of sheep and camels, the region of settled life and the region of nomadism.' The area of the former measures some 300×75 km. and is controlled not merely by the greater rainfall and occurrence of scrub, but also by the presence of perennial springs which are found within the same zone. Both goats and cattle feed readily off scrub forest, but require regular watering at frequent intervals.

Sheep and camels on the other hand can go for much longer periods without water, a fact which greatly extends their radius of grazing from a given source of water, and at the same time they are better adapted to exploit the specialized vegetation of the desert. When feeding on the young growth of the littoral and inland steppe after the short period of the winter rains, they are able to do without watering of any kind. These creatures then supply a means of existence and foster a type of society that extends up to 75 km. inland from the Gebel and along the whole coast to the east and to the west.

It is interesting to conjecture how far the population estimates for men and beasts immediately prior to the Italian occupation, may have compared with the corresponding populations in the second and third millennia B.C. of which we learn something from ancient Egyptian texts. The maximum values of the recent populations are given by Evans-Pritchard as in the neighbourhood of a quarter of a million souls plus one and a quarter million sheep and goats, 23,500 cattle, and over 100,000 horses and camels. Bates concludes from contemporary records that the number of Libyan fighting men and their allies taking part in the battles against Merneptah and Rameses III in the thirteenth and twelfth centuries B.C. must have been in the order of 20,000–25,000 and 30,000 respectively. In the former battle 1300 cattle are mentioned as having been captured by the Pharaoh.

While the total population of Cyrenaica and Marmarica can of course only be estimated in the vaguest fashion from these figures, after deduction of an unknown number of foreign allies, it is difficult to believe that the requisite number of young fighting men can have been produced by a population of less than, say, 100,000. Again, the number of cattle mentioned shows that these were already a mainstay of the Cyrenaican economy.

Generalized though these conclusions are, they serve nevertheless to illustrate a point of some interest to the archaeological evidence, namely that despite the relatively low density of population of a pastoral as opposed to an agricultural society, the former still represents an enormous advance in this respect on any hunting society living in a comparable environment.

The maximum density of the population of Cyrenaica and the surrounding districts under a hunter-gatherer economy can of course only be guessed in the most arbitrary fashion, but a comparison with some modern primitive peoples (Australian aborigines, Hottentots, Bushmen, and Paiute Indians), suggests that it may not improbably have been in the order of 10 sq. km. per man for the plateau, and 26 sq. km. per man for the steppe to the south and along the coast, i.e. a total population of just over 2000 for the whole Gebel. Even at this low density, however, it is important to remember that some diffusion of ideas and even trade of a kind can take place over surprising distances, as shown by recent studies in Australia.[1]

In conclusion therefore, a word concerning the pattern of outside economic and cultural relations of the Beduin of Cyrenaica in recent times may not be irrelevant. Their main commercial relations are overland with Egypt, to whom until recently they supplied considerable quantities of foodstuffs. Such overflow of population as took place from Cyrenaica seems also to have been regularly directed in an easterly rather than a westerly direction. Sirtica to the west has indeed been described as 'without dispute one of the most decided frontiers, natural or human, to be found anywhere in the world'.[2] 'The people of Cyrenaica are linked to the classical Arab world to the east. More particularly they are linked to Egypt, with which their country has had political ties from the earliest times, and which is their natural market.'[3]

The particular cultural connexions seen in recent times are of course almost entirely the result of a specific historical accident—the place and mode of origin of the eleventh-century Arab conquerors of the territory. On the other hand such historical accidents, where partly conditioned by geographical factors, have

[1] Thomson (1949), also Mountford (1951).
[2] Despois, quoted in Evans-Pritchard (1949), p. 47.
[3] Evans-Pritchard (1949), p. 47.

13

a tendency to repeat themselves, even at widely separated periods and levels of culture. It may be recalled in this connexion that the Hamitic languages spoken by the pre-Islamic population along the whole North African coast are assigned by some philologists ultimately to the same group of languages as the Semitic stock indigenous to Western Asia. As will be seen from the archaeological section of this report, the distributions of some prehistoric culture traits are not without analogy to those of the present day.

Future investigations will no doubt show how far such analogies are illusory. At the present early stage of the investigations it would seem none the less a sound policy to bear in mind the existing structure of relationships as illustrating one form at least which the connexions may have assumed in the past, particularly towards the end of the prehistoric epoch.

PART I
PLEISTOCENE GEOLOGY

A. SHORELINES AND MARINE DEPOSITS

INTRODUCTION

It was by no means certain before the 1947 expedition that the coast of Cyrenaica would necessarily be suitable for the study of ancient shorelines. Scarcely a single reference to the existence of any such feature could be found in the published literature. Nevertheless, there were several reports, to be referred to later, describing the occurrence of Pleistocene marine deposits at various points, more or less elevated, in the western coastal plain. It had also been suggested by Marinelli,[1] Ahlmann[2] and Stefanini,[3] that the escarpments might themselves be the products of comparatively recent marine erosion. This view had been opposed by Desio, who considered that they could more probably be ascribed to faulting or abrupt folding, but even Desio had admitted the possibility of marine erosion at relatively low levels.[4] It thus seemed clear enough that much of the Cyrenaican coast, if not all, must have undergone submergence during the Pleistocene period. Hence it could reasonably be expected that traces of Plcistocene shorelines would in fact be present, in some form or another, in most parts of the coastal region.

Moreover, there were reasons for believing that these traces might be present in a form suitable for easy and rapid examination; that is to say, that they might be accompanied by well-marked erosional features. From accounts of other regions, it appeared that such features were most often developed on coastlines where the topography was steep, and this could certainly be said of the entire north coast of Cyrenaica. As for their chances of preservation, almost the whole of Cyrenaica was known to be made of hard limestone, a material which, owing to the formation of underground drainage systems, is particularly resistant to subaerial erosion.

Since time was limited, a detailed examination of the entire coastline was obviously out of the question. It was decided, therefore, that the coast between Tolmeita and Ras et Tin should, by reason of its steepness, be examined first, and should receive the greater amount of attention. This decision proved to be well justified. Almost all points which were visited on this part of the coast showed abundant signs of marine erosion, at many different levels. These signs were also clear and easy to study, and could thus provide much of the informa-

[1] Marinelli (1920).   [2] Ahlmann (1928).
[3] Stefanini (1930), pp. 23–4.   [4] Desio (1939), pp. 63–7.

tion which had been hoped for. On the west coast, by contrast, the evidence proved to be not only difficult to find and interpret, but also relatively scanty. Even here, however, enough evidence could be collected to suggest a number of interesting conclusions; moreover, the differences between the nature of this evidence and of that found on the north coast were so great as to provide in themselves an additional source of interest.

The information obtained from these two parts of the coast will be discussed in the two following chapters. In Chapter III, various chronological questions will be dealt with. Finally, Chapter IV will contain a summary of the conclusions reached, followed by a discussion of the more important implications of the investigation as a whole.

CHAPTER I

TOLMEITA TO RAS ET TIN

1. TOPOGRAPHY AND GEOLOGY

The greater part of the coast between Tolmeita and Ras et Tin possesses a more or less uniform topography. The dominant feature throughout is the coastal escarpment, which is steep, often precipitous, and notched at frequent intervals by deep wadis (Pl. 1). The crest of the escarpment is generally well defined, and shows great variations of altitude when followed along the coast. At Tolmeita it stands about 300 m. above sea-level; eastwards, it declines almost to zero at Wadi Giargiarummach, rises to a maximum of 540 m. near Ras el Hillal, and drops finally to sea-level near Ras et Tin.

The foot of the escarpment, on the other hand, is quite indefinite. In most places, indeed, a narrow plain appears to separate the escarpment from the sea; this at any rate is the impression obtained both from an examination of the 1:100,000 maps and from a distant view on the ground. Even without a close inspection, however, it is everywhere obvious that the plain merges more or less gradually into the main part of the escarpment, and, moreover, that it has a decided slope towards the sea. The real nature of this feature will be discussed more fully below. For the moment, it will be best to refer to it by its Arab name, the Sahel.

The rocks which compose the escarpment and the Sahel show little variety, at any rate as far east as Derna. Everywhere they consist of hard, even-bedded limestones, often containing bands of flint. For the most part, they are of Eocene and Upper Cretaceous age, with Oligocene and Miocene beds confined to the highest parts of the escarpment; it is only between Derna and Ras et Tin that these younger beds descend to lower levels. Except in one disturbed zone, between Apollonia and El Atrun, dips are generally low and faulting probably absent.

2. PLAN OF INVESTIGATION

The investigation was conducted throughout with two immediate objects in view. First, it was intended not only to record the positions of as many fragments of shoreline as possible, but also to measure their altitudes at frequent intervals along their lengths. If these altitudes should remain constant, they might provide a means of establishing the relative ages of fragments widely separated from one another. If not, they might at any rate throw some light upon

2-2

the recent tectonic history of the coastal regions. Secondly, it was intended to make a careful search for associated geological and archaeological remains.

It soon became obvious that an investigation of this kind would be slow and laborious, and various devices were used, as described below, in order to increase the speed of operation. Progress, however, always remained relatively slow. Hence, it was only possible during the 1947 expedition to examine certain stretches of coast easily accessible by road or track. These lay respectively between Ras Aamer and Apollonia, between Derna and a point about 15 km. to the west, and in the vicinity of Ras el Hillal. In all three cases, the evidence was both abundant and clear, but could hardly be said to lead to any very interesting conclusions. The three stretches of coast were too far apart for correlations to be attempted, and each one by itself was too short to yield much information, either geological or archaeological.[1]

Nevertheless, it now seemed likely that only a few weeks' additional field-work, in some future expedition, might well produce results of real interest. It would clearly not take long to examine the gaps between the three original stretches of coast, and thus to complete the survey of a continuous and very considerable section of the north coast of Cyrenaica. At the same time, there was no reason to suppose that the shorelines would not be as well developed within the gaps as elsewhere. If so, an examination of these gaps might well reveal, first, how many individual shorelines were represented on this particular section of the coast, and secondly, what were the present altitudes of each.

Results of this kind, obtained from so great a length of coast, might then enable definite estimates to be made of the extent to which the present elevations of the shorelines were due on the one hand to changes of sea-level, or on the other to movements of the land. Such estimates, in turn, should finally provide the kind of information which had been hoped for with regard to local tectonics and general movements of sea-level; they might also provide a means of correlating the Cyrenaican shorelines, not only with one another, but also with similar features elsewhere.

One of the main objects of the 1948 expedition, therefore, was to fill in the two gaps. This object was almost completely achieved; by the end of the expedition there were only 5 km. of coast, a little to the east of El Atrun, which had not yet been visited. In addition, the survey was extended to the east of Derna. Thus, during the two expeditions, a detailed survey was made of almost the whole of the coast between Ras Aamer and a point 12 km. east of Derna. The total length of this stretch of coast is nearly 110 km.

[1] A summary of these results has been published in a preliminary report: McBurney, Hey and Watson (1948), pp. 34–6.

The results obtained will be discussed under four separate headings. First, the nature of the evidence will be discussed in general terms, without reference either to altitudes or numbers of shorelines. Secondly, a description will be given of the methods used during the detailed survey, and of the uses which were subsequently made of air-photographs. Thirdly, the results themselves will be presented. Fourthly, an attempt will be made to interpret these results, and to consider their implications.

Some observations were also made on certain other parts of the north coast of the Gebel, both to the east of Derna and to the west of Ras Aamer. These observations, though not made according to any definite system, are all of interest in various different respects. They are discussed in a final section, under the heading 'Miscellaneous Observations'.

3. NATURE OF EVIDENCE FOR MARINE ACTION

DETAILS OF COASTAL TOPOGRAPHY

Almost all parts of the coast between Tolmeita and Derna appeared, when seen from a distance, to have a simple topography, the escarpment and the Sahel together forming a smooth curved surface, concave upwards. To some extent, this impression was confirmed by a closer examination of the ground. At most localities, the areas nearest the sea were found to be partly covered with accumulations of alluvial deposits, and these did indeed possess concave upper surfaces.

Such accumulations, however, were seldom present on the higher slopes of the escarpment, while even at the lower levels they were always discontinuous and sometimes absent over large areas. Where this was so, surfaces were laid bare which were generally composed mainly of bare rock, and whose detailed topography was usually found to be anything but simple.

Except where they had obviously been affected by stream erosion, these surfaces almost invariably consisted of alternate steep and gentle slopes, arranged in the form of a staircase with the 'treads' running parallel to the modern coastline. The gradients of the steep slopes, though seldom approaching the vertical, were generally between 1 in 4 and 1 in 2 (15° and 30°); those of the gentle slopes were normally less than 1 in 30. It will therefore be justifiable to refer to the two types of surface, for the sake of convenience, as 'cliffs' and 'terraces' respectively. Examples of these features are shown in Pl. 1 b.

Both the terraces and the cliffs showed a considerable range of size. Each individual terrace, when followed along the coast, was found to vary continually in width. The greatest width attained was about one kilometre, but each was found, sooner or later, to dwindle to the point of vanishing; for this reason, even in the absence of alluvium or subaerial erosion, it was seldom that one terrace, as a single morphological feature, could be traced laterally for more than a few kilometres. As for the cliffs, their range of heights was virtually unlimited, and particularly so since the disappearance of a terrace inevitably resulted in the merging of the cliffs immediately above and below.

It was noticeable, however, that the heights of the cliffs showed a general tendency to increase in proportion to the altitudes at which the features stood above sea-level. As a result of this tendency, the average slopes of the ground were usually less steep near the sea than further inland; this, in turn, was evidently the sole reason for the apparent differentiation of the coastal topography into Sahel and escarpment. Hence, it was clear that the Sahel could no longer be regarded as a distinct morphological feature.[1]

PLEISTOCENE MARINE DEPOSITS

For the most part, the terrace surfaces presented bare limestone pavements, often showing some degree of *karst* erosion. In many places, however, local patches of terra rossa had accumulated, permitting the growth of scrub, or even of crops. In addition, the terraces were found to carry certain patches of deposits which were of Pleistocene age and of marine origin. These deposits were of two types. The majority of the patches had evidently been laid down by the wind, but there were others which showed clear signs of having been laid down by the sea itself.

Deposits of the latter type occupied very small areas, being almost entirely confined to the landward margins of the terraces. In this position, nevertheless, they were regularly present, in the form of narrow, somewhat discontinuous bands, usually lying a little below the cliff-bottoms but sometimes rising just above them. Normally, these deposits consisted almost entirely of calcareous organic debris, with a firm calcite cement; the most common constituents were shell fragments, foraminifera, echinoid spines and pieces of calcareous algae. Pebbles of flint and limestone often occurred as well, sometimes so abundantly as to form a true conglomerate. Large unbroken fossils were scarcely ever found, and when found were extremely difficult to extract. The few which were collected came mainly from the younger and less firmly cemented deposits, and could all be identified as forms still living in the Mediterranean; details, when not given elsewhere in the text, will be found in Appendix B.

Sections of these truly marine deposits were rare, and it was therefore difficult

---

[1] The term has nevertheless been retained in the present work, as a convenient name for the flatter and more habitable parts of the coastal belt.

to gather much information about their thicknesses. In general, the thickness seldom seemed to exceed half a metre. Exceptions were noted chiefly in places where a storm-beach had evidently been piled up against a cliff; the deposit could then be as much as 3 m. thick.

Deposits laid down by the wind were much more widespread. Many of the occurrences nearest the modern shore had assumed definite topographical forms; these will be described later under the name of 'fossil dunes'. More generally the deposits occurred as sheets, or banks, along the landward edges of the terraces, hiding not only the bedrock but also any marine deposits that might be present; very often, they also extended for considerable distances up the lower slopes of the associated cliffs. Lithologically, these wind-blown deposits were composed of much the same materials as the finer grades of the marine deposits. The chief points of distinction were the greater degree of rounding and sorting of individual grains, the rarity of large fossils and pebbles, and the occasional presence of land-snails. In practice, since all but the last of these criteria were questions of degree, many occurrences were found which could not be definitely assigned to either category, even after a subsequent examination of specimens under the microscope.

It should be added that both the water-laid deposits and the dune deposits were usually extremely hard, and firmly cemented to the surface of the bedrock. Moreover, both materials, and the bedrock itself, tended to present identical appearances on weathering. Their recognition in the field, therefore, was not always easy.

ORIGIN OF TERRACES

On morphological grounds alone, it seemed highly probable that the terraces were products of marine erosion, and the degree of probability was raised almost to the point of certainty by their association with marine Pleistocene deposits. Nevertheless, it was always just conceivable that the deposits might in some cases have been laid down on surfaces already formed by other means. Two possibilities in particular had to be considered: first, that some of the terraces might be structural surfaces, formed by the weathering of alternate hard and soft layers within the bedrock, and, secondly, that they might represent a series of pediments, formed entirely by the action of streams.

For the terraces on the highly disturbed beds to the east of Apollonia, the first possibility could be dismissed immediately. On other parts of the coast, however, owing to the generally low dips of the bedrock, no terrace could be shown to be independent of the underlying geological structures until it had been traced and levelled over a distance of several kilometres. In practice, this meant that the detection of structural terraces, if any were present, might in some cases have to be delayed until the final collation of results.

The second possibility presented a similar problem. This possibility, admittedly, was never strong; the landward edges of the terraces appeared to be too well defined and too regular in plan to represent the landward edges of pediments, as described by Johnson.[1] According to Johnson, however, the most distinctive feature of a pediment is that the surface of the bedrock along its landward margin takes the form of a series of 'fans', or low half-cones. Since the height of these features is often small compared with their lateral extent, it was clear that pediments also would have to remain undetected until the end of the investigation.

Strictly speaking, therefore, no general statement on the origin of the terraces should be made until after the presentation of the final results of the survey. Nevertheless, it was always obvious in the field that the vast majority of the terraces must be of marine origin, and it was in fact decided during the investigation that this origin should be assumed in all cases as a working hypothesis. It will therefore be convenient at this point to anticipate, and to say that none of the terraces, according to the final results, showed any signs of having been formed otherwise than by marine erosion. Each one, in other words, represented a fragment of a wave-cut platform; similarly, each cliff-bottom marked the approximate position of an ancient shoreline.

One general question remains to be mentioned. It has often been assumed that each member of a series of marine terraces must necessarily be younger than that which lies immediately above it. Since, however, this rule is not universally accepted, it was felt that some independent evidence should be sought as to whether or not it could be applied to the terraces of Cyrenaica. Such evidence was found in the observation that the surfaces of the terraces showed in general an increasing degree of subaerial erosion in proportion to their altitudes above sea-level; at the same time, the aragonite fossils in their associated deposits showed to an increasing extent the effects of solution. Evidently, therefore, this was a case in which the rule was valid.

It thus became possible to ascribe the varying widths of the terraces, and their limited lateral extent, to the encroachment of each new shoreline upon its predecessors. The same explanation could also be used to account for the condition of certain exceptional lengths of coast from which the staircase topography was altogether absent. In every such case, where this absence was not due merely to subaerial erosion, the escarpment was found to present a single slope of unusually steep gradient, with perhaps one narrow terrace at its foot (e.g., on parts

[1] Johnson (1932), and (1944) p. 796.

of the coast between Apollonia and Ras Aamer, and between Ras el Hillal and Ras ben Gebara). Very probably, these were localities at which each terrace in turn had been completely destroyed by the formation of its successors, so that the whole escarpment could itself be regarded as one single marine cliff.

4. COLLECTION OF EVIDENCE

An account will now be given of the methods which were used during the detailed survey in the collection of evidence in the field. This will be followed by a short account of the uses made of air-photographs.

FIELD-WORK

Estimation of mean sea-levels

Whenever a fragment of high shoreline was encountered, the first necessity was to locate, as closely as possible, the position of the line corresponding to its contemporary mean sea-level. In theory, of course, this was a problem to be dealt with separately in each individual case. At first, therefore, levels were taken on all features that might conceivably provide some information, and these levels, together with copious descriptive notes, were preserved for later consideration. It was soon realized, however, that the problem, owing to the uniform manner in which the shorelines were developed, was virtually the same in every case, and that it would thus be possible to save much time by the adoption of some general rules, suitable for use in the field.

Two possible sources of information were available: marine deposits and erosional features. Of the two sources, the former was clearly the less reliable, since any particular patch of marine deposits might have been laid down several metres either above or below sea-level; in any case, such deposits were often absent. From a study of the erosional features at any one locality, on the other hand, it should be possible to obtain a fairly exact idea of the highest level reached by the sea during their formation. Moreover, the present maximum tidal range on the Cyrenaican coast was known to be less than one metre,[1] and there was no reason to suppose that it had been any greater in Pleistocene times; in each instance, therefore, the former *maximum* sea-level should not be greatly different from the former *mean* sea-level.

Even with the use of erosional features, a certain margin of error was inevitable. Marine caves and undercut notches were rarely present, and no definite signs were found of the holes of boring molluscs. Thus, it was almost always necessary in practice to base all deductions upon the positions of the landward

[1] *Med. Pilot*, vol. v.

limits of the terraces. This introduced two kinds of uncertainty. First, the limits were usually ill-defined; the profiles of the junctions between cliffs and terraces were almost always curves rather than sharp angles. Secondly, the relationship between these limits and the contemporary sea-levels could never be exactly known. Nor could any guidance be obtained in this respect from observations made on the modern shores of the Gebel; the sea almost everywhere was eroding Pleistocene materials whose physical properties were very different from those of the bedrock.

It seemed reasonable to assume, nevertheless, that the mean sea-level in most instances would have stood within 1 m. above or below the limit of normal marine erosion; also that the position of this limit would be roughly denoted by the line of most rapid change of slope, or, in other words, by the line of greatest curvature. The location of this line by eye involved a further possible error, but it generally seemed that this was unlikely to exceed ± 1 m. It was therefore concluded that the former positions of mean sea-level could probably be located, by this method, with a total possible error which should normally be less than ± 2 m.

The measurement of altitudes was carried out as accurately as possible. Since spot-heights on the maps were rare, the sea itself had to be used as a datum throughout. For altitudes less than 100 m., a hand-level was used almost exclusively. By means of various checks, the possible error involved in this method of levelling was found to be about ± 3 per cent; for a shoreline at 100 m., the estimation of the relative displacement of sea-level might thus involve a maximum total error of about ± 5 m. An aneroid barometer was also used in certain cases, though seldom for shorelines below 100 m. With this method, the possible error was found to be about ± 10 per cent.

Procedure in the field

It has already been made clear, in the description of the nature of the evidence, that no shoreline was physically continuous throughout the section of coast examined. On the contrary, as a result of erosion both subaerial and marine, and of concealment by alluvium, the shorelines were represented only by fragments, of which none were more than a few kilometres long and the majority much less.

It was suspected from the start that such a condition might be general, and the whole survey was planned on this assumption. The most obvious danger of the situation was that it might be impossible at the end of the survey to decide which fragments had originally belonged to the same shorelines. There could be no question of correlation by marine faunas. Nor would it be possible to depend upon altitudes alone, for there was no reason to assume that any shoreline would have remained horizontal since its formation. Strictly speaking, in fact, the establishment of wholly reliable correlations would be out of the question. Nevertheless, the uncertainties might still be reduced to reasonable proportions, provided that every effort was made in the field to locate and level the largest possible number of fragments. It would therefore not be enough merely to record the more conspicuous fragments of shoreline; an important part of the investigation would consist in searching for those which were ill-developed and obscure.

A completely exhaustive search could scarcely be attempted, for time was limited and it was essential that the survey should cover a really substantial length of coast. Some compromise was therefore necessary. Eventually it was decided that the investigation should be based upon a series of reconnaissances carried out at frequent intervals along lines at right-angles to the coast. During each reconnaissance, records should be made of the levels of all shorelines that might be encountered, and notes should also be made of the existence of any visible connexions between the segments of shoreline observed on adjacent reconnaissances. The position of each line of reconnaissance should then be marked on the 1:100,000 map. For the intervals between reconnaissances, a standard distance of 2 km. was decided upon; it was considered that this would be small enough for the purpose, yet large enough to allow a reasonable rate of progress.

This, in fact, was the system used throughout the 110 km. of coast. The only real departure from the system was in the matter of the 2 km. interval; more often than not, the sites of the reconnaissances were determined in practice by local variations in the state of the evidence. It should also be mentioned that levels were taken on many features other than the landward limits of the terraces, even after it had been decided that these latter features alone should be used for the estimation of former mean sea-levels. Thus, records were often made of the levels of any associated patches of beach or consolidated dune, also of all cliff-tops that were encountered; information of this kind would clearly be of considerable general interest.

INFORMATION FROM AIR-PHOTOGRAPHS

Air-photographs eventually became available for most of the areas examined, though unfortunately not until after the 1948 expedition. With the permission of the Air Ministry, the writer was able to inspect these photographs at the R.A.F. Central Print Library, and later to buy certain prints for detailed study at Cambridge.

It was found that the photographs, when examined stereoscopically, showed the details of the staircase topography remarkably clearly. This was especially so in the case of prints of 6 in. focal length, with which the disadvantages of a small scale were amply made up for by the great vertical exaggeration. By the use of such photographs, it was possible to locate many lengths of shoreline whose existence had been completely overlooked in the field, as a result either of poor preservation, poor exposure or simple inaccessibility. Apart from their own interest, discoveries of this kind were especially valuable as a means of establishing correlations between various fragments of shorelines which had previously appeared to be isolated.

In a few cases, when the photographs were suitable, attempts were made to find additional levels, on lengths of shoreline either inadequately surveyed or newly discovered, by means of an Abrams contour-finder. According to checks carried out upon points of known altitude, levels obtained in this way appeared to involve a possible error of about ±10 per cent.

5. DETAILED SURVEY (RAS AAMER—DERNA): RESULTS OBTAINED

The essential features of the results obtained, whether in the field or from air-photographs, are shown graphically in Fig. 3. In effect this diagram represents an attempt to show the position and extent of each fragment of shoreline that was located, and its altitudes above present sea-level (or, more precisely, the altitudes of the line corresponding to its contemporary mean sea-level).

The two axes represent respectively distances along the coast, measured from an arbitrarily chosen point 5 km. west of Ras Aamer, and altitudes above present mean sea-level. Points at which levels were taken are shown either as crosses or as circles; the crosses indicate those points whose altitudes are considered to be the more reliable, having been taken with a hand-level, while altitudes obtained with an aneroid or from air-photographs are shown as circles. The full lines indicate fragments of shoreline which were actually visible, either on the ground or on air-photographs; the significance of the other lines will be discussed later.

Where a fragment of shoreline terminates at a point whose position is known but whose altitude is unknown, it has generally been assumed, as a matter of convention, that the altitude is the same as that of the nearest levelled point on the same shoreline. The position of the termination has been indicated in such cases by means of a short vertical line. It should be emphasized that the higher fragments of shoreline, as a result of the use of this convention, appear on the diagram as showing a constancy of levels which almost certainly they do not in reality possess.

Before proceeding to a further discussion of the diagram, it will be useful to give a certain amount of supplementary information in verbal form. This in-

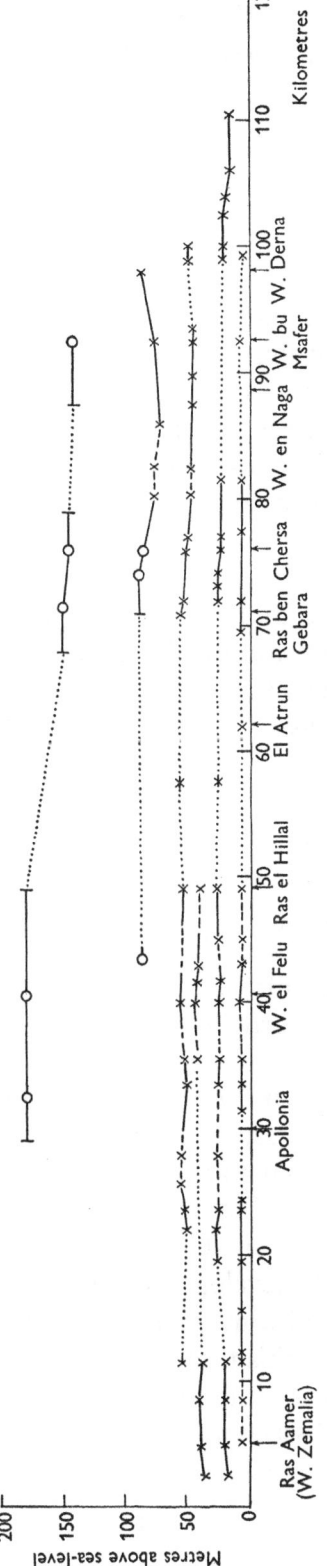

Fig. 3. Results of detailed survey of ancient shorelines between Ras Aamer and Derna

Note. The levels of the higher fragments of shoreline are only very approximately known. In all probability they are far less constant than would appear from the diagram.

formation will be mainly concerned with three matters: the nature of the ter-
minations of the various fragments of shoreline, the exact locations of the more
isolated and obscure fragments, and details of any particularly unusual features.
The evidence will be considered in the order of its occurrence, from west to east.

RAS AAMER—APOLLONIA

At Ras Aamer, only the two higher shorelines were associated with terraces.
The evidence for the third and lowest was derived solely from exposures of its
deposits, largely buried by alluvium, in the banks of Wadi Zemalia at a distance
of a few metres from the sea.

The stretch of coast between 7 and 15 km. east of Ras Aamer was one of those
from which terraces were almost entirely absent; the slope of the escarpment
was here steep and unbroken, and its foot was separated from the sea only by a
single terrace. A curious feature of this part of the escarpment was that its face
was found to be very largely covered by thin sheets of consolidated dune, from
about 20 m. to at least 100 m. above sea-level.

The scarcity of evidence for about 5 km. to the west of Apollonia could be
ascribed to heavy subaerial erosion, assisted by the presence of large masses of
alluvium at the lower levels.

APOLLONIA—RAS EL HILLAL

Between Apollonia and Ras el Hillal fragments of terrace were numerous but,
with one exception, all were rather short. This could be ascribed to some extent
to subaerial erosion; in many cases, however, it was clearly due to encroachment
by younger terraces. Alluvium again became troublesome in the last 5 km. before
Ras el Hillal. Near Ras el Hillal also the terraces themselves became wider and of
less pronounced relief. Nevertheless, a reasonably clear succession of shorelines
was found on the headland itself, along the road leading to the lighthouse.

It has been mentioned that there was one fragment of shoreline on this part
of the coast which was not especially short. This was the highest of all. The
existence of this feature was not at first recognized in the field. The maps of the
area, however, as Desio had observed,[1] indicated the presence of some kind of
terrace between 100 and 200 m. above sea-level. An examination of air-photographs showed that a well-developed terrace within this range of levels could
indeed be traced from Wadi Susa, due south of Apollonia, as far eastwards as
Ras el Hillal. The width of this terrace was remarkably great, being seldom less
than 1 km., except towards either end, where it rapidly dwindled to zero. In
spite of a high degree of dissection, it was not difficult to see that the terrace had

[1] Desio (1939), pp. 59 and 61.

originally possessed a surface with a regular and gentle seaward slope, ending abruptly towards the land against a small-scale but well-defined escarpment. All these observations were confirmed when the area was revisited in 1952.

On morphological grounds alone, there could be little doubt that this terrace also was of marine origin, and it may be added that a patch of consolidated dune was found by chance on its surface, to the south-east of Apollonia, in 1947. The landward edge of the terrace therefore represented almost certainly yet another shoreline. When plotted from the photographs on to the 1:100,000 maps, this line was everywhere found to lie between 150 and 200 m.; two points whose levels could be measured by means of the contour-finder showed an altitude of approximately 180 m., and this altitude was subsequently checked for the more westerly point by means of the aneroid.

RAS EL HILLAL—RAS BEN GEBARA

At Ras el Hillal, as can be seen from the maps, the coastal topography undergoes an abrupt change. On the east side of the headland, the present shoreline turns sharply to the south-west and south, forming the west side of the large bay of Marsa el Hillal. At the same time, it approaches more closely to the crest of the escarpment, whose average gradient is thereby very greatly increased. After a short distance, the coastline resumes its easterly direction; the escarpment, however, remains equally steep as far as Ras ben Gebara.

On close inspection, it was found that certain other and more detailed changes took place at this point. In spite of a heavy development of alluvial fans, it could be seen that the escarpment was largely free from marine terraces from here at least as far east as El Atrun; air-photographs indicated that this was also the case between El Atrun and Ras ben Gebara. Almost certainly, this state of affairs could again be ascribed mainly to the destruction of each shoreline by its successors. On the whole of this stretch of coast, indeed, only one point was found at which the staircase topography was clearly visible; this was a little to the west of Wadi Merghes. Elsewhere, the only obvious sign of ancient marine erosion was in the form of a narrow shelf, often covered with consolidated beach-deposits, which was persistently present at a height of 3–5 m. above sea-level. The shelf was almost invariably accompanied by a steep, high cliff which, being close to the present shore, had a deceptively modern appearance.

RAS BEN GEBARA—DERNA

To the east of Ras ben Gebara, the escarpment soon became less precipitous, and terraces were found to appear once more. From this point onwards, however, much trouble was caused by alluvial fans. Thus, even though many of the shorelines could be followed for great distances by visual means, it was only rarely that they were well enough exposed for accurate levels to be obtained. It was mainly for this same reason, indeed, that the two lowest shore-lines could be traced for no more than 5 km. to the east of Chersa. The discontinuities of the higher shorelines, on the other hand, were clearly due to encroachment.

For some 5 km. to the west of Derna only two terraces were present, and of these the lower was entirely hidden by alluvium. Thus there was only one shore-line which could be continuously traced along this part of the coast; its shore-deposits, at 86 m., were admirably exposed in sections on either side of Wadi Derna. To the east of Derna the situation became still more obscure. As on the coast to the east of Ras el Hillal, the crest of the escarpment again approached the modern shore, and the escarpment as a result became generally steeper; along the modern shore itself a line of cliffs developed, often of great height and always dropping straight into the sea. At the same time alluvium once more became abundant, even on the higher slopes of the escarpment.

Thus, as the combined result of encroachment and the deposition of alluvium, the coast for many kilometres on either side of Derna was found to carry few obvious traces of marine shorelines. When, however, the question arose of dating the terrestrial deposits of Wadi Derna, it was decided that special efforts must be made to locate any fragments of shoreline in this area that might still be preserved and exposed. Largely owing to the fortunate fact that many sections were available, both natural and artificial, these efforts were surprisingly successful. Since many of the fragments were found in very obscure places, they will now be described in some detail.

To the west of Wadi Derna, only one additional piece of evidence was found, in the form of an exposure on the east bank of Wadi bu Msafer (6 km. west of Derna), just above the bridge which carries the main road. The exposure itself was unusual, in that it revealed a section of what must have been either a low, short cave or a deeply undercut notch, at the foot of a fairly steep slope of bed-rock. Whatever its origin, the cavity had eventually been filled with beach-material almost up to its roof, which lay at a height of 5·5 m. above present sea-level; the small remaining space had then been filled with dune-sand, and the same material had been banked up against the rock slope at the entrance. The height of the contemporary sea-level, in this one unique case, was taken as being the height of the roof of the cavity.

The evidence obtained to the east of Wadi Derna was much more abundant. One kilometre east of the wadi, in artificial exposures on the east side of the Tobruk road, sections of two different shorelines were found. Beach-deposits were present in both cases, and both features were deeply buried beneath alluvium. Their respective sea-levels were 20 and 48 m.

The higher of the two shorelines could be traced eastwards, in various excavations, for about another kilometre. The lower had a much greater extension. Westwards, it could be traced as far as a garden on the southern edge of Derna, only 500 m. east of the wadi; still further west, it had evidently been removed by encroachment. Eastwards, it could be followed along the cliffs until, 5 km. east of the wadi, it developed into a narrow ledge, never more than 100 m. wide and usually much less, near the top of the modern cliffs. The ledge continued as far as a point 13 km. east of Derna, where it was finally obliterated by a large landslip. It was seldom found to carry beach-deposits, but was frequently covered with a layer of consolidated dune. It may be added that the levelling of this feature presented much difficulty, since its inner edge was nearly always hidden by terrestrial deposits; for the most part, also, it was inaccessible from the sea, so that the first stage of the operation required the use of a weighted tape.

Finally, a second and lower ledge was found, at a single locality, about 1 km. east of Derna, where the cliffs made their first appearance behind the modern shore. The ledge was only 2 or 3 m. wide, and was heavily obscured by slope-deposits; nevertheless, both beach-deposits and consolidated dune were found on its surface. Its height above sea-level was about 3·5 m.

6. DETAILED SURVEY: INTERPRETATION OF RESULTS, AND CONCLUSIONS

CONCLUSIONS NOT BASED UPON CORRELATIONS

As a result of its fragmentary state, the evidence available between Ras Aamer and Derna is obviously very far from perfect. Considered as it stands, it permits only the most general statements to be made concerning the number and dispositions of the shorelines represented. If more detailed conclusions are to be put forward, some attempt must be made to correlate the various fragments of shoreline. Yet, as has already been pointed out, any such attempt must necessarily introduce some degree of uncertainty.

Nevertheless, even if no correlations are made, at least two conclusions can be reached which, in spite of their very general nature, have a certain interest. First, although many of the fragments are sufficiently long to show their inclinations, if any, from the horizontal, these inclinations are in every case either very slight or non-existent. This suggests that the whole of this length of coast has been free from drastic earth-movements, unless of a strictly vertical nature, during and since the formation of the entire series of shorelines. Secondly, successions of four or even five shorelines are of widespread occurrence, whereas successions of six or more are apparently absent. In view of this and of the previous conclusion, it seems very probable that the total number of shorelines represented may not greatly exceed five.

SUGGESTED SCHEME OF CORRELATION

To the present writer, it appears that a series of correlations can in fact be proposed which stands a very fair chance of being correct. Since the fragments of shoreline themselves show no signs of violent crumpling, and since they are always fairly widely spaced in a vertical direction, it seems almost certain that any two which lie at nearly the same level, and whose horizontal separation is not more than a few kilometres, may be regarded as parts of the same shoreline. In Fig. 3, accordingly, all pairs of fragments which appear to satisfy these conditions have been joined by means of broken lines, signifying correlations which are 'almost certain'. For a correlation to be included within this category, the greatest permissible distance of separation has been fixed, somewhat arbitrarily, at 5 km. Yet even with this limitation it is found, as can be seen from the figure, that the number of such correlations that can be made is still fairly large.

These particular correlations are still too few to enable any definite conclusions to be drawn. Nevertheless, they provide very strong support for the two conclusions already reached, and this in turn suggests that further altimetric correlations may still be proposed, with a reasonable degree of confidence, even where the fragments concerned lie considerably more than 5 km. apart.

On these grounds, a second group of proposed correlations, regarded as 'very probable', has been indicated in Fig. 3 by means of dotted lines. In this case, no account has been taken of distances between fragments. Some latitude has been allowed, also, in the matter of correspondences of level; even the slight degree of warping observed on the fragments themselves might give rise to appreciable differences of level as between widely separated points on the same shoreline.

As can be seen from Fig. 3, a complete scheme of correlation has thus been arrived at. It must again be emphasized that none of the individual correlations suggested is regarded as having been proved. None, on the other hand, is regarded as being less than 'very probable'. The scheme, therefore, may perhaps be used as a reasonably secure foundation upon which to base further conclusions.

CONCLUSIONS BASED UPON SUGGESTED CORRELATIONS

Assuming that this scheme of correlation is correct, it may be said that the total number of ancient shorelines represented on this stretch of coast is six, or at most seven. Certain fairly definite statements may also be made concerning the present levels of these reconstituted shorelines. In general, as is clear from Fig. 3, there is none which departs very far from the horizontal, though all appear to show some minor undulations. To some extent, however, a false impression is given by the levels on the diagram, since all involve a certain margin of error. For this reason, a brief statement will now be given of all that can justifiably be said, in the writer's opinion, regarding the levels of each individual shoreline. A separate paragraph will be devoted to each one; the order in which they will be considered will be the order of their altitudes, from the lowest to the highest.

(i) It is only on two parts of the coast that the lowest shoreline can be suspected to depart from its normal level of 6 m.: to the east of Derna, and between Ras el Hillal and Ras ben Gebara. In both cases, the landward limit of the associated terrace stands at a level scarcely higher than 3 m. Wherever this was so, however, it was noticed that the terrace was always accompanied by an unusually steep cliff. At all such places, therefore, the conditions under which the shoreline developed must in any case have been somewhat abnormal. It is thus believed that the low levels of the terraces also may well be original but abnormal characteristics, and that this shoreline after all is in reality completely unwarped.

(ii) The second lowest shoreline stands a little below 20 m. at Ras Aamer, rises as high as 25 m. between Apollonia and Ras ben Gebara, and then sinks once more, towards Derna, to levels below 20 m. East of Derna, levels were both difficult to obtain, for reasons already given, and difficult to interpret, since the steepness of the associated cliff raises the suspicion that here again the entire platform may have been formed at an unusually great depth below sea-level. Nevertheless, its landward limit certainly drops as low as 14 m., and this would suggest a sea-level no higher than 17 m.

(iii) This shoreline, of which no fragments have been found to the east of Ras el Hillal, rises from 35 m. near Ras Aamer to 40 m. to the east of Apollonia. There is some evidence of a subsequent fall towards Ras el Hillal.

(iv) This shoreline stands at about 55 m. in the neighbourhood of Apollonia. It shows a definite drop from 53 m. at Ras ben Gebara to 44 m. near Derna, and just east of Derna a slight rise to 48 m.

(v) Levels for this shoreline are reliable only in the Derna area; elsewhere, they have been taken with the aneroid. The westward rise in the vicinity of Chersa may thus be only apparent. The rise from 70 m., 12 km. west of Derna, to 86 m. at Derna itself, is, however, well authenticated; excellent sections of the shoreline were available at both points, in a road-cutting at Derna West Pass and in the sides of Wadi Derna respectively; all altitudes were obtained by hand-level, with subsequent independent checks. This average lateral inclination of 16 m. in 12 km. (1 in 700) was the steepest recorded on any of the Cyrenaican shorelines.

(vi) Levels for the eastern section of this shoreline were obtained with the aneroid, for the western section with the contour-finder. The least margin of error is therefore ± 15 m. throughout. Hence, no statement can be made regarding the changes of level shown by either section; even the correlation between the two must be looked upon as particularly uncertain. The most that can be said is that all the fragments probably lie between 140 and 190 m., and that if all are parts of the same shoreline it is probable, though not certain, that this shoreline rises towards the west.

If it is still assumed that the shorelines have been correctly pieced together, something may now be said concerning the manner of their formation, and their subsequent history. In the first place, since every shoreline is associated at some point or other with a terrace whose width amounts to hundreds of metres, and since the bedrock must everywhere be highly resistant to marine erosion, it is clear that each shoreline must represent a very long period of stability, both of land and of sea. The present terraces, moreover, are no more than the stubs of the original wave-cut platforms, and their present widths can only be fractions of their original widths. These original widths, however, are purely a matter for speculation; for this reason, it is impossible to make any estimates of the durations, either absolute or relative, of the periods of stability.

For the present state of elevation of each individual shoreline, two possible explanations can be offered. On the one hand, the entire shoreline may have been raised as a result of tectonic processes. On the other hand, the sea-level itself may have dropped, the effects of earth-movements being confined to the production of small-scale undulations. There appears to be no final argument against either of these alternatives. In the writer's opinion, however, it is scarcely conceivable that even 10 km. of coast, let alone 110 km., could be shifted bodily in a vertical direction without itself being very considerably warped or tilted. Yet no shoreline appears to show signs of any such treatment, except in a very minor degree. It is therefore concluded that the latter alternative is correct. In other words, it is believed that the shorelines represent a series of stages, each one lower than the last, at which the sea-level halted before reaching its present position.

Since warping, though slight, was not altogether absent, it is impossible in most cases to make exact estimates of the levels at which the six (or seven) reconstituted shorelines might originally have stood. Nevertheless, there can be little doubt that the lowest represents a mean sea-level very near 6 m. As for the remainder, it seems most probable that each would originally have stood at a level intermediate between its present extreme limits of variation; these limits are 15–25, 35–40, 44–55, 70–90 and 140–190 m. respectively. As has already been said, the existence of the highest of these shorelines is regarded as less certain than that of the others; there may in reality be two separate shorelines between the limits given.

Finally, some mention must be made of the nature of the warping, such as it was. There can at least be no doubt that it affected all parts of the coast, from Ras Aamer to far beyond Derna. If the suggested correlations are accepted, it then appears that the amplitude of distortion increases with the age of the shore-lines; in other words, that the movements were more or less continuous until after the second lowest of these features had been formed. It appears also that the general tendency of the movements was towards a slight but persistent up-ward warping between Ras Aamer and a point about 10 km. east of Chersa, the shorelines being at their highest levels between Apollonia and Ras ben Gebara. Further east, in the Derna area, the movements seem to have been more con-siderable but less systematic; warping seems to have occurred in different direc-tions at different times.

7. MISCELLANEOUS OBSERVATIONS

A certain amount of information has been gathered concerning the remainder of the north Cyrenaican coast. It is only to a small extent based on first-hand observations; most of it has been obtained from air-photographs and maps. Nevertheless, in the light of the information already discussed, it suggests some interesting conclusions.

DERNA—RAS ET TIN

The most westerly part of this stretch of coast has already been described. Mention has been made of the existence of a large landslip, at a point some 13 km. from Derna. According to air-photographs, landslips are continuously present for a distance of many kilometres to the east of this point. Beyond the land-slips, cliffs begin once more, themselves representing the whole of the coastal escarpment. At first they are over 100 m. high, but they gradually decline eastwards, finally dying out at Marsa Belaghigh (45 km. east of Derna). As far as can be seen from the photographs, the cliffs are almost sheer, with no associated marine terraces, and plunge straight into the sea. The headland of Ras et Tin shows nothing but dunes, probably for the most part consolidated.

The most interesting feature of this stretch of coast, and one which appears to be unique, emerges from a study of the Admiralty chart[1] of the adjacent sea-floor. Off all other parts of the Cyrenaican coast, depths of water show a gentle

[1] No. 3355.

and more or less regular increase over the whole of the distance between the
modern shore and the edge of the continental shelf. Between Derna and Ras et
Tin, on the other hand, great depths of water are shown even by the soundings
closest to the land. As little as 9 km. east of Derna, a depth of 13 fathoms (24 m.)
is shown immediately offshore. 35 km. from Derna, the depth in a similar
position is given as 21 fathoms (38 m.), and depths of this order, with a maximum
value of 40 m., continue as far as Ras et Tin. Along the whole of this stretch of

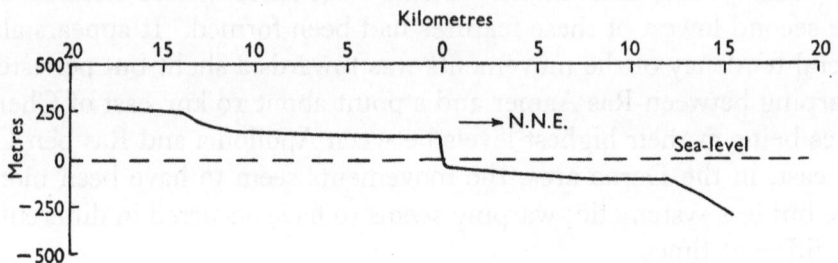

FIG. 4a. Coastal profile 35 km. east-south-east of Derna

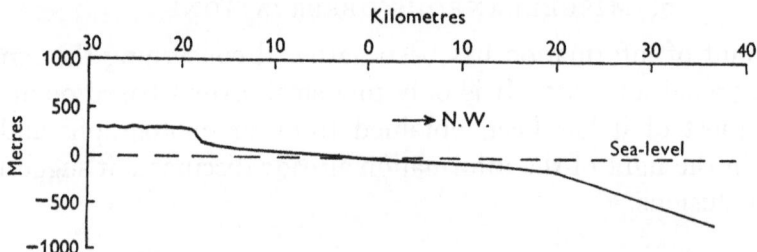

FIG. 4b. Coastal profile 20 km. north-east of Benghazi

Levels taken from Admiralty Charts Nos. 3354 and 3355, and 1:100,000 maps.
Vertical exaggeration: × 10.

coast, in other words, the escarpment itself is partly submerged. The increase in
depth does not, indeed, continue at the same rate for any great distance from the
present shore; the initial drop gives place almost immediately to a broad continental shelf, as shown in Fig. 4a. Nevertheless, it can be said that the submergence of the escarpment extends, in some places, to about one-third of its
total height.

On the assumption that the escarpment has remained as stable in this area as
it has further to the west, its present condition could hardly be explained except
as the result of marine erosion during some period of eustatic low sea-level. This,
however, is almost inconceivable. It appears, therefore, that this is the one part of
the coast where considerable Pleistocene earth-movements must have occurred.

At present there is little evidence regarding their exact mode of operation.

There seem to be two possibilities: a simple downward tilting towards the east, or the formation of a fault-scarp which, on account of its steepness, has been able to resist extensive marine erosion. It is not yet possible to decide between the two.

As regards the dating of these movements, it can at least be said that they must have preceded the second youngest shoreline. The level of this feature is not, indeed, perfectly steady. Nevertheless, even though it continues well into the area where the escarpment is partially submerged, it is not known to drop by more than a few metres (see Fig. 3).

RAS AAMER—TOLMEITA

For 15 km. to the west of Ras Aamer, the coast has been examined only on air-photographs. Little could be distinguished in the way of marine terraces; the escarpment, besides becoming lower and less steep, seems to have suffered considerably from subaerial erosion.

Still further to the south-west, a personal reconnaissance has been made in the vicinity of Wadi Giargiarummach. This is the area in which the top of the escarpment descends almost to sea-level. It was found, in fact, to be reduced to a single steep slope, perhaps 20 m. high. Near its foot was a discontinuous band of beach-conglomerate which, in sections in the sides of the wadi, could be seen to rest upon a narrow platform whose inland edge stood at 6·5 m.

The remainder of the coast, almost as far as Tolmeita, is known only from air-photographs. 6 km. west of Wadi Giargiarummach a line of cliffs begins, dropping straight into the sea; this again appears to represent the whole of the escarpment. Just before Meghiunes (19 km. west of the wadi) there is another belt of landslips. West of Meghiunes, the escarpment resumes its normal form. At first it is extremely steep and apparently without terraces. Then, within 15 km. of Tolmeita, the gradient begins to slacken and terraces are once more visible.

From here onwards, the bedrock is largely hidden by alluvial fans. At Tolmeita itself, however, a certain amount of information could be obtained in the course of a personal visit. 1 km. east of the town, in the banks of Wadi Zuiana, a wave-cut platform with beach-deposits could be traced upstream from the modern shore as far as its termination, 6·4 m. above sea-level, at the foot of a low cliff; all of these features were completely buried beneath alluvium and consolidated dune. Further upstream, a second beach-deposit was found, buried beneath alluvium but resting on bedrock, at an altitude of 33 m.; within this deposit were remains of *Clanculus*, *Bittium reticulatum* and abundant *Patella caerulea*. To the west of the town, fragments of two well-developed terraces could be distinguished. Both were encumbered with consolidated dune, but the lower of the two indicated a shoreline at an approximate level of 20 m.

39

CHAPTER II

TOLMEITA TO ANTELAT

I. TOPOGRAPHY AND GEOLOGY

Immediately to the south-west of Tolmeita, as has already been stated, the coastal escarpment begins gradually to retreat further inland, while at the same time a coastal plain comes into existence between the escarpment and the sea. A general idea of the nature of this new kind of topography can be obtained for the northern part of the area in question from the 1:100,000 maps. These maps extend southwards as far as a line about 15 km. south of Benghazi; their contours, in this particular region, are drawn at intervals of 10 m.

Within this part of the area, it can be seen that the escarpment itself has an average gradient which is often steeper than at most points to the east of Tolmeita. This gradient, however, now slackens off abruptly along a well-defined line; the escarpment, in other words, can here be said to possess a definite foot. The altitude of this line, at any rate from Tocra southwards, remains always between 150 and 200 m. above sea-level. The levels of the crest of the escarpment are somewhat more variable, but generally lie between 250 and 300 m. As on the north coast, the continuity of the escarpment is still broken at frequent intervals by deep gorges.

The strip of country between the foot of the escarpment and the sea is, of course, an area whose average gradients are gentle. From the maps they appear also to be generally regular. Almost anywhere in this area the contours show a remarkably even spacing, scarcely varied except by a progressive increase in width as the escarpment withdraws further and further inland.

There is in fact only one locality between Tolmeita and the southern limit of the area covered by the 1:100,000 maps where the regularity of the contour-lines is seriously disturbed. This lies 18 km. to the east of Benghazi, immediately to the west of the village of Benina. The gentle seaward slope of the ground is here interrupted by a step, running from north to south. Though hardly a conspicuous feature, the step can be traced on the map for a distance of about 10 km.; towards either end it loses its steepness and merges into the regular slopes of the surrounding country. The foot of the step maintains an altitude of about 90 m. above sea-level, while its crest lies between 100 and 120 m.

It is obvious, in spite of the presence of this step, that the entire strip of country, unlike the Sahel of the north coast, must be regarded as a distinct morphological feature. It is obvious also that its average gradients are sufficiently low to justify its being described as a 'coastal plain'. A typical section across the escarpment and the coastal plain, together with the adjacent sea-floor, is shown in Fig. 4b.

No detailed survey is available of the area to the south of that covered by the 1:100,000 maps. The small amount of information which exists concerning its topography indicates, however, that no great changes take place. The escarpment appears to remain steep and well defined; according to levels taken by Desio, its foot continues to lie at altitudes between 150 and 180 m. above sea-level, while its crest declines gradually towards the south until in the vicinity of Antelat the whole feature ceases to exist.[1] The average gradient of the plain, also, appears to retain its regularity, while, at the same time, the width of the plain continues to increase until it reaches a maximum of 50 km., at a distance of 50–60 km. to the south of Benghazi.

The solid geology of the west coast of Cyrenaica, and of its hinterland, is in many respects very similar to that of the north coast. The rocks are still of Tertiary age, and dips remain low, except that a faulted monocline, with down-throw to the north-west, follows the course of the escarpment between Tolmeita and Benina. There is also one other notable difference. A few kilometres to the south-west of Tocra, Miocene rocks descend to sea-level; from this vicinity southwards the rocks of both the plain and the escarpment are entirely of Middle Miocene age, and, though still mainly limestones, are much less uniform and compact than the Eocene and Cretaceous rocks of the north coast.

2. EVIDENCE FOR MARINE ACTION

Various patches of Pleistocene marine deposits have long been known to exist in the coastal plain of western Cyrenaica, indicating at least some degree of submergence. A certain amount of evidence for marine erosion has also been generally admitted. So far, however, no definite conclusions have been reached as to the number of times the sea advanced, or the heights which it attained. There has certainly been no agreement over the suggestion, already mentioned, that both the plain and the escarpment may themselves be wholly the products of marine erosion.

The present writer has only been able to visit a small part of this very large region, and has, indeed, confined his activities almost entirely to the area north of Benghazi. Nor has any attempt been made, as on the north coast, to carry out a systematic investigation. Nevertheless, the information obtained, taken in

[1] Desio (1939), p. 65, fig. 13.

conjunction with that which has already been published, has led to a number of conclusions regarding the physiographical history of the region.

In view of the relatively haphazard way in which the evidence was collected, and of its varied nature, it would scarcely be possible for it to be discussed according to the same plan as was used in the previous chapter. It has been decided that the subject can most conveniently be dealt with under three headings. First, the whole of the existing geological evidence, both old and new, will be presented in full, with the exception of that from the immediate vicinity of Benghazi; evidence of this type is given priority because its significance, though often very limited, is at least unequivocal. Secondly, the morphology of the coastal region will be discussed, so far as it is known, with a view to deciding the extent to which it can be ascribed to marine erosion, and the state of the sea-level while such erosion was in progress. Thirdly, a separate section will be devoted to the Pleistocene geology of the Benghazi area, since this is not only complicated but also highly abnormal.

GEOLOGICAL EVIDENCE

The geological evidence will be considered in the order of its geographical position, from north to south.

Tocra Area

5 km. to the south-west of Tocra, the writer found many sections on the modern shore which revealed the presence of a bed of coarse shell-limestone, lying between the coastal consolidated dunes and the underlying Eocene bed-rock. It was never more than 1 or 2 m. thick, and lay very near sea-level.

Marine Pleistocene deposits have also been reported as extending in a continuous sheet from Bersis (9 km. south-west of Tocra) to Driana.[1] It is suspected, however, that some of these deposits, though undoubtedly of marine origin, were in fact laid down by the wind. The area is characterized by large numbers of hillocks, clearly marked on the 1:100,000 maps, and many of these, if not all, appear to be fossil dunes.

Coefia

At Coefia (14 km. north-east of Benghazi), the ground is pitted by a group of *dolinas*, or large natural depressions, in which excellent sections are available of both the Miocene bedrock and its overlying deposits. This is one of five Cyrenaican localities where Stefanini was able to collect a marine Pleistocene fauna, later to be identified and published by A. C. Blanc.[2]

[1] Stefanini (1935), p. lxxii.   [2] Blanc, A. C. (1936), pp. 279–82.

In this case, the fauna was obtained from an exposure in a *dolina* 'to the south-east of Ain Zeiana'; twenty-eight species of mollusca were identified, all still living in the Mediterranean. Further stratigraphical details were obtained by the present writer, who noted that the deposit in question was a shell-limestone, not more than 0·5 m. thick, resting on the bedrock and covered by about 1 m. of consolidated dune. Near the main road the bedrock platform lay at about 3 m. above sea-level; traced inland, it rose to 6 m., while the shell-limestone at the same time became less and less distinguishable from the overlying dune. No shoreline was definitely located, but it seemed very probable that the altitude of 6 m. indicated the maximum height of this particular transgression.

South of Benghazi

South of the latitude of Benghazi, Desio's geological map shows a coastal strip of Pleistocene limestone, narrow towards the north but broadening southwards until it includes the whole of the coastal plain. Verbal descriptions of these deposits are, however, very scanty. Pleistocene marine deposits are found, according to Desio,[1] in wells and excavations along the road from Benghazi to Ghemines (52 km. to the south). 3 km. south of Sidi Magrun (75 km. south of Benghazi) is one of the fossiliferous deposits mentioned by Blanc;[2] as far as can be gathered from the maps, the altitude here would be about 20 m. above sea-level. It is conceivable, however, that this deposit, also, may be wind-blown; the present writer has noticed the almost continuous occurrence of dune-lime-stones along the road from Sidi Magrun southwards to Agedabia; they are easily recognizable as such by their bedding, which is visible in road-cuttings, and by their undulating surfaces.

East of Ghemines, Desio has reported a limestone, with recent Mediterranean molluscs, extending as far east as a point half-way between Tilimun and Soluch.[3] This deposit must reach an altitude of over 50 m. above sea-level.

Molluscan faunas have been collected from marine Pleistocene beds in the immediate vicinity of Agedabia.[4] According to Desio, the altitude of the beds is 36 m.; the maps, however, give the altitude of Agedabia itself as 5 m., and the surrounding country is notably flat.

Finally, a Pleistocene deposit a little to the east of Mersa Brega is mentioned by Marchetti as being of brackish-water facies, with *Cardium* and *Cerithium*.[5] A Pleistocene fauna from the same vicinity is also mentioned by Blanc.[6] No

---

[1] Desio (1935), p. 87.   [2] Blanc, A. C. (1936), pp. 279–82.
[3] Desio (1935), p. 85.
[4] *Ibid.* pp. 87–9, 347–8; Blanc, A. C., *op. cit.*
[5] Marchetti (1934), p. 320.   [6] Blanc, A. C., *op. cit.*

altitudes are given in either case, but the whole area, except for its consolidated dunes, is very low-lying.

MORPHOLOGICAL EVIDENCE

Form of bedrock surfaces

For two reasons, a mere examination of the 1:100,000 maps could not by itself enable any definite conclusions to be drawn regarding the effects of marine erosion in the western coastal regions. First, small but important irregularities of gradient might exist which, owing to the spacing of the contour lines, might yet have remained unrecorded; this was especially likely to have been the case on the face of the escarpment itself. Secondly, even if the surface of the ground should be more or less as shown on the maps, its form might still have been controlled to some extent by the accumulation of material rather than by erosion.

The writer had many opportunities of investigating these two possibilities in the course of various journeys in the northern part of the coastal plain. As regards the first possibility, it soon became clear that the simplicity of the coastal topography was by no means only apparent. The escarpment, wherever it was inspected, was always found to be free from subsidiary terraces. As for the plain, only one interruption of its gradient was found which did not appear on the 1:100,000 maps. This took the form of yet another step, cut in the solid rock, running parallel to and about 1 km. from the modern shore in the vicinity of Tocra. Though not steep, and scarcely more than 2 m. high, it could be traced for 8 km. to the north-east of the town, and for at least 10 km. to the south-west. The whole of this feature lay on the seaward side of the 10 m. contour-line; the altitude of its foot was measured at two widely separated points, and was found in each case to be 5 m. above sea-level.

It was already fairly certain, from the reports of previous writers, that the surface of the coastal plain was in fact to a large extent composed of bedrock, and that it was mainly erosional in origin. This impression was amply confirmed. The cover of marine Pleistocene deposits was usually very thin, and often altogether absent. Continental Pleistocene deposits had, indeed, produced a noticeable effect on the topography in certain places. Thus, consolidated dune-deposits sometimes formed gently undulating surfaces of great extent, and also formed chains of hills rising to considerable heights; such hills, however, were mostly confined to a narrow belt adjacent to the modern shore. Alluvial fans, also, covered certain large areas of ground, and could attain great thicknesses, particularly at the foot of the escarpment; for this reason, indeed, the bedrock was completely invisible over the greater part of the coastal plain between Tolmeita and Tocra. Nevertheless, most of the areas examined showed a complete absence of Pleistocene deposits of any kind.

As a result of these observations, it thus appeared, on the one hand, that most of the essential features of the northern part of the coastal plain were in fact perfectly adequately shown on the 1:100,000 maps; on the other hand, that these features were for the most part cut in the bedrock. Generally speaking, therefore, only two surfaces had to be considered: the steep, unbroken slope of the escarpment, and the very gentle but almost equally unbroken slope of the coastal plain. Additional surfaces were present, so far as was known, in two areas only, where the slope of the plain was broken by the local development of more or less abrupt steps.

No new field-observations were made to the south of the area covered by the 1:100,000 maps, and little can be added to the general topographical information already given. In this region, however, certain parts of the escarpment have been examined on air-photographs, and have been found to be as free from terraces as the part between Tolmeita and Benina. It should also be mentioned that the only known topographical irregularities in the southern part of the plain are in the form of occasional rock-cut steps, a metre or so in height.[1] Finally, as regards the distribution of superficial deposits, it appears from various previous accounts that much of the surface of the plain is still composed of bedrock, and that Pleistocene deposits, where present, still generally occur only in thin layers. Thus it seems that any general conclusions reached concerning the physiographical history of the mapped area can justifiably be extended, with due reservations, to the area which is still relatively unknown.

Origins of bedrock surfaces

The nature of the coastal topography having been described so far as it is known, it will now be possible to discuss its origin. It will be convenient first of all to discuss the origin of the Tocra and Benina steps.

These features can only be explained either as products of differential subaerial erosion, in which case the areas between each step and the sea must be structural surfaces, or as products of marine erosion, in which case the same areas must be wave-cut platforms. For the Tocra step, the first alternative is ruled out by the fact that the underlying bedrock has a steep seaward dip; confirmation that the feature is a marine cliff is provided by the constancy of its levels, and by the presence of the marine limestone exposed on the modern shore. It is not so easy to eliminate the possibility that the Benina cliff may have been formed by subaerial erosion, since the bedrock on its seaward

[1] Desio (1939), pp. 46–7.

side appears to have roughly the same inclination as the surface of the ground.[1] Moreover, it is not known to be associated with any marine Pleistocene deposits. Nevertheless, the constant levels of its foot, so far as they can be determined, indicate that this also is a marine cliff. Such a conclusion has in fact already been generally accepted.[2] These two features therefore provide the first real indication that the sea not only invaded this region, but also that its level remained stationary for considerable periods of time at various positions higher than that which it occupies at present; in other words, that the relative levels of land and sea may have behaved in much the same way here as on the north coast of the Gebel.

The remainder of the plain, excluding the areas between the steps and the sea, can now be considered as a whole, together with the escarpment. The first question to be discussed is whether the escarpment really marks the original limit of the plain, or whether, in view of its association with the Tolmeita-Benina monocline, it may not merely represent a line along which a far larger surface has been tectonically dislocated. The latter possibility was suggested by Desio, the main evidence for the theory being the undoubted correspondence which exists in the area east of Benghazi between the slopes of the ground and the dips of the Miocene bedrock, both above and below the escarpment.[3]

An obvious objection to this theory is that the plain, at least from Tolmeita to Driana, actually cuts across the edges of steeply-dipping beds involved in the monocline. This fact is ascribed by Desio to subsidiary marine erosion,[4] but two other objections can also be raised. First, it seems incredible that the level of the foot of the escarpment, if this is a tectonic feature, should not vary by more than 50 m. in a distance of 200 km., from Tocra to Antelat. Secondly, no evidence has yet been produced to show that the faulted monocline continues for any distance beyond Benina.

On the whole, therefore, it seems far more likely that both the escarpment and the plain in their present forms are wholly due to some erosional process, and that the escarpment is associated with the monocline simply because the latter structure presented some kind of increased resistance. If so, only three possible origins can be suggested for the plain itself. It may be largely a structural surface, exposed as a result of differential subaerial erosion; it may also be either a pediment, or wholly a product of marine erosion.

Since so little is known of the geology of the coastal plain, the first alternative cannot be finally excluded; on the other hand, the only known evidence in its favour, the correspondence of gradients to the east of Benghazi, comes from a very small area. The second alternative can definitely be dismissed, in view of

[1] See Marchetti (1938), p. 207, fig. 40 and p. 209, fig. 41.   [2] Desio (1939), p. 65.
[3] *Ibid.* pp. 63–7.   [4] *Ibid.* p. 67.

the apparent absence of 'rock-cut fans', the regularity of the plan of the escarpment, and the abruptness of its foot.

By far the most likely theory is that the entire plain was formed by marine erosion. All the available morphological evidence suggests, in fact, that the plain and the escarpment are respectively a wave-cut platform and its associated cliff, both features having been formed almost in their present state during a single period of high sea-level. For this theory, the constant levels of the foot of the escarpment provide very strong supporting evidence.

If the theory is correct, most of the plain must be older than the Benina and Tocra steps, which thus represent the total effects of subsequent marine erosion. It remains to inquire why these steps should be of such small extent and so few in number. The most reasonable answer seems to be provided by the great width and low gradients of the plain itself. Any waves attacking such a surface would already have had to travel across a wide expanse of relatively shallow water, and their remaining energy would be low. Under these circumstances, however long the sea-level remained stable at any one altitude, marine erosion might well be completely inhibited. As will be seen in the following section, there are indeed many indications, in the Benghazi area, of the existence of shore-lines unaccompanied by any morphological feature.

3. THE PLEISTOCENE GEOLOGY OF THE BENGHAZI AREA
HISTORICAL

Topographically, the Benghazi area is no different from any other part of the coastal plain, except perhaps that its 10 m. contour-line lies unusually far from the sea. For a long time it appeared as though its geology also was perfectly normal. It was recognized by Gregory[1] that Miocene limestones were exposed in the vicinity, and that these were covered by a younger limestone to the east of the town; the fauna of the latter, according to R. B. Newton,[2] was post-Pliocene and marine, a hint of brackish conditions being provided by the presence of *Paludestrina*. At Benghazi itself, a third and still younger limestone was found, the other deposits having here sunk below sea-level;[3] this was evidently the limestone forming the main coastal consolidated dunes, features which are well developed in this area next to the modern shore. Subsequently, details of a single fossiliferous exposure of a marine Pleistocene bed in a quarry 5 km. east-south-east of Benghazi were provided by Desio. The bed consisted of about 20 cm. of sandy limestone, resting on the Miocene floor; nine species of molluscs were found, all probably still living in the Mediterranean.[4]

[1] Gregory (1911), p. 592.   [2] Newton (1911), pp. 622–5.
[3] Gregory (1911), p. 593.   [4] Desio (1935), pp. 79–80.

Complications were first revealed when a series of boreholes was sunk in connexion with the water-supply of Benghazi. It was then found that the Pleistocene limestones in certain places were underlain, not by the usual Miocene limestones, but by a series of deposits in which clays and marls predominated; one

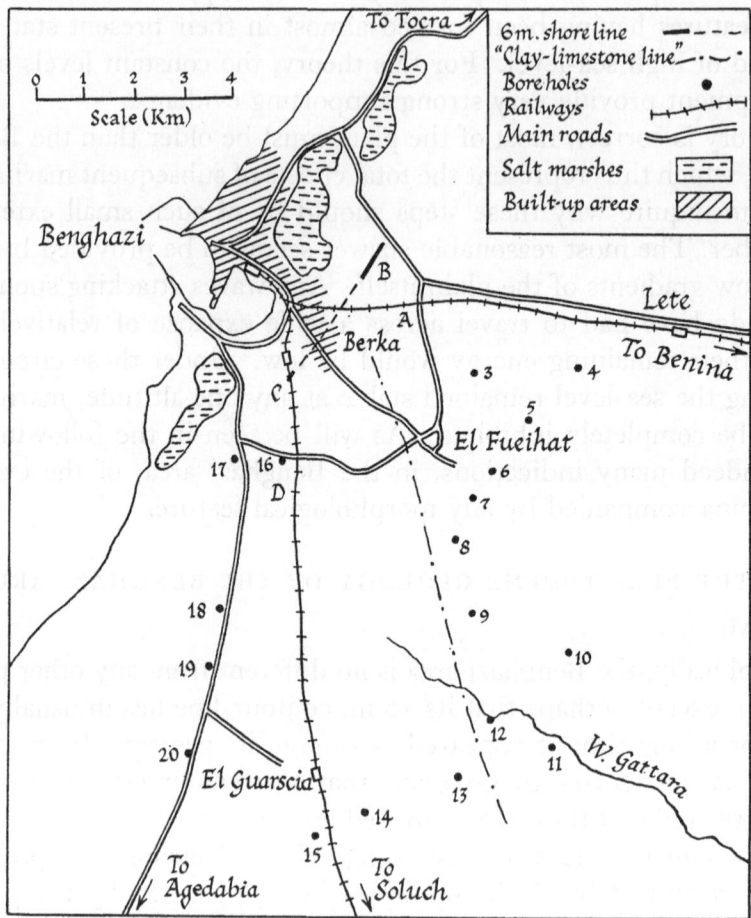

FIG. 5. Map of Benghazi area, showing positions of boreholes mentioned in
Marchetti (1938) and of open sections examined by the present writer

of the boreholes was still in clay at a depth of 140 m. below sea-level. These argillaceous deposits were at first assigned to the Pliocene and Pleistocene.[1] Later, however, after a re-examination of much of the material, Marchetti was able to show that the real situation was both different and more complicated.[2]

[1] Amato, A., in Romano (1933), pp. 66–8 (quoted in Marchetti (1938), p. 216, n.).
[2] Marchetti (1938), pp. 205–27.

48

In Marchetti's work, details are given of the contents of the boreholes and of their positions; the latter are shown here in Fig. 5.[1] Bore-holes Nos. 3–12 reveal a normal state of affairs, with a thin and discontinuous sheet of Pleistocene lime-stone resting on a floor of limestone of Langhian (Lower Miocene) age. No. 20 did not penetrate below the superficial Pleistocene limestone. It was in Nos. 13–19 that the anomalous deposits were found. The records of these boreholes are as follows; the first column gives translations of Marchetti's own lithological descriptions (with certain minor omissions), the second column the altitude above sea-level, in metres, of the upper surface of each bed:

LITHOLOGY	Max. alt. (m.)	LITHOLOGY	Max. alt. (m.)
No. 13:		*Sandy Quaternary limestone	0
Terra rossa	27	with marine molluscs	
Sandy Quaternary limestone	21	White earthy clay and limestone	−2
Yellowish clay	18	Greenish clay	−5
Pebbles of coarse limestone	15·5	Compact bluish clay with sandy	−80
Green clay	14	cemented lenses (lowest level:	
Sandy Quaternary limestone	12	−141 m.)	
*Greenish sandy marl with	9		
carbonaceous remains		*No. 17:*	
Green clay	0	*Terra rossa	13
Green sandy clay (lowest	−2	Greenish-yellow clay with	12·5
level: −9 m.)		*H. costata* (lowest level: +3 m.)	
No. 14:		*No. 18:*	
Terra rossa	21	Terra rossa	13
Coarse limestone (Quaternary	20	*Sandy Quaternary limestone	12
sand)		Greenish-yellow clay with	2·5
*Yellow clay	16	*H. Costata*	
Greenish-blue clay, marine	−2	Green clay, with crystals of	−4
(lowest level: −24·5 m.)		celestine at base (lowest level:	
		−9 m.)	
No. 15:		*No. 19:*	
Calcareous Quaternary sand	18·5	Terra rossa	13
*Yellowish sandy clay	17·5	Sandy Quaternary dune-	11·5
Greenish-blue clay (lowest	2·5	limestone	
level: −13·5 m.)		* ? Pebbles (Miocene limestone)	8·5
		Greenish clay with *H. Costata*	4
No. 16:		(lowest level: −1·5 m.)	
Terra rossa	12		
Yellowish-green clay	8	*No. 20:*	
Earthy greenish-yellow and	3	(Sample missing)	13
reddish clay with fragments		*Sandy Quaternary limestone	6
of *Cardium*		(lowest level: +5 m.)	

[1] Based mainly on Marchetti's plan (*op. cit.* fig. 39). In this plan, however, boreholes 17–20 are scattered widely over the area, whereas in the text they are assigned to definite positions along the Agedabia road (p. 214). The latter positions are evidently correct, since they agree with these bore-holes being included under the heading 'Zona di Er Rahaba-Castellaccio'. If so, their levels, which in these four cases are not mentioned, would be about 13 m.

In boreholes Nos. 13–16, according to Marchetti, the lower parts of the argillaceous deposits were entirely without fossils, but the upper parts were certainly of Pleistocene or Recent age. The lowest beds which he appears to have included within this last category have been marked in each case with an asterisk. It will be noticed that they are of rather mixed facies, conglomerates and even lime-stones being represented in addition to clays; Marchetti considered, moreover, that the yellow clays of Nos. 14 and 15 were either fine alluvium or, more likely, brackish lagoon deposits, and that the limestone pebbles of No. 13 were probably alluvial.[1]

In the clays of Nos. 17, 18 and 19, however, Marchetti found the foraminifer *Heterostegina costata*. This species had already been identified from a single locality 4–5 km. south-south-west of Benghazi by Silvestri, who regarded it as indicative of a Middle Miocene (Helvetian) age.[2] This, therefore, was the age of the clays in Nos. 17–19, and in all probability the age of some of the unfossiliferous clays of Nos. 13–16.

As can be seen from Fig. 5, the boreholes in which the argillaceous deposits were found are entirely confined to the western side of the area; moreover, the apparent disappearance of the Lower Miocene limestones is abrupt, and takes place across a fairly straight line running south-south-east from Benghazi. In Marchetti's view, this line represents a fault throwing down to the west.[3]

As regards the relative ages of the various post-Miocene deposits, there was little that could be said for certain. It was at least clear that the terra rossa in each borehole was part of one single bed, representing the latest sediments laid down in the area by wadis issuing from the Gebel. For the remainder of the deposits, Marchetti could only suggest a tentative scheme, based upon lithology. Thus, on the assumption that all the Pleistocene limestones were equivalent in age, he arrived at the inference that the *Cardium* clays of borehole No. 16 were younger than the limestones of the other boreholes, and younger even than the coastal consolidated dunes.[4]

NEW OBSERVATIONS

In view of Marchetti's results, the present writer decided in 1948 to search the Benghazi area for open sections, from which it was hoped to obtain still further information about the local Pleistocene succession. In particular, it was hoped that some connexion might be found with the situation revealed on other parts of the coast of Cyrenaica. The localities visited are described below, together with the conclusions which they suggest; their positions are shown in Fig. 5. The order of arrangement is purely one of convenience.

[1] Marchetti, *op. cit.* p. 224. [2] Silvestri (1929).
[3] Marchetti, *op. cit.* pp. 222–3. [4] *Ibid.* p. 224.

Locality A

This site comprises a group of quarries in the south-west angle between the Benina railway and the by-pass road round Benghazi. It was in one of these that Desio found his fossiliferous Pleistocene limestone, the fauna of which was as follows: *Cardium lamarcki, Loripes lacteus, Fragilia fragilis, Mytilus minimus, Ostrea* sp., *Cerithium* cfr. *vulgatum, Nassa costulata, Rissoa parva* and *Bittium reticulatum*. To this list the present writer can add *Conus mediterraneus*.

A similar deposit can be seen in each of the quarries, its thickness varying up to about 1 m. The most interesting new observation is that it contains in some places abundant rounded pebbles. This fact, together with the absence of the deposit on the landward side of the quarries, suggests that a shoreline was not far away. The nearest break of slope, however, is at the Benina step, 14 km. to the east. This shoreline, therefore, must be one of those at which the sea was unable to carry out any erosion. Strictly speaking, its exact position cannot be determined, since the position of the landward boundary of the deposit may well have been partly determined by subaerial erosion. All that can be said is that the shoreline stood a little above 15 m., the level of the surface of the ground at this locality.

It may be mentioned that Pleistocene limestones are found in the boreholes further to the south up to a level of 20 m.[1] According to the writer's own observations, however, there is a strong possibility that these beds may be partly of aeolian origin. This would certainly account for their remarkable thickness of 15 m. in borehole No. 3, only 2 km. to the south-east of the present locality.

Locality B

The exposures included under this heading were provided by a line of small excavations, running south-westwards from the Benina road for about 1 km. Here again the Miocene limestone was overlain by a fossiliferous Pleistocene deposit, similar in many ways to that which was seen at Locality A; it was clear, in fact, that the two deposits had once been continuous, since numerous patches of a similar material were found on the surface of the ground between the two localities.

Certain changes, however, had occurred. The material had now become a pale grey, greenish or yellowish marl. The bivalve shells, among which *Cardium lamarcki* and *Fragilia fragilis* now predominated, were usually in a perfect state of preservation, a fact which, together with the lithology, indicated deposition in very quiet water. In addition, abundant ostracods and shells of *Paludestrina*

[1] Marchetti (1938), p. 223.

were found, suggesting brackish-water conditions. Finally, the thickness was now slightly greater, up to 1·5 m., but the general level was much lower, the Miocene limestone floor being only 4 m. above sea-level in the most south-westerly excavation.

An entirely new feature was the presence of a second Pleistocene deposit, overlying the marl. This was a coarse, buff-coloured conglomerate, with numerous shells of *Glycimeris glycimeris*; other shells were represented mainly by rolled fragments, but it has been possible to identify the following among the specimens collected: *Arca noae*, *Cardium edule*, *Dosinia lupinus*, *Loripes lacteus*, *Murex* sp., *Natica* sp. This fauna was quite different from that of the marl, and the lithological difference was very marked also. The deposit, in fact, was evidently a beach-conglomerate, belonging to some later transgression of the sea. Its average thickness here was about ·5 m., and its highest level about 6 m. Since the deposit scarcely extended further east than the line of excavations, it was considered that this level must represent the approximate maximum height of the transgression.

Locality C

About 400 m. south of the ruins of Berka station, on the Benghazi-Soluch railway, the following section was obtained on the west side of a shallow cutting:

LITHOLOGY	Altitude of upper surfaces (m.)
3. Consolidated dune (continuous with the main coastal fossil dunes)	7·0
2. Conglomerate with pebbles of Miocene limestone, up to 20 cm. in diameter, some with worm-holes; also numerous rolled shells of *Glycimeris*. Matrix of buff-coloured calcareous sand, partly aeolian. Petered out a few metres to the south	6·0
1. Grey and green marl with bands of shells, many with both valves united. Upper surface hardened in some places to a depth of up to 3 cm. by development of a calcite cement	5·5

In a nearby excavation in the same cutting, layer 1 could be traced downwards to an altitude of no more than 3 m. above sea-level; its base was not seen. At this low level it was found to be interrupted by layers of finely laminated, pale brown silt. These contained no fossils, except for ferruginous impressions of plant-remains, unfortunately unidentifiable, and innumerable impressions of rootlets at right-angles to the bedding.

Except for the silt and the consolidated dune, the succession was identical, faunally and lithologically, with that of Locality B. It thus appeared, first, that all of the Pleistocene marine deposits so far observed in the area must be older than the main coastal fossil dunes; secondly, that the lower part of the grey marls must here have been laid down in water which from time to time was virtually fresh; and thirdly, that this water, in view of the presence of rootlets, must have been very shallow, which implies a sea-level no more than 3 m. higher than at present.

Locality D

The exposure at this locality was provided by a trench, 150 m. long from east to west, situated about 200 m. west of the Soluch railway and 500 m. south of the by-pass road. The deposits in the sides of the trench showed perfectly even bedding, apparently horizontal; a measured section was as follows:

LITHOLOGY	Altitude of upper surfaces (m.)
5. Terra rossa	12
4. Fine yellow-green marl with numerous shells; partly cemented	9·8
3. Similar to 4, but fewer shells and less cementation	9·6
2. As 4	9·2
1. As 3, with lenticles of brown and grey clay	9·1

The base of the section was at 8·0 m.; no bedrock was visible.

The fauna comprised both molluscs and ostracods.[1] The molluscan fauna was collected from layers 1–4; except for the presence of *Paludestrina* and the absence of *Bittium* and *Rissoa*, it was identical with that which Desio obtained at Locality A. It therefore provides evidence that this is still the same series of deposits. The ostracods have been examined by Mr P. C. Sylvester-Bradley, who has very kindly submitted the following report:

There appear to be only two species in your collection from Benghazi: (1) *Cyprideis littoralis* Brady, in great numbers and with many juvenile moult stages; (2) *Loxoconcha* sp. which may perhaps be referable to *L. elliptica* Brady (*L. elliptica* has been regarded as a synonym of other species, probably in error). This species bears a strong resemblance to one figured by G. W. Müller (1894) under the name *Cythereis teres* Brady, from the Gulf of Naples. Both these species suggest a brackish-water or estuarine environment rather than a truly marine one.

Once again, both the lithology and the state of the shells indicated deposition in quiet water. Once again also, changes in the lithology and in the distribution

[1] Sedg. Mus. reg. nos. D. 5269–325.

of the fauna indicated rapid changes of facies; most of the section, indeed, showed brackish conditions, but marine influence was definitely stronger in layers 2 and 4 than elsewhere.

CONCLUSIONS

The examination of the open sections revealed the existence of two distinct series of Pleistocene deposits in the Benghazi area, other than those which were purely terrestrial. The younger series consisted merely of typical beach-material, indicating a shoreline at about 6 m., which, however, was accompanied by no erosional feature. The older series, on the other hand, was more complicated, and will need further discussion.

Although the point cannot be proved, there can be very little doubt that the clays, marls and silts at each locality are the products of one continuous process of deposition, and that they have not subsequently been tilted by earth-movements. In other words, their age decreases as their altitude above sea-level increases. If so, their deposition must have been accompanied by a continuous rise of sea-level, in view of the presence of rootlets at 3 m. (Locality C), and of beach-deposits at 15 m. (Locality A). In addition, it must also have been accompanied by certain other, more peculiar, conditions. First, until the deposits had accumulated to a height of at least 10 m. above present sea-level (Locality D), they must all have been laid down in very quiet water; rolled pebbles were found only at the highest levels. Secondly, during this same time, the salinity of the water was undergoing constant changes.

The ultimate change in the degree of agitation of the water cannot be ascribed to a decrease of depth; this would conflict with the evidence of the rootlets, and would in any case be unlikely in view of the small vertical distance involved. It can only be suggested that the water in which the deposits were laid down was initially sheltered from the sea by some kind of barrier, rising to a height of at least 10 m.; eventually, as the sea-level rose, this was either overwhelmed or partially destroyed.

The changes of salinity show that a fairly large stream must have entered the sea nearby; except for their regular bedding, the deposits might be called typically estuarine. The true estuarine environment, however, requires a large tidal range, whereas the tidal range of the Pleistocene Mediterranean is believed to have been much the same as at present. In this case, the changes of salinity might have been caused either by seasonal floods, or by competition over a long period between a perennial stream and the rising sea-level, fresh water being retained in either case behind the presumed barrier. The latter theory is perhaps the more likely, in view of the considerable thickness of the layers of each facies.

It is now necessary to discuss the relationship between the deposits just described and the argillaceous Pleistocene beds in boreholes Nos. 13–16. Only one borehole, however, is near enough to an open section for a definite correlation to be possible. This is No. 16, which lies only 500 m. to the north of Locality D. It can thus be said with certainty that the yellow-green clays of this borehole are the same as the marls of Locality D. This shows that the marls are not, as Marchetti suggested, a recent deposit younger than the coastal consolidated dunes. It shows also that the marls themselves descend as low as present sea-level.

There can be no such certainty in the case of boreholes Nos. 14–16, all of which lie at least 6 km. to the south of the nearest open section. It is quite conceivable that all or most of their Pleistocene deposits belong to older transgressions than the two which have been mentioned; this may also be true of the sandy Quaternary limestone in No. 16.

Nevertheless, the conditions under which these deposits were laid down must have been remarkably similar to those deduced for the older Pleistocene deposits in the open sections; in every instance there appears to have been a continual alternation of marine and fresh-water conditions, the extreme example being provided by borehole No. 13. It thus seems very probable that these sediments also were laid down during the transgression which culminated in a shoreline at a little over 15 m.

On the assumption that this is true, it is interesting to consider the nature of the floor over which the transgression would have taken place in this area. Everywhere to the east of the 'clay-limestone line', the Miocene appears to rise well above present sea-level. This is also the case in the most westerly boreholes, along the Agedabia road; in Nos. 17, 18 and 19 the top of the Miocene appears to stand at levels of $+12$ m., $+2\cdot5$ m. and $+4$ m. respectively, though it must be admitted that the first of these figures, differing so widely from the others, may perhaps be erroneous. Between the clay-limestone line and the Agedabia road, on the other hand, there are two boreholes (Nos. 14 and 16) in which deposits believed to belong to the '15 m.' transgression are known to descend below sea-level, and not one in which the top of the Miocene has ever been located for certain.

Marchetti himself has suggested, in fact, that all the clays of borehole No. 13, and even those of No. 16, down to -141 m., may be early Pleistocene rather than Miocene, representing perhaps a phase of sedimentation within a drowned valley belonging to Wadi Gattara.[1] This theory, of course, is tentative, and the evidence is still insufficient for it to be accepted in its entirety. Nevertheless, it

[1] Marchetti (1938), pp. 225–6.

is at any rate extremely probable that the base of the sediments of the '15 m.' transgression, in the area east of the Agedabia road, lies well below the present level of the sea. If so, it would certainly be true to say that these sediments, at least, were laid down upon a floor which was interrupted by a trough, only 3–4 km. wide, running due south from Benghazi for a distance of at least 11 km. A trough of such proportions could scarcely have originated except as the valley of a small river, cutting into the extreme eastern edge of the area of Miocene clays; as for the river itself, this could scarcely have been any other than Wadi Gattara, by whose waters the whole area is regularly flooded at the present day.

Provided that these conclusions are correct, it is possible to make certain further suggestions about the conditions under which the deposits of the '15 m.' transgression were laid down. During the early stages of the transgression, there would have been a time when, although the trough itself was submerged, there remained a narrow tongue of land on its western side, separating it from the open sea. The trough itself would then have formed a long, narrow inlet, presumably communicating with the sea on the present site of Benghazi. Such an inlet would always have been protected from the action of the waves; if at the same time the flow of the wadi was maintained, the sea-water of the inlet might also have been diluted to a very considerable extent. In this way, conditions would have arisen exactly similar to those which have been deduced from the nature of the deposits themselves. There is, admittedly, no reliable evidence that the tongue of land ever rose to the height of 10 m., which is the maximum level attained by the brackish-water deposits at Locality D. Being composed of soft clays, however, it might in any case have been largely destroyed by marine erosion when the level of the sea had risen sufficiently high.

No signs of the existence of a similar channel are shown by the deposits of the 6 m. transgression. This fact could have two possible explanations. Either the mouth of Wadi Gattara had shifted into another area, or the volume of its waters had decreased until their erosive power on the seaward edge of the coastal plain was as nearly negligible as it appears to be at the present day.

4. CONCLUSIONS

Although the evidence from the western coastal plain is still relatively scanty, it does at least imply a recent geological history which involved something more than one simple submergence. In spite of a very different topography, the region in fact resembles the north coast in that here also there are traces of a descending series of marine shorelines. In this case, however, it is only rarely that the shorelines appear as erosional features; much of the evidence for their existence is purely geological, and, owing to imperfections of preservation and exposure, is difficult both to find and to interpret.

The following is a list of levels and localities at which ancient shorelines are thought to have been recognized, in this region, with a fair degree of certainty:

150–200 m.	Tocra-Antelat (still somewhat doubtful)
90 m.	Benina
15 + m.	Benghazi
5–6 m.	Tocra, ? Coefia, Benghazi

It is obvious that this evidence is quite insufficient to lead to conclusions as detailed or as definite as those drawn from the evidence obtained between Ras Aamer and Derna. In particular, the majority of the more authentic fragments are much too short to provide by themselves any direct indications of the extent to which they have been tectonically disturbed since the time of their original formation. Nevertheless, some conclusions may be suggested which are not so uncertain as to be entirely without interest.

In the first place, it seems very probable that the escarpment between Tocra and Antelat does indeed represent the cliff of a single marine shoreline, whose altitudes now lie between 150 and 200 m. If so, it can only be concluded that the feature was formed at a time when the sea itself stood between 150 and 200 m. above its present level, and that it has subsequently undergone no important tectonic disturbances. Under such conditions of high sea-level, it may be added, most of the low-lying region to the south of the Gebel would certainly have been submerged. The absence of escarpments on the southern slopes of the Gebel, specifically commented upon by Marinelli,[1] does not form a valid objection to this idea; this might well be another case in which marine erosion was inhibited by the low initial gradient of the ground. On the other hand, it cannot necessarily be assumed that the Gebel at this time would have formed an island; both Marmarica and the eastern end of the Gebel are suspected to have been tectonically unstable in comparatively recent times.[2]

According to the same theory, this would also be the time at which the coastal plain was first brought into existence, almost as it is at present, by the action of marine erosion. Whether or not this is true, it is at least virtually certain that the plain, by some means, had already received its present form before the appearance of the lower and least doubtful shorelines. This being so, the gentle slopes of the plain could themselves provide a sufficient explanation for the undeniable fact that these latter shorelines are so rarely accompanied by erosional features.

[1] Marinelli (1920), p. 74.
[2] See above, p. 38, and Desio (1939), pp. 25–33.

For reasons already given, there is little more that can be said for certain at this stage concerning the origins and histories of these lower fragments of shore-line. It should be pointed out, moreover, that the problem still remains, even if it be agreed that the foot of the escarpment represents one single shoreline; in spite of the uncertainty as to the levels of this feature, there is little doubt that it shows undulations which, though relatively slight, are none the less real. Nevertheless, it can be said with almost complete confidence that the Tocra step, with its constant level of 5 m. maintained over a distance of at least 18 km., must owe its present elevation entirely to a shift of the sea-level. It is only slightly less certain that this is also true of the other fragments of shoreline which were found to lie at the same altitude.

Finally, there is evidence to show that the sea-level in the Benghazi area, immediately before rising to its shoreline at 15 + m., had sunk at least as low as it is at present. During this regression, some water-course, presumably Wadi Gattara, appears to have cut itself a channel in a strip of Miocene clay which, probably as a result of faulting, was present at ground-level in the area to the south of Benghazi. As the sea-level rose, the channel was then filled with sedi-ments which are largely of fresh and brackish-water facies; evidently the sea-water in the drowned portion of the channel continued for some time to be diluted by the waters of the wadi.

CHAPTER III

CHRONOLOGY

I. INTERNAL CHRONOLOGY

All the available evidence has now been presented concerning the nature and dispositions of the high shorelines of the Cyrenaican coast. For a part of the north coast, between Ras Aamer and Derna, an attempt has been made to explain the present elevations of these features in terms of large movements of the sea-level itself in recent geological times, and of small but perceptible movements of the land. For the remainder of the coast, no such definite conclusions can yet be drawn. Nevertheless, certain individual shorelines on the west coast have been attributed to times when the sea stood at various heights above its present level. It should therefore be possible to suggest correlations, however tentative, between these shorelines and those on the north coast. Such correlations would in themselves be of interest, both geological and archaeological; in addition, they should provide some information about the recent tectonic behaviour of the Cyrenaican coast as a whole.

Only one further set of correlations can be made which could fairly be described as definite. It has already been concluded that the 6 m. shorelines at Benghazi, Coefia and Tocra were formed at a time when the sea-level actually stood about 6 m. higher than at present. Exactly the same conclusion has also been reached for the lowest fragments of shoreline between Ras Aamer and Derna. It can scarcely be doubted, therefore, if these conclusions are correct, that all these fragments of shoreline were formed during one and the same period of high sea-level.

As will be seen later, these correlations are of importance in connexion with the dating of continental deposits. At this stage, however, it will only be necessary to point out their tectonic implications. Each of the fragments in question, as has already been made clear, is believed to have remained completely undisturbed by earth-movements since the time of its formation. Whatever the absolute date of the 6 m. high sea-level, therefore, it appears that the subsequent period has been one of complete tectonic stability for a large part of the 300 km. of coast between Benghazi and Derna. It is virtually certain, indeed, in view of the presence of shorelines at 6 m. near both Tolmeita and Wadi Giargiarummach, that the whole of this stretch of coast has remained similarly stable.

A conclusion only slightly less definite can be arrived at in relation to the older

59

Pleistocene deposits of the Benghazi area. As has been seen, these deposits, often of brackish or even fresh-water facies, appear to have been laid down during a marine transgression whose highest level is indicated by the remains of beach-conglomerates at an altitude of about 15 m. There is no direct evidence to show whether this transgression, and the subsequent regression, were in any way assisted by earth-movements. One of the unique features of these deposits, however, is the fact that they are immediately overlain by deposits of a later transgression. Since the later deposits are those of the 6 m. shoreline, and since no intervening deposits have been found, it seems fair to conclude that the beach-conglomerates at 15 m. are contemporary with the shoreline which, between Ras Aamer and Derna, is thought to lie immediately above the shoreline at 6 m. As has already been pointed out, this second lowest shoreline appears to have stood originally at some altitude between 15 and 25 m. This, in itself, tends to confirm the correlation; it also implies that the Benghazi area has not been tectonically disturbed since the deposition of the 15 m. conglomerates.

It is unfortunate that the most interesting group of conclusions is suggested by the least firmly established evidence. Between Apollonia and Derna there are undoubted signs of shorelines between 140 and 190 m.; it is possible that two separate shorelines are present, but it is suspected that there is only one. It is strongly suspected also that the foot of the coastal escarpment, for the whole of the distance between Tocra and Antelat, represents one single shoreline whose levels now stand between 150 and 200 m. If this latter suspicion could be finally confirmed, there would then be very little doubt that the formation of the western coastal plain, together with the length of escarpment by which it is bounded, took place at the same time as the first known signs of marine erosion were making their appearance on the coastal escarpment between Apollonia and Derna. It would also be possible to say that the entire coastal region, from Derna to the head of the Gulf of Sirte, had been free from all but the slightest tectonic movements since some time when the sea-level was still 150–200 m. higher than at present.

To this there is one obvious objection. The greatest width of the western coastal plain is over 50 km., yet there is no known terrace between Apollonia and Derna which is more than 1 km. wide. The disproportion remains, moreover, even if it is assumed that the original wave-cut platforms on both parts of the coast extended as far as the present edge of the continental shelf. If the conclusions are accepted, it must therefore be supposed that the rate of marine erosion during one particular period was incomparably greater on the west coast of Cyrenaica than on the north coast. A simple explanation can be offered, however, for a difference of this kind: the rocks of the north coast, as has already been stated, are known to be far more resistant than those on the coast from Tocra southwards.

It may be added that this theory of long-continued stability is confirmed, rather than contradicted, by the presence of the Benina step (90 m.) and the higher shorelines found at Tolmeita (33 and 20 m.); all of these levels lie within the supposed ranges of various shorelines between Ras Aamer and Derna. The evidence in these cases, however, is still too scanty to allow the establishment of correlations.

2. EXTERNAL CORRELATIONS

An attempt will now be made to relate the Cyrenaican shorelines to geological time-scales established elsewhere in the world. For this purpose, little use can be made of the associated marine faunas. Of the species so far identified, whether by Newton, Blanc, Desio or the present writer, every one has been living in the Mediterranean from Pliocene times until the present day. Nor can any help be obtained from the Pleistocene continental deposits of the region. Although, as will be seen, these have in fact yielded abundant remains, both organic and archaeological, nothing can be said on independent evidence regarding the dates at which the various animals, plants and industries first entered the region; the discovery of these dates was, indeed, one of the main objects of the whole investigation. The continental deposits have also provided evidence for a climatic succession, but even this by itself is of little use; the climatic phases represented are such as might have occurred again and again during the Pleistocene period.

At present, indeed, only one possible means of correlation appears to be available. Each of the Cyrenaican shorelines is believed to have been formed during a period in which the sea-level, in the course of a general descent from an altitude a little below 200 m. to that at which it stands at the present day, halted temporarily at some intermediate stage. In many cases it is also believed that an approximate estimate can be made of the height of the sea-level at the time when a particular shoreline was formed. This descent of the sea-level, however, together with its interruptions, must have been of worldwide occurrence. Since each interruption appears to have lasted for a considerable time, each one must have left its mark in parts of the world other than Cyrenaica; in other words, series of shorelines similar to those of Cyrenaica should be widely distributed elsewhere. In any such series, provided it has never been disturbed, it should be possible to recognize the counterparts of individual Cyrenaican shorelines merely by similarities of levels. Thus, if any undisturbed series should exist whose component shorelines had been dated by independent means, it should then be possible to transfer these dates to the Cyrenaican shorelines themselves.

Series of high shorelines have, in fact, been reported from all the coasts of the world. There are many cases, moreover, in which it has been claimed, as in Cyrenaica, that a series of this kind has been scarcely disturbed since its formation, and various theories have consequently been put forward dealing with the manner in which the sea-level has changed in recent geological times. Much has already been written also with regard to the chronology of these suggested sequences of changes.

It cannot be said, however, that any of these conclusions have been generally accepted. As far as the chronological arguments are concerned, many of them can be criticized on the grounds of being excessively indirect; it is only rarely, indeed, that they have been based upon direct geological evidence. A more fundamental objection arises from the suspicion that the various series of shorelines from which the behaviour of the sea-level has been deduced may not after all have remained entirely unmoved. This point will be discussed more fully in the next chapter. For the moment, it need only be said that there are very few individual shorelines, let alone complete series, which have been shown beyond any reasonable doubt to remain horizontal over long distances.

At present, therefore, even this method of correlation is of very limited value. It is nevertheless found, as a result of two fortunate circumstances, that it will serve well enough for the present purpose. In the first place, there is no real need to discuss the dates of any but the lowest and the highest of the Cyrenaican shorelines; the former, as will appear later, will be of interest in connexion with the Pleistocene continental deposits, the latter in connexion with tectonics and general physiographical history. In the second place, reliable evidence for undisturbed shorelines, or for features closely associated with them, appears in other parts of the world to be far more abundant both above 150 m. and below 30 m. than at any intermediate levels. Above 150 m., the successive stages seem in general to be widely spaced, so that the evidence for one stage is often found to have been damaged relatively little by the erosion carried out during the stages which followed. At lower altitudes, the effects of this kind of damage become increasingly intense, but the lowest shorelines of all have often remained conspicuous owing to the relatively short time for which they have been exposed to subaerial erosion.

The chronology of the 6 m. shoreline of Cyrenaica and of the shoreline or shorelines between 140 and 200 m. will now be dealt with in turn.

6 *m. shoreline*

Isolated fragments of shoreline at levels between 6 and 9 m. have been reported from all over the world, and several good examples are known of shorelines which remain between these levels for great distances. Thus, the position of the lowest distinct strandline on the coast of the eastern United States is marked by a feature known as the Suffolk Scarp, which can be traced in spite of some large gaps for a distance of about 800 miles (1300 km.), between New Jersey and Florida. The foot of the scarp lies at 20–30 ft. (6–9 m.) above sea-level.[1] Similarly, on the coast of South Africa abundant traces of a shoreline at 6 m. have been found over a distance of 1500 km. from Port Nolloth to Durban.[2] There is therefore much evidence, in addition to that from Cyrenaica, to show that 6 m. was in fact the altitude of the sea-level during its last period of prolonged stability.

Since this stage of high sea-level is evidently comparatively recent, it will be necessary to relate it, if possible, to some Pleistocene time-scale. There are three regions in which a Pleistocene time-scale has been worked out that is generally regarded as 'standard'; these are the glaciated areas of North America, the Alps and North-West Europe, together with their immediate surroundings. In each case, however, there can only be a very small area in which marine shorelines, undisturbed by isostatic adjustments of the land, could ever be found in direct contact with glacial or periglacial deposits. For this reason, the dating of the 6 m. high sea-level is still not entirely secure; it is largely for this same reason, indeed, that the dating of Pleistocene shorelines in general remains in a state of uncertainty.

On the east coast of North America, the Suffolk Scarp cannot yet be correlated with any glacial deposits,[3] while the shorelines on the remainder of the North American coast have generally been disturbed by later tectonic movements.[4] In the case of the Alps, it should certainly be possible by a study of river terraces to correlate the glacial succession with the shorelines of the Riviera; unfortunately, these shorelines also are suspected of having been disturbed.[5]

Circumstances in North-West Europe should, in theory, be more favourable. Fragments of 'raised beach' at levels below 30 ft. have been reported from numerous localities on the southern coasts of England, Wales and Ireland, and, for many of the English and Welsh examples, J. F. N. Green has deduced contemporary mean sea-levels between 25 and 30 ft. (8–9 m.) above O.D.[6] In South Wales and Southern Ireland, moreover, the beaches are actually overlain by moraines, while in England, also, they are often covered by 'head', indicative of a cold climate.[7] As Wright points out, however, no boulder-clay has yet been

---

[1] Flint (1947), pp. 438–9.
[2] Krige (1927).
[3] Flint (1947), p. 440.
[4] *Ibid.* p. 442.
[5] Denizot (1935), pp. 568–9.
[6] Green (1943), pp. 129–31.
[7] Wright (1937), pp. 113–17.

shown to be older than any one of these fragments of shoreline;[1] in any case, the correlation of the British drifts with those of the Continent is still in a state of uncertainty. Even in this region, therefore, it is still impossible to obtain direct stratigraphical evidence.

Thus it is necessary after all to fall back upon less direct arguments. One fact which emerges from the British evidence is that the 6–9 m. high sea-level was earlier than one or more glacial stages. In the British Isles, however, no indication has been found, in the deposits overlying the corresponding shoreline, of any climatic phase which could be interpreted from its high temperatures and long duration as being interglacial.[2] As will be seen later, the same can be said of the continental deposits which overlie the 6 m. shoreline of Cyrenaica. It can also be said of continental deposits in similar positions elsewhere in the Mediterranean, although in these cases, admittedly, the stability of the associated shorelines has not always been proved. Notable examples are the cave deposits of Grotta Romanelli (Southern Italy), and the deposits of the Pontine Marshes, southeast of Rome.[3] In spite of the negative nature of the evidence, it therefore seems certain that the 6–9 m. high sea-level, although not post-glacial, cannot have been earlier than the Last (Riss-Würm) Interglacial.

The possibilities are still further reduced by the evidence regarding the temperature of the 6 m. sea. The faunas of the 6–9 m. shoreline of Britain suggest, according to Zeuner, a climate which was probably temperate and certainly not glacial.[4] In the eastern United States the Suffolk Scarp is associated with a series of marine deposits known as the Pamlico formation. The fauna of these deposits was studied by Richards, who concluded that the contemporary water-temperatures at all points along the scarp were at least as warm as those of today, and in many cases somewhat warmer.[5] On the shores of Italy and Southern France, numerous fragments of beaches have been recorded, as Zeuner points out,[6] at levels of 7–8 m., and many of these contain marine fossils belonging to the 'Strombus fauna', the components of which at the present day are found no further north than the coast of Senegal. In spite of the absence of real proof, some of these Mediterranean beaches at least must surely date from the time of the 6–9 m. high sea-level.

It cannot be doubted, therefore, that this high sea-level occurred at a time when the sea was warmer than it is at present. This means that it can only be placed either in the Last Interglacial or in one of the interstadials of the Last

---

[1] Wright (1937), p. 124. [2] But see footnote on p. 65.
[3] For summary of climatic successions, see Zeuner (1945), p. 203, fig. 64.
[4] Zeuner (1945), pp. 235–40. [5] Richards (1936), p. 1644.
[6] Zeuner, op. cit. pp. 232–3.

(Würm) Glaciation. Beyond this statement, in the writer's opinion, it is impossible to go with complete certainty. Nevertheless, the great width of some of the associated wave-cut platforms suggests a period of very considerable duration, and this in turn suggests that the first alternative is the more likely to be correct. The evidence of the water-temperatures, also, points to the same conclusion.[1]

140–200 m. shorelines

As has already been emphasized, there is no real certainty that those Cyrenaican shorelines which now lie between 140 and 200 m. were ever in continuity with each other. Even if there were, it would still be impossible on the evidence so far available to make any but the roughest estimate of the position of the sea-level during their formation. All that can safely be said, in fact, is that they were formed at a time or times when the sea-level stood at some altitude or altitudes intermediate between these two limits. The present problem, therefore, must be stated in very general terms: to establish the date at which the sea last stood at or near 200 m., and the date at which it last stood at or near 140 m.

It must first be admitted that there appears to be no undisturbed shoreline within this range of altitudes which has yet been directly correlated with any of the faunal stages recognized in the Mediterranean region. Thus, it is again necessary to rely upon indirect arguments, and in particular, to consider the chronology of changes of sea-level which occurred both before and after the period in question.

There are many areas where evidence has been found, though not necessarily in the form of shorelines, showing that the sea-level, before arriving at 200 m., had already halted at a series of still higher altitudes. A well-known example is provided by the southern slopes of the Massif Central of France. In this region, certain widespread erosional surfaces are found which, according to Baulig, are fluvial rather than marine in origin, but which nevertheless are considered to have been formed while the sea stood at certain very high levels. Of these levels, the three most important are at 380, 280 and 180 m. respectively,[2] and there is good reason to believe that all these surfaces were formed after the beginning of Pontian (Late Miocene) times.[3]

From the chronological point of view, however, the most interesting published evidence is that which has been obtained from a mass of hills, 25–30 km. from

---

[1] These, indeed, are the views most commonly held at the present day. Since this was written, however, Mr G. F. Mitchell has told the writer that there is a deposit near Wexford, S.E. Ireland, which, while probably younger than the lowest (so-called 'Preglacial') raised beach on the nearby coast, may yet represent an Interglacial period. This raises the possibility that the 6–9 m. shorelines of Cyrenaica and elsewhere may, after all, have to be assigned to the Second-Last (Mindel-Riss) Interglacial.   [2] Baulig (1935), pp. 18–21.   [3] *Ibid.* p. 21.

north-east to south-west, which lies behind the city of Algiers. On all sides of these hills occur well-developed series of terraces, originally interpreted by de Lamothe as being of marine origin and as denoting the positions of eight shore-lines, of which the four highest stood at 325, 265, 204 and 148 m.[1] This view has, indeed, been opposed by Baulig, who considers that the four highest terraces, at least, are products of fluvial erosion; he still derives from them, however, a series of sea-levels not greatly different from Lamothe's: 280, 250, 180 and 140 m.[2] As regards the age of the terraces, there is no suggestion that any have been warped, yet the rocks of the area include tilted Lower Pliocene beds.[3] Thus, in spite of the difference of opinion as to the sea-levels which they represent, these terraces show fairly conclusively that the last halt of the sea-level at a height near 200 m. took place long after the beginning of Pliocene times.

Confirmation for this view is provided by a certain series of deposits which lies within the Nile Valley. From Luxor to the Delta, a distance of over 500 km., the maximum levels of these deposits have been found to remain remarkably constant, at about 180 m. above present sea-level. The uppermost beds of the series are everywhere estuarine or fresh-water; marine deposits are found in the lower beds only, and in Upper Egypt are altogether absent. It is thought, never-theless, that the maximum levels represent a position occupied by the sea-level itself for a considerable length of time, while the highest beds of the series were being laid down.[4] It has not yet been possible to suggest a date for these highest beds on the basis of their own faunas, which are composed entirely of fresh and brackish-water molluscs with the gastropod *Melanopsis* as the characteristic form.[5] The underlying marine beds in Middle and Lower Egypt have, however, been assigned to the Astian, that is, to a part of the Pliocene.[6]

At this point it should be mentioned that it is impossible, according to Gignoux, to establish any chronological subdivisions in that part of the Mediterranean Pliocene which lies beneath the Calabrian; the Astian represents no more than the sandy facies of this 'Lower Pliocene', the argillaceous facies being the Placentian.[7] It is for this reason that the Egyptian evidence must be regarded as purely confirmatory, and not as a source of new information.

As far as the writer is aware, there is in fact no published evidence which would enable any shoreline between 200 and 140 m. to be assigned, with any certainty, to a date more recent than 'Lower Pliocene'. The evidence which,

---

[1] De Lamothe (1911), pp. 57–178.   [2] Baulig (1928), pp. 523–34.
[3] Baulig (1935), p. 21.   [4] Sandford and Arkell (1933), p. 6.
[5] Blanckenhorn (1921), pp. 151 *et seq.*   [6] Blanckenhorn, *op. cit.* pp. 130–42.
[7] Gignoux (1950), pp. 626–7. According to a decision of the International Geological Congress of 1948, the Calabrian itself has now been transferred to the Pleistocene.

perhaps, comes nearest to being helpful in this respect is that provided by the Netley Heath Beds in the south of England. These deposits rest upon a surface which is believed on geomorphological grounds to be an ancient wave-cut platform, associated with a more or less undisturbed shoreline at about 200 m.;[1] the deposits themselves contain detached blocks of a material which, in turn, contains a Red Crag fauna.[2] According to a correlation recently put forward, the lowest Red Crag is now thought to be equivalent to the Calabrian,[3] in which case the Netley Heath Beds would be of Calabrian age or slightly later. Unfortunately, it cannot be assumed that these were the beach-deposits of the 200 m. shoreline; they could be considerably younger. Even this evidence must therefore be regarded as inconclusive.

Nevertheless, the Netley Heath Beds do at least show that the last halt of the sea-level at or near 200 m. cannot have occurred long after the end of Calabrian times. It now remains to inquire whether or not this can also be said of the last halt at or near 140 m.

There appears to be no undisturbed shoreline in the Mediterranean below the height of 140 m., or indeed at any other level, which is known to be directly associated with Calabrian deposits. Hence, it is necessary once again to rely upon indirect evidence. In this case, the evidence is supplied by the deposits which contain the succeeding Mediterranean fauna, the Sicilian. Localities in which these latter deposits have been connected with a particular shoreline are certainly not numerous; nor is this surprising, since the true Sicilian fauna occurs only in beds of deep-water facies. These localities, moreover, are entirely confined to Sicily and Southern Italy, regions noted for their tectonic instability. Nevertheless, Gignoux has described several cases in which a shoreline of this age can be traced for a considerable distance, and has found that the altitudes of such features seldom vary beyond the limits of 80–100 m.[4] Shorelines suspected to be Sicilian do, indeed, occur as high as 250 m. on the Italian side of the Straits of Messina,[5] but this is known to be a particularly unstable area. On the whole, therefore, it seems likely that the sea-level in Sicilian times did not stand more than 100 m. higher than at present.

Thus the higher Cyrenaican shorelines may all be regarded as pre-Sicilian; yet, on the evidence from Algiers and the Nile Valley, even the oldest must have been formed long after the beginning of the Pliocene. Strictly speaking, there is no more that can be said. In the writer's opinion, however, it is most likely that all these shorelines are in fact of Calabrian age.

[1] Wooldridge and Linton (1939), pp. 48–63.
[2] Dines, Edmunds and Chatwin (1929), pp. 110–19.
[3] Lagaaij (1952).
[4] Gignoux (1913), pp. 628–30.
[5] *Ibid.* p. 230.

5-2

CHAPTER IV

SUMMARY AND FINAL DISCUSSION

I. SUMMARY OF RESULTS

(i) Almost everywhere on the northern side of the Gebel Akhdar, the surface of the ground between the sea and the crest of the lower escarpment was found to possess a 'staircase topography', composed of alternate cliffs and terraces. This was also true of the western side of the Gebel, though here in general there was only one cliff, the escarpment itself, and one terrace, the plain which lies on its seaward side. Partly on morphological grounds, partly on account of the presence of patches of marine deposits on their surfaces, the terraces are considered to be ancient wave-cut platforms, and the cliffs also to be of marine origin.

(ii) The total number of different shorelines to which the terraces correspond does not appear to be more than six or seven. So far as they are preserved, none of these shorelines has been found to show signs of violent warping. Hence, two deductions have been made regarding their manner of formation. First, they are thought to represent a series of successively lower altitudes at which the sea-level halted, for a considerable length of time in each case, before reaching its present position; approximate estimates of these altitudes are as follows: 140–200 m. (? two separate stages), 70–90, 44–55, 35–40, 15–25 and 6 m. Secondly, the period in which they were formed is thought to have been one of almost complete tectonic stability for the whole of the west coast of Cyrenaica, and for the north coast as far east as Derna; it is only to the east of Derna that definite signs of instability have been found.

(iii) Certain deposits near Benghazi, mainly marls and clays and mainly of brackish-water facies, are interpreted as having been laid down within an ancient drowned channel of Wadi Gattara, during a period in which the sea-level was rising from some altitude below 0 m. to an altitude a little over 15 m. It is believed that this rise of sea-level took place immediately before the last but one of the series of halts referred to above.

(iv) Except in the case of the Benghazi marls, few identifiable marine fossils have been obtained from deposits associated with the terraces, and these few are of little obvious interest, either chronological or otherwise.

(v) On evidence provided by similar features in other parts of the world, the highest shorelines of Cyrenaica, below the crest of the lower escarpment, have been assigned to the Calabrian. On similar evidence, the lowest of all, at 6 m., has been assigned to the Last Interglacial (but see footnote on p. 65).

2. FINAL DISCUSSION

It now remains to consider the more interesting implications of these results. Their local implications are obvious; it need only be pointed out that they provide not only some new information concerning the recent geological history of Cyrenaica but also a possible means of assigning dates to archaeological and geological remains found on certain parts of the coast.

From a more general point of view, the results are chiefly of interest in so far as they throw light upon the behaviour of the sea-level itself in Pleistocene times. On this point some further discussion is needed.

Early in the present century Lamothe made a detailed study of the high shorelines of Algeria and Tunisia. His conclusion, already referred to, was that the shorelines in these countries from 325 m. downwards were only eight in number, and that each remained horizontal over great distances. Hence, he arrived at the further conclusion that the present levels of these features represented successively lower altitudes at which the sea-level had stood before reaching its present position.[1]

Lamothe's conclusions received strong support from Depéret, who considered that the four lowest levels of Lamothe could be recognized in many other parts of the Mediterranean. To these four levels, Depéret assigned the following limits: 90–100, 55–60, 28–30 and 18–20 m.; he also gave them the respective names Sicilian, Milazzian, Tyrrhenian and Monastirian. In addition, Depéret claimed to have recognized one further level at an altitude of 6–8 m., but this he considered to be too unimportant to deserve a name.[2] The reality of these five levels has since been accepted by many authors, and there has even been some agreement as to their chronology. They have in fact already been used, in many cases, as an argument for supplying absolute dates for both geological and archaeological material.

There has, however, been an increasing tendency to doubt whether this 'classical' version of the recent behaviour of the sea-level is really correct. Severe criticisms have been made of the methods and arguments used both by Lamothe and Depéret, and by those who have attempted to find confirmatory evidence in other areas. As a result, the suspicion has grown that the stability of

[1] De Lamothe (1911).
[2] Depéret (1918). The terms 'Sicilian' and 'Tyrrhenian' had previously been assigned to two of the Pleistocene marine faunas of the Mediterranean.

the shorelines described by these writers has not been sufficiently proved, and that it is therefore impossible to accept unreservedly the estimates which have been made of their original levels.[1]

In the present writer's view, as already stated, the existing evidence, especially that from North America and South Africa, appears to leave little doubt about the existence of a level at or near 6 m. It seems certain also that this level represents the last occasion on which the sea-level remained stable for any considerable length of time. These conclusions are confirmed by the evidence from Cyrenaica; they also agree with the conclusions of Depéret. For the higher levels of Depéret, however, adequate independent confirmation does indeed appear to be lacking, and it is in this respect that the results obtained between Ras Aamer and Derna are of particular interest.

For reasons already mentioned, it is certainly not claimed that these results can provide a conclusive detailed history of the behaviour of the sea-level in recent geological times. In the first place, the correlations suggested between the various fragments of shorelines must all be regarded as conjectural, to some extent at least. In the second place, even if these correlations are accepted, it is still impossible, owing to a slight but unmistakable degree of warping, to make any exact estimates of the original levels of the reconstructed shorelines, except in the case of the lowest of all.

Nevertheless, even if no correlations are attempted, the evidence from northern Cyrenaica is sufficient as it stands to throw some light on the question. In particular, it provides a very strong indication that the descent of the sea-level from the high altitudes at which it stood in Pliocene times has indeed been interrupted by a series of halts, each of considerable duration, of which the 6 m. shoreline represents only the latest example. It suggests, in addition, that the number of such halts, from 100 m. downwards, was about half a dozen. So far as they go, all these conclusions tend to confirm the theory of Lamothe and Depéret, in outline if not in detail.

Moreover, it is at least very probable that the proposed correlations between the fragments of shoreline are correct, and that the original levels of these features lay between, or not far outside, the limits given. Since these limits are generally wide, and should perhaps in some cases be even wider, it would hardly be fair to use the Cyrenaican figures for the purpose of criticizing the accuracy of those obtained by Depéret. A comparison between the two sets of figures shows, indeed, that it is only between the two lowest members of each series that any kind of correspondence exists:

[1] See, for example, Baulig (1935), p. 7; Blanc, A. C. (1937), pp. 622–5; Gignoux (1950), p. 658; Johnson (1931), pp. 43–50.

CYRENAICA (in metres)	DEPÉRET (in metres)	
140–200 (? two shorelines)	—	
70–90	90–100	(Sicilian)
44–55	55–60	(Milazzian)
35–40	28–30	(Tyrrhenian)
15–25	18–20	(Monastirian)
6	6–8	—

Provided the correlations are correct, however, it is at least safe to say that the Cyrenaican results indicate that the number of levels below 100 m. is five. This is also the case with the series of Depéret. On the basis of these results, it is therefore concluded that the Depéret-Lamothe theory, whatever the faults of the means by which it was arrived at, may well be substantially correct.

Finally, mention should be made of one other view held by Lamothe: that the sea-level did not descend regularly from one stage to the next, but that it first sank in each case to an altitude lower than that of the succeeding stage, subsequently returning to that stage from below.[1] The results obtained from the Pleistocene deposits around Benghazi suggest that this did indeed occur during the descent to the second lowest stage.

[1] De Lamothe, *op. cit.* pp. 220–6.

B. CONTINENTAL DEPOSITS

INTRODUCTION

Previous publications contain many references to the existence of continental Pleistocene deposits in the coastal regions of Cyrenaica. For the most part, however, such references are brief and generalized, and have provided little new information regarding the geological history of the territory. During the writer's examination of the marine shorelines, much attention was therefore given to any deposits of this kind found either along the coast or in the lower reaches of the wadis. Attempts were made in particular to determine their relationship to the succession of shorelines, and to deduce the conditions under which they were laid down.

The results obtained will be described in the following chapters. For this purpose, it has been found convenient to divide the deposits into three categories, as follows: deposits composed largely of alluvial material, deposits of consolidated dune, and deposits which include much tufaceous material. The deposits of each category will be dealt with in a separate chapter.

To a very great extent the occurrences of continental deposits, including even some of consolidated dune, were found to be associated with individual wadis. It will be useful, therefore, to give a short preliminary account of those wadis which run from the northern and western flanks of the Gebel into or towards the Mediterranean.

The number of such wadis is very large. Each one, as it approaches the coast, becomes more and more deeply entrenched within a narrow gorge, which, owing to the hardness and homogeneity of the bedrock, may be very steep-sided. To the south-west of Tolmeita, none of these gorges appears to continue beyond the coastal escarpment. From Tocra to the head of the Gulf of Sirte, indeed, there is probably no wadi which maintains a permanent course of any description across the full width of the coastal plain. On the north coast, however, where no true coastal plain exists, many of the gorges extend as far as the present shore. In such cases, it is often evident that the bedrock floor, though now hidden by alluvium, sinks well below sea-level.

The main watershed of the Gebel lies nowhere far from the north coast. The wadis of this part of the coast, therefore, with which the following account is chiefly concerned, are for the most part very short. The great majority are less than 20 km. long; the longest of all, Wadi Maallegh, attains its length of 120 km.

only by running parallel to the watershed for the whole of its course, finally reaching the sea at the extreme eastern end of the Gebel.

Perennial springs are confined almost entirely to the eastern half of the Gebel. Within this area they occur at four well-defined geological horizons, of which one is in the Middle Eocene, one near the base of the Oligocene, and two in the Lower Miocene.[1] Although numerous, these springs are generally of very small discharge. As a result, very few of the wadis contain a perennial stream of any appreciable length, and no wadi contains such a stream for the whole of its course. Almost all the wadis are in fact completely dry at all times, except during a few days in each year when violent flooding takes place.

[1] Marchetti (1938), p. 89. The Aquitanian, to which one of the four horizons belongs, is here included in the Miocene.

CHAPTER V

ALLUVIAL DEPOSITS

I. DESCRIPTION

Various references have already been made, in connexion with the marine shore-lines, to the presence of large masses of alluvium both on the lower slopes of the escarpment and along the inland edge of the western coastal plain. It is these deposits which form the subject of the present chapter.

The existence of such deposits has been noted by many former writers on the geology or geography of the Gebel. It is not considered, however, that any purpose would be served by giving a list of previous references. All are very short, and there is none which sets out to provide a detailed description of the deposits; certainly, no serious discussion either of their age or of their conditions of deposition has yet been attempted.

It should be mentioned at this point that some of the material dealt with in the following account is now believed to have been transported by means other than running water; it will nevertheless be retained within this chapter, in view of its close association with material which is genuinely water-borne. On the other hand, all deposits of alluvium containing contemporaneous developments of tufa have been excluded. Deposits of this kind were found to show many striking differences from the remainder of the alluvium, not only in their lithology but also in their mode of occurrence and their included organic and archaeological remains; since their total volume is in any case insignificant, it has been found convenient to consider them under the heading 'Tufaceous Deposits'.

NATURE OF DEPOSITS

The two main constituents of the deposits in question are fragments of rock of local derivation, and a matrix which generally consists almost entirely of terra rossa. Owing to the presence of the latter material, the deposits nearly always have a characteristic red colour. The relative proportions of the two constituents are exceedingly variable. In some places the whole deposit is composed of terra rossa; occasionally, the matrix is altogether absent. The rock-fragments also vary greatly in size, ranging from boulders to small grains, and sorting is generally poor. As a rule, the fragments are well rounded, but they may also be so angular as to be indistinguishable from scree; this is one case in which running water does not appear to have been the main agent of transportation.

Stratification is normally well marked, and frequently shows cross-bedding. Certain occurrences were found, however, in which bedding could scarcely be seen; the material then consisted of a mass of terra rossa in which fragments of rock, angular or subangular, were dispersed widely and apparently at random. The contrast between this kind of material and the normal alluvium is illustrated by Pl. 2. It is thought that these occurrences of unstratified material almost certainly represent ancient mud-flows, as described by Blackwelder.[1] Here again, therefore, the material cannot strictly be called alluvial. Unfortunately, it was only towards the end of the investigation that the peculiar nature of these unstratified deposits was first recognized, and no special note was made of their distribution. A general impression was gained, however, that they were widespread, but far less abundant than the undoubtedly water-borne deposits with which they were usually interbedded.

In many places, the alluvial deposits are entirely unconsolidated. Elsewhere they have been converted into an extremely hard conglomerate by the formation of a calcite cement within the terra rossa matrix. It was often found that hard and soft layers occurred alternately within the same small exposure. This showed that the progress of cementation had depended mainly upon very local conditions, such as movements of ground-water, rather than upon the passage of time.

MODE OF OCCURRENCE

To some extent, and particularly upon the higher marine terraces, the deposits occurred as isolated patches. By far the greater part of the material, however, formed more or less continuous sheets, extending seawards from the lower slopes of the escarpment and often reaching the modern shore. It was eventually realized, moreover, that these sheets, although at first giving an impression of complete topographical confusion, were in fact composed of well-defined 'conoidal forms', either deltas or alluvial fans, overlapping one another laterally and each one originating at the mouth of the gorge of one particular wadi (Pl. 1a).

The upper surfaces of these forms, when seen in profile against the sky, invariably appeared as gentle curves, concave upwards but never becoming horizontal. Where the seaward edge lay entirely on land, as was the case not only on the plain to the south-west of Tocra but also at many points on the north coast of the Gebel, the upper surface was always seen to converge gradually towards that of the underlying bedrock, with no abrupt change of slope; a feature of this kind was clearly an alluvial fan, not a delta.

[1] Blackwelder (1928).

Between Tocra and Derna, however, many of the 'conoidal forms', perhaps the majority, extended to the modern shore, and had been partly destroyed by the sea, often with the formation of a considerable cliff (Pl. 3*b*). Where this was so, there could be no strict proof of their non-deltaic nature. Nevertheless, several were observed which, although they reached the shore, had for some reason resisted marine erosion. In all such cases the upper surfaces of the features could then be seen to maintain their gradients down to the water's edge, sometimes actually passing beneath the sea. It was concluded, therefore, that all the 'conoidal forms', in so far as they were visible above the sea, could be classified as alluvial fans. This conclusion was confirmed by the fact that none of the gravels showed traces either of marine horizons or of foreset bedding; often, indeed, they contained shells of land-snails at or near sea-level. It should be added that every fan which was observed was of the 'simple' kind, 'compound' fans being apparently absent.

A certain amount of alluvium occurred, in addition, within the lower reaches of many of the larger wadis. In almost every such case the upper surface of the deposits, where preserved, appeared as a single, well-defined terrace (Pl. 3*a*), lying above the modern stream-bed at a height which invariably increased in the up-stream direction; in some instances, the difference in altitudes exceeded 20 m. If an alluvial fan was present outside the gorge of the wadi, it was generally obvious that the terrace and the upper surface of the fan were in continuity with each other.

The bases of these wadi deposits were often hidden by recent accumulations of loose pebbles and boulders. In most wadis, however, there were many points at which the modern stream-bed had reached the bare rock. This fact indicated that the thickness of recent alluvium was generally very small. It also made it possible to see that the ancient alluvium almost invariably rested directly upon the original floor of the wadi.

Only two wadis were visited—Wadi Derna and Wadi en Naga—in which a more complicated situation was found to exist. In these wadis, two distinct deposits were present, separated from each other by a marked unconformity, and each surmounted in certain places by its own terrace. In both cases, it was only the upper deposit which was continued downstream as an alluvial fan; the lower deposit was found only within the gorge. The lower deposits also showed certain lithological peculiarities, since, although partly of detrital origin, they were composed very largely of tufa. These two instances will be described in detail under the heading 'Tufaceous Deposits'.

ORGANIC REMAINS

The only organic remains found within the alluvial deposits (always excepting those associated with tufa) were the shells of land-snails. These were often present in great abundance. Notes were taken of the species observed in many different sections, but unfortunately little systematic collecting was undertaken. This may partly explain the fact that the number of species recorded was only six: *Rumina decollata* (L.), *Helicella variabilis* (Drap.), *Xerophila cyrenaica* Mart., *X. icmalea* (West.), *Albea candidissima* (Drap.) and *Helix melanostoma* Drap.[1] It should be added also that all these are forms with particularly substantial shells; the more fragile members of the contemporary faunas might well have been destroyed before deposition.

2. CHRONOLOGY

For reasons already given in Chapter III, no *direct* correlation can yet be attempted between any one of the Cyrenaican continental deposits and any Pleistocene successions in other parts of the world. At the present stage, therefore, only two problems can be discussed in connexion with the chronology of the alluvial deposits: the age of the various occurrences with relation to each other, and their ages with relation to the marine shorelines.

As regards those patches of alluvium which are completely isolated, it can only be said that each one is younger than the terrace upon which it rests. In some cases, indeed, the differences of age appears to be very small. To the east of Wadi Derna, for example, both the 20 and the 48 m. terrace were found to be overlain by several metres of weathered, well-rounded pebbles, already mentioned by Pfalz under the name of the 'higher Quaternary gravels'.[2] The two sets of gravel were quite distinct, and it seemed very likely that each had been separately laid down by the waters of Wadi Derna soon after the emergence of its respective marine terrace. Even in these cases, however, it was impossible to be sure that the deposits were not in fact of some more recent date.

It was at first expected that the dating of the alluvial fans and terraces, in spite of their more definite morphological forms, would prove to be equally inconclusive. After a time, it was realized that the actual situation was surprisingly different.

First, as has already been stated, compound alluvial fans seemed to be altogether absent; also, it was only very rarely that more than one terrace was present within the gorges. This could only mean that all the fans and almost all the terrace-deposits had been laid down more or less simultaneously, a conclusion

---

[1] D. 5189–5221.   [2] Pfalz (1931), pp. 16–17.

which was strongly supported by the nature of the archaeological remains found within the deposits.

Secondly, although many of the fans descended to sea-level, none was found which carried any traces of the 6 m. shoreline. On the contrary, the alluvium could often be seen resting directly either upon the shoreline itself or upon its wave-cut platform. This relationship was observed at innumerable points on the north coast of the Gebel, from Derna to Wadi Sleib, 7 km. south-west of Tocra; further to the south-west, the fans retreated too far inland for any such observations to be made.

It was concluded, therefore, not only that all the fans and most of the wadi-terraces were contemporary, at any rate on the north and north-west coasts of the Gebel, but that the whole series was younger than the 6 m. shoreline. According to the conclusions already reached in Chapter III, this means that all these features are almost certainly younger than some part of the Last Inter-glacial. Since they are presumably younger also than the majority of the isolated patches of alluvium, the material of which they are composed will henceforth be referred to for convenience as the 'Younger Gravels'.

3. CONDITIONS OF DEPOSITION

Nothing of interest can be said as regards the conditions under which the isolated patches of alluvium were deposited; their chronology is too uncertain and their geographical extent too limited. This discussion will therefore be confined to the Younger Gravels. The evidence will be considered under three headings: organic remains, morphology and lithology.

ORGANIC REMAINS

Of the six species of snails recorded from the Younger Gravels, every one is still living in the Gebel and none is found far beyond the limits of the Mediter-ranean region.[1] On this evidence, it might be thought that the climate of the Younger Gravels was very similar to that of the present day.

The dispersion of snails, however, must proceed at a very slow rate, and must be almost completely halted by physical barriers such as seas and deserts. For this reason, the mere fact that a snail is now confined to the Mediterranean can scarcely be used as an indication of its climatic preferences, still less of its range of toleration. Thus, although some conclusions might yet be legitimately drawn if only the number of species were larger, it is considered that the fauna of the Younger Gravels, as it is known at present, cannot be used as a basis for climatic deductions.

[1] Zavattari (1934), pp. 906–8.

MORPHOLOGY

As has already been seen, the Younger Gravels, in so far as they lie outside the gorges and above the surface of the sea, form a series of topographical features which are believed to be genuine alluvial fans. From this fact alone it is already possible to reach two conclusions about the conditions under which the deposits were laid down. First, the level of the sea can have been no higher than it is at present. Secondly, deposition must have been initiated by some change of conditions which caused the wadis to steepen their gradients; this is clearly shown also by the forms of the terraces within the gorges.

On the evidence of the marine shorelines, this change cannot have been caused by earth-movements. The only possible alternative is that the wadis suddenly became incapable of carrying the load of material which was supplied to them. Either the water-supply decreased, or the amount of available rock-waste increased, or both changes occurred at once. Unfortunately there is no means of deciding which process took place in the present instance, and in any case nothing is known as to the exact nature of the previously existing conditions. From the morphology of the fans and terraces alone, therefore, no further conclusions can be drawn.

Some additional information can be obtained, however, by comparing the slopes of these features with those of the modern stream-beds. It has already been seen that the stream-beds and the upper surfaces of the alluvium showed, within the gorges, a tendency to converge in the downstream direction. Outside the gorges, this same tendency was noticeable as the stream-beds crossed the alluvial fans. The wadis, in fact, since the time when the gravels were laid down, had once more reduced their gradients. Moreover, in view of the large expanses of bare rock exposed in all but the lowest reaches of every stream-bed, it appeared that this reduction must still be in progress. On these grounds, although it is still impossible to draw any very definite conclusions, it can at least be said that the gravels were laid down at a time when conditions differed from those of today in one, or both, of two respects. On the one hand, the supply of rock-waste may have been greater than at present; on the other hand, the volume of water in the wadis may always have remained below its present maximum values.

LITHOLOGY

The nature of the deposits in itself is scarcely more informative. In rugged country a certain amount of rock-waste, however small, is formed under most conditions. Special climatic conditions are undoubtedly required for the formation of terra rossa, though their exact nature is still controversial. In this instance, however, the material is in any case obviously derived, having presumably originated on the broad and level limestone surfaces of the higher Gebel, where it still occurs, thinly and unevenly distributed, at the present day. It is quite possible, therefore, that the terra rossa of the Younger Gravels was formed long before its final deposition.

Certain deductions can, indeed, be made from the presence of the mud-flows. The formation of a mud-flow, according to Blackwelder, is favoured by four conditions, of which the first two are a steep topography and an abundance of some substance which becomes slippery when wet; the latter condition would be fulfilled in the present case by the terra rossa. The other two conditions are a plentiful supply of water, and insufficient protection of the ground by vegetation. The water-supply, however, if too continuous, will encourage the growth of vegetation; if too plentiful, it will rapidly remove the mud by normal means of transportation.[1] Thus the presence of the Cyrenaican mud-flows implies a rainfall which was heavy but highly intermittent, perhaps similar to that of the present day. Although, so far as is known, mud-flows do not now occur on the northern slopes of the Gebel, this may only be due to the absence of large enough quantities of terra rossa in the catchment areas of the wadis.

So far, no account has been taken of the length of time available for the deposition of the Younger Gravels. If, as seems most likely, the process represents a single Pleistocene climatic phase, its duration can hardly have been more than a few score thousands of years. This is confirmed by the fact that flint implements of one and the same Middle Palaeolithic industry were found not only throughout the gravels but also on their surfaces (see Chapter XI). Yet the gravels of each wadi contain a very large volume of rock-fragments, most of which must have been detached from the bedrock while deposition was in progress. This points to a relatively high rate of disintegration.

An exact estimate of this rate would obviously be out of the question. Nevertheless, there can be little doubt whether it was greater or less than that of the present day. Nowadays in this region the bare rock is widely exposed, often on steep or precipitous slopes, and the degree of exposure has undoubtedly increased in historic times as a result of deforestation. Yet very few active screes were observed, and even these were of small size; their presence could easily be ascribed to such agencies as earthquakes and the growth of tree-roots. As for the loose boulders in the wadi-beds, the majority had obviously been derived from older alluvial deposits. It was clear, in fact, that the present rate of disintegration was far too slow to have produced within the time available the great volume of rock-waste which is represented by the Younger Gravels.

[1] Blackwelder (1928), pp. 478–9.

During the deposition of the gravels, conditions must therefore have been very much more favourable to disintegration than they are at present. As a matter of observation, there are two types of climate in which disintegration is now proceeding at an especially high rate: cold and wet, and hot and dry. In the first case, it is believed that the rocks are shattered by the alternate freezing and thawing of water in fissures; in the second case, the effect has generally been supposed to be produced by differential expansion, resulting from large daily changes of temperature.

In regions whose present climates are relatively hot and dry, former phases of aggradation or of scree-formation have in consequence usually been interpreted as indicating periods of still greater heat and aridity. Recently, however, Blackwelder has pointed out that certain types of rock under modern desert conditions show little or no sign of disintegration. According to his own observations, a high degree of crumbling and exfoliation may be shown by coarse-grained rocks with complex silicates; marbles and quartzites, on the other hand, remain completely untouched, while fine-grained lavas also may be only slightly affected.[1] For this and for other reasons Blackwelder concludes that the disintegration observed in modern deserts should be ascribed, not to thermal expansion, but to some kind of chemical action, probably induced by the presence of very small quantities of moisture.[2]

Whether or not this last conclusion is accepted, Blackwelder's field-observations do at least imply that no phase of disintegration should be attributed to desert conditions until it has first been proved that the rocks concerned are among the types which would in fact be broken up under these conditions. Massive, unmetamorphosed limestones such as occur in the Gebel are not included among the rocks whose behaviour is described by Blackwelder. The writer has therefore attempted to discover from existing publications whether the Tertiary limestones of North Africa are anywhere undergoing rapid disintegration at the present day, and, if so, under what conditions. There is, indeed, some evidence which might suggest that the process is taking place in areas to the south and east of the Gebel, where the climate is both hotter and drier than in the Gebel itself. First, in addition to the normal exfoliation of igneous rocks, in certain localities the limestone itself is reported to have broken into flakes and slabs, which now lie upon the surface of the ground. This effect has been described by Hume from the Cairo area and the central Libyan desert,[3] and by Desio from southern Marmarica and the Harug (27° N., 17° E.).[4] Secondly, enormous slopes of talus are found at the foot of limestone hills in many parts of Egypt and Libya.

[1] Blackwelder (1933), p. 109.   [2] Ibid. pp. 110–11.
[3] Hume (1925), pp. 24–5.   [4] Desio (1939), p. 18.

Neither the slabs nor the talus-slopes are of interest in this connexion, however, unless it can be proved that they are still forming at the present day. In the case of the slabs, such a proof would scarcely be possible. As regards the talus slopes, there is little doubt that such features are still growing, both in Egypt and in Libya, in the vicinity of igneous rocks. On the other hand, the writer has failed to find a single specific reference to the continued growth of a talus-slope at the expense of a limestone bedrock in either country. It can, in fact, be said that no positive evidence has been found which shows that any of the sedimentary rocks of these regions are now producing large volumes of rock-waste from any cause whatsoever.

It is concluded, therefore, that the Younger Gravels of the north coast of the Gebel owe their origin neither to thermal expansion of the rocks nor to chemical effects; on the contrary, it is believed that they were formed at a time when frost was a frequent occurrence during a relatively large part of the year. To the present writer, indeed, it seems very probable that frost may have been responsible for the bulk of the screes and alluvial fans which now exist in Egypt and Libya, and that these are not modern but of Pleistocene age.

It is interesting to note that the same conclusion has been reached by Black-welder himself with regard to the talus-slopes of the Basin-and-Range region of the western United States, a region which for the most part is far hotter and drier than the Gebel. The talus-slopes of this region are abundant, but the majority are 'dead' and many are now being eroded away. As regards their distribution, Blackwelder has noted that they increase in number and volume both towards the north and towards areas of higher altitude. Evidently, therefore, their formation was favoured by cold, and indeed by humidity.[1]

At present, there seems to be little precise information available concerning the limits of the conditions under which rocks can be extensively shattered by frost, but Professor Gordon Manley has made certain suggestions on this point in a personal communication to the writer. In his opinion, the first requirement is an abundance of moisture. Provided this is available, and provided the rocks are sufficiently well jointed, considerable disintegration should result with a daily temperature range of, say, 24–44° F., and a mean temperature of 35° F., even if these conditions were maintained for only one month in each year. The process would also be greatly assisted by intermittent snow showers.

Since the Younger Gravels of Cyrenaica are found in many of the smallest coastal wadis, for example to the west of Derna, they must have been laid down at a time when heavy disintegration took place at altitudes not very far above modern sea-level. At the present day, the January mean temperature at Derna,

[1] Blackwelder (1935).

near sea-level, is 56° F.[1] During the formation of the gravels, the January mean temperature on the present site of Derna may therefore have been as much as 20° F. lower than it is today. It must be admitted, however, that this difference seems excessively great, even if allowance be made for probable changes of sea-level.

CONCLUSIONS

The Younger Gravels were laid down at a time when the sea-level was certainly no higher than it is today. Their deposition was initiated by some climatic change which resulted in a climate with a January mean temperature very much lower than at present. On the evidence of the mud-flows, the rainfall was intermittent but at times heavy. It should also be mentioned that the present tendency of the wadis to increase their gradients can now almost certainly be wholly ascribed to a decrease in the supply of rock-waste; a recent increase of the water-supply, in fact, need not be assumed. During the deposition of the gravels, therefore, the showers of rain, when they occurred, may have been no less heavy than those of the present day.

[1] Information kindly supplied by the Air Ministry Meteorological Office.

CHAPTER VI

CONSOLIDATED DUNE-DEPOSITS

I. DESCRIPTION

Deposits of consolidated dune are found on all the marine terraces of the north coast of the Gebel, and on many parts of the western coastal plain. Very often, as has already been mentioned, they occur as discontinuous patches with no particular morphological form except in so far as they tend to soften the topography of the underlying bedrock surfaces; occurrences of this kind require no further description. Many of these deposits, however, have themselves given rise to definite topographical features. Such features, which are very characteristic of all the flatter parts of the North African coast, may properly be referred to as 'fossil dunes'. It is with the genuine fossil dunes of the Cyrenaican coast that the following account is mainly concerned.

When present, these features occupy a belt immediately adjacent to the modern shore, and of very variable width. In their general form they are similar to modern coastal dunes, except that their outlines have lost all sharp angles and are now gently rounded. Sometimes they occur as continuous ridges, but more often as elongated hillocks, either single or arranged in chains.

Their material, as already stated in Chapter I, consists almost entirely of minute shell-fragments, more or less firmly cemented with calcite. It may be added that the fossil dunes of Cyrenaica have been quarried for building-stone since classical times. As regards their faunas, these, apart from microscopic remains, were again found to be composed almost entirely of land-snails; such collections as have been made by the writer and others are described or mentioned in Appendix C. In addition, small marine molluscs, of species common in the Mediterranean at the present day, were sometimes found.

Desio has already given a certain amount of information regarding the varying development of the fossil dunes on different parts of the Cyrenaican coast.[1] In the following account the subject is dealt with in considerably greater detail. For the west coast, the account is based largely upon Desio's work and upon existing maps; for the north coast, it is based almost entirely upon the present writer's own observations.

[1] Desio (1939), pp. 38–45.

HEAD OF GULF OF SIRTE—TOLMEITA

Fossil dunes are believed to be present around the entire coastline of the Gulf of Sirte. On the south-western and southern shore, the dune-belt appears to be comparatively narrow and compact. Towards the head of the Gulf, as the coast turns north-eastwards, the belt expands. Although its full width in this area is not known, the dunes certainly extend at least as far as the main coast road, which in places lies more than 20 km. from the sea. This fact alone means that the dune-belt is wider here than on any other part of the Cyrenaican coast.

To the north of Agedabia, the width of the dune-belt begins once more to decrease. Along the eastern shore of the Gulf, its landward edge seems to be ill-defined, but the presence of fossil dunes at many points along the coast road indicates that it must here extend about 5 km. inland. The landward edge continues to be ragged as far as the vicinity of Driana, where the presence of numerous outlying fossil dunes has already been mentioned (p. 42). A few kilometres beyond Driana, however, the dune-belt becomes reduced to a single chain, adjacent to the modern shore. About 8 km. north-east of Tocra even this one chain finally disappears. From here as far as Tolmeita the belt is represented only by a few small hillocks partly buried in alluvium.

Two points of interest must be mentioned regarding the fossil dunes of the whole of this stretch of coast. First, wherever the dune-belt is more than a kilometre or so in width, its component dunes show a remarkable difference of development according to their distance from the sea. Those which lie nearest to the sea tend to form a continuous and well-marked chain, or sometimes two or three parallel chains; those further inland, on the other hand, are generally much more widely distributed and less regularly arranged. Moreover, at any point on the coast, the dunes of the marginal chains are almost always higher than those further inland. It may be of interest, incidentally, to give some idea of the heights which these marginal chains can reach. At the head of the Gulf they reach an altitude of over 60 m.; further to the north they decline to about 20 m. in the area south of Benghazi; between Benghazi and Tocra they are seldom more than 10 m. high.

Secondly, the outermost chains of dunes are often separated by shallow salt-water lagoons, known as *sabakh* (sing. *sebkha*); these are also found on the landward side of the whole dune-belt, in places where its total width is small. Good examples occur in the immediate hinterland of Benghazi, which is itself built upon a ridge of consolidated dune, and further to the north-east between Driana and Tocra.

Finally, mention must be made of an unusual feature shown by certain fossil dunes at the head of the Gulf. These dunes are crossed by the main coast road, between Mersa Brega and a point about 20 km. west of Agedabia, and their aeolian nature is clearly revealed by the appearance of their bedding as exposed in numerous road-cuttings. Their material is marine sand, presumably derived from the shores of the Gulf, yet the orientation of each dune is roughly from north-west to south-east, that is, at right-angles to the line of the coast. It seems very probable, therefore, that they represent a series of genuine longitudinal (*seif*) dunes.

TOLMEITA—RAS EL HILLAL

On the coast between Tolmeita and Ras et Tin, where the escarpment runs close to the sea, fossil dunes are only intermittently present, and are always confined to a very narrow belt adjacent to the modern shore. As a rule the belt comprises simply one well-marked chain, or sometimes two.

The first large development beyond Tolmeita begins at the mouth of Wadi Giargiarummach, and extends for 25 km. towards the north-east. It consists of two separate chains of hillocks, accompanied in many places by *sabakh*. For the first 10 km. beyond the wadi, as far as El Hania, the dunes are not impressive. From this place onwards, however, the hillocks of the landward chain reach unusually great heights; almost all approach 50 m. above sea-level, and one rises as high as 62 m. With their steep sides, and a tendency to assume sugar-loaf forms, the fossil dunes of this stretch of coast present a most remarkable appearance.

Between Ras Aamer and Ras el Hillal fossil dunes are confined to two main localities. The first of these is at Apollonia, a part of which town is built upon a single chain of dunes, 7 km. long and rising to a height of 35 m.; the chain is interrupted at many points, a fact which to some extent at least is due to its having been partially buried beneath the alluvial fans of a series of wadis. The second locality lies 6 km. to the west of Ras el Hillal, where a short but well-developed fossil dune, known as Ras el Aslab, lies along the seaward border of the alluvial fan of Wadi bu Sahela.

RAS EL HILLAL—RAS ET TIN

On the steep coast immediately to the east of Ras el Hillal, consolidated dune-deposits are rare. When they occur at all, they take the form of steep banks piled up against the cliffs which overlook the modern shore; at some points these banks can be seen to descend below sea-level.

Between Ras ben Gebara and Derna, however, there are five distinct groups of genuine fossil dunes, of particular interest owing to their possession of certain features in common. First, each group is situated near the mouth of a large wadi, and, apart from one exception, always on its western side. Secondly, although most of this stretch of coast has a remarkably regular orientation from west-north-west to east-south-east, the dunes occur as chains which run from west to east. Lastly, each group of dunes projects to some extent beyond the general line of the coast; thus each one forms a well-marked headland, either by itself or in conjunction with the alluvial fan of the associated wadi.

From west to east, the names of the headlands and of their associated wadis are as follows:

Ras ben Gebara	Wadi ben Gebara
Chersa	Wadi el Angil
Ras bu Meddad	Wadi en Naga
(Name unknown)	Wadi bu Msafer
Ras bu Azza	Wadi Derna

The one headland which lies to the east of the mouth of its wadi is Chersa. Even in this case a deserted channel still exists which shows that Wadi el Angil in the not very distant past entered the sea to the east of the headland.

The dune-chains of Ras ben Gebara and Chersa are short and compact, and compose the entire headland at both localities. At each of the three remaining headlands, however, the chains are considerably longer, and the acute angles between them and the original line of the coast are occupied by expanses of red alluvium, connected in each case with a normal aggradation terrace within the gorge of the wadi. The structure of the largest headland of all, Ras bu Azza,[1] is illustrated in Fig. 6. Exactly the same structure and the same triangular form is shown by Ras bu Meddad and the unnamed headland at the mouth of Wadi bu Msafer. Only the scales are different, Ras bu Meddad being smaller than Ras bu Azza, and the unnamed headland being smaller still.

Between Derna and Ras et Tin, the coast once more becomes precipitous, and consolidated dune-deposits, as to the east of Ras el Hillal, are again represented only by occasional banks. From air-photographs, it appears that genuine fossil dunes are abundant in the vicinity of Ras et Tin itself, where they occur in chains running from north-west to south-east.

2. EVIDENCE FOR SUBMERGED FOSSIL DUNES

Everywhere on the Cyrenaican coast the fossil dunes are now being attacked by the sea. In some places fragments of chains have been almost detached from the shore to form peninsulas; in other places they have been completely detached to

---

[1] The name is here used to designate the whole of the promontory upon which Derna and its suburbs are built.

form islands, either single or in chains. It is obvious, however, that marine erosion is not the sole cause of this disruption. Wherever the fossil dunes are in contact with the sea, they can always be observed to pass beneath its surface; evidently there has also been some degree of submergence. The same conclusion is suggested by the salt-water lagoons which in so many places lie between or behind the dunes, and which often have a very restricted outlet to the sea.

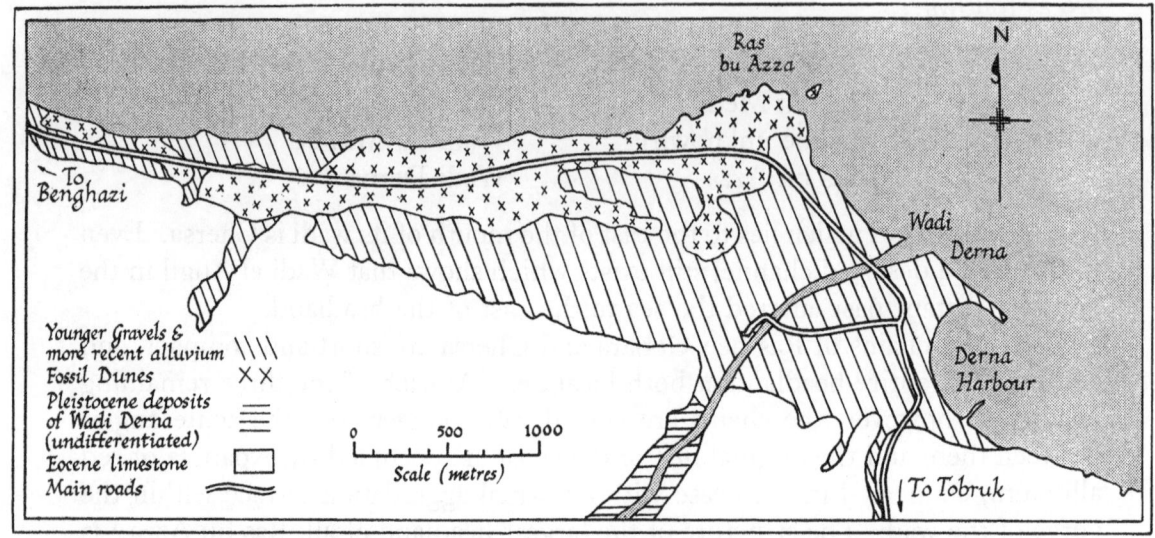

FIG. 6. Deposits near mouth of Wadi Derna

Geological boundaries determined partly by personal reconnaissance, partly from air-photographs. The seaward boundary of the Eocene is everywhere marked by a sharp fall of level.

It might therefore be expected that charts of the sea-bottom, if of sufficiently large scale, would show traces of fossil dunes which are now completely submerged. It is found in fact, as has already been noted by Desio,[1] that many such traces can be located, often in an excellent state of preservation.

In several of the large-scale insets on the Admiralty charts, areas are covered in which chains of fossil dunes have been partly overwhelmed by the sea. One of the clearest examples is at Apollonia, where the chain upon which the town is built has been deeply indented by the formation of two small bays, separated by a peninsula. About 700 m. to the east-north-east of the peninsula are two islets, both composed of consolidated dune. From the Admiralty chart,[2] it can be seen that the islets are joined both to each other and to the peninsula by a line of shoals, only just submerged, and that the water on their landward side has a

---

[1] Desio (1939), pp. 38–45.　　　　[2] No. 3355; inset Apollonia, scale 1:12,500.

maximum depth of about 4 m. These features can only represent an offshoot of the main chain of dunes, and, unless some local scour has taken place, the figure of 4 m. must denote the minimum depth to which they have been submerged.

A minimum submergence of the same order is indicated at Derna, where the main chain of fossil dunes is continued offshore for some 600 m. to the east of Ras bu Azza by an islet and an almost completely submerged shoal. At Tolmeita, a chain of shoals runs north-eastwards from the peninsula of consolidated dune which carries the lighthouse; in this case, the implied submergence is somewhat greater.[1] The greatest submergence of all is indicated on the chart of Mersa Brega, near the head of the Gulf of Sirte.[2] From the peninsula upon which the fort of Mersa Brega is built, a line of shoals runs out to sea and continues towards the north-east, parallel to the coast, for about 10 km. According to Desio, this again represents the remains of a chain of dunes,[3] in which case the submergence, always assuming the absence of scour, cannot have been less than 20 m.

In all the cases so far mentioned, the nature of the various shoals and islets is betrayed by their alignment with chains of fossil dunes on shore. On the Admiralty charts, however, many groups of shoals which nowhere approach dry land can be seen to exist off the coast of Cyrenaica. A minor example occurs on the Apollonia inset, which shows a line of shallows running roughly from east to west at a distance of 500–600 m. from the shore, with depths of water up to about 11 m. on its landward side. Many similar features exist which are large enough to be visible on the 1 : 500,000 charts; these appear to be especially common from Tocra southwards, where they often take the form of long ridges, running parallel to and some distance from the shore. The following table gives some details of the more important examples, the second and third of which have already been mentioned by Desio.[4] Depths and distances have been obtained from the Admiralty chart,[5] and, in view of its small scale, must be regarded as approximate.

Locality	Distance offshore (km.)	Length (km.)	Maximum depth on landward side (m.)	Named points
North of Driana	3·5	15	17	Secche di Driana
North and south of Zuetina	1·5–6·0	35 +	16	Elfie Rock, Tre Scoglie
South-west of Zuetina	6–8	30 +	35	Garah Island, Hericha Rock, North Lamaresk Reef
North of El Agheila	5	13 +	22	Bu Sheefa Rock

[1] Adm. chart, No. 3355; insets Derna and Tolmeita, scale 1:25,000.
[2] Adm. chart, No. 3354; inset El Brega, scale 1:50,000.
[3] Desio (1939), p. 45.   [4] *Ibid.* pp. 44–5.   [5] No. 3354.

The first of these ridges rises to within 4 m. of the surface; each of the others is exposed at one or more points.

Largely on the grounds of their general dimensions, Desio has already suggested that all such ridges, in spite of their complete separation from the land, may be submerged fossil dunes.[1] Although proof could be supplied only by the collection of rock-specimens, all the available evidence does in fact tend to support this view.

In the first place, there is no obvious alternative explanation. All of the features are much too high to be beach-ridges. They can hardly be coral-reefs, since none are known to have been formed in the Mediterranean on a large scale in recent geological times. Neither is it likely that they are submarine scarps, in view of the very gentle dips of the beds on the mainland.

In the second place, not only their dimensions but also their topographical details are remarkably similar to those of the known submerged dunes, and, in some cases, to those of the dunes on dry land. These details can be studied in the two cases of the El Agheila reef and the inner reef off Zuetina, both of which are partially covered by large-scale insets on the Admiralty chart.[2] The latter is seen to be composed of large numbers of hillocks irregularly distributed over a belt 2–3 km. wide. Both in scale and topography, it closely resembles the reef off Mersa Brega, which, as already mentioned, is almost certainly a submerged chain of dunes. The reef off El Agheila is even more remarkable. It consists of two distinct parallel ridges, whose crests are about 1·5 km. apart but whose combined width is not more than 2 km. Both ridges can be traced across the entire width of the inset, that is, for 13 km. Although the outer ridge rises as high as − 5·5 m. while the inner ridge breaks surface at the Bu Sheefa Rock, the intervening trough is as much as 26 m. deep. These two ridges can only be compared with the fossil dunes on the mainland, and, moreover, with those which have been especially well developed and preserved.

It thus appears that Desio's view is almost certainly correct. If so, submerged dunes must be fairly common on the inner edge at least of the continental shelf off the coast of Cyrenaica. It is likely, moreover, that the ridges so far located comprise only a small proportion of those which actually exist, since the soundings on the Admiralty charts become increasingly widely spaced according to their distance from the shore.

[1] Desio (1939), p. 44.
[2] No. 3354; insets Ez Zuetina, El Agheila, scale 1:50,000.

3. CHRONOLOGY

The dating of the consolidated dune-deposits presented problems similar in many respects to those which arose in the case of the alluvial deposits. Once again, it would only be possible to say, in general, that each occurrence was younger than the marine terrace upon which it rested, a conclusion which would be of very little interest if the terrace itself were of any great age. An additional difficulty might arise, moreover, from the fact that the uses of morphological evidence would in this case be strictly limited. First, most marine dunes occupy smaller areas than most alluvial fans. Secondly, marine dunes are quickly and easily formed, and once consolidated are not easily destroyed; dunes of many different ages may thus be preserved intact within one small area.

In spite of all theoretical possibilities, however, the actual situation again proved to be both simpler and more conclusive than had been expected. In the first place, of all the occurrences of consolidated dune which were examined on the coast between Benghazi and Derna, and which lay partly or wholly below 6 m., not a single one was found to carry any signs of the 6 m. shoreline. All such occurrences must therefore have been younger than the 6 m. shoreline; in many cases, indeed, they could be seen to rest either upon the shoreline or upon its associated terrace.

In the second place, it was found that those occurrences whose relative ages could be demonstrated in this way included all the genuine fossil dunes on the coast between Tocra and Derna, with a very few minor exceptions, and the whole of the high and continuous 'marginal' chain between Benghazi and Tocra. Yet these features in themselves comprised the vast bulk of all the consolidated dune-deposits on this particular part of the coast. Here, at any rate, it could thus be said that almost all the consolidated dune-deposits, in addition to almost all those of alluvial origin, must have been laid down after the regression of the sea from the 6 m. shoreline. This was clearly true also of all those fossil dunes lying offshore between Benghazi and Derna in a state of partial or complete submergence; although they appeared to have survived their present submergence remarkably well, they could hardly be supposed to have survived in addition a previous period of submergence during which wide platforms were eroded, in many places, in solid Tertiary limestones.

By analogy with the alluvial deposits, all consolidated dune-deposits which were shown to be of this age will henceforth be referred to as 'Younger'.

Of those few occurrences which could not be shown to belong to this category, the majority were in the form of isolated sheets, generally of small area, lying on the higher marine terraces. In addition, certain genuine fossil dunes on the coast to the south-west of Tocra were found to lie wholly on the landward side of the 6 m. shore-line, as for example in the hinterlands of Driana and Benghazi, but these were few, low, and irregularly developed in comparison with those nearer the sea. Even these occurrences could not be finally excluded from the 'Younger' category, though a greater age was in fact suggested in some instances by their more advanced state of weathering and cementation. Perhaps in most cases the deposit was laid down soon after the final regression of the sea from the terrace upon which it rested.

It now remains to consider the relationship between the Younger Dunes and the Younger Gravels. At most localities where both were present, it was clear that much of the gravel had been laid down since the formation of the dunes. Moreover, no indications were found that any of the fossil dunes had been formed after all deposition of gravel had ceased. It was at first believed, there-fore, that these two terrestrial deposits must represent two distinct periods of sedimentation.

Subsequently, however, evidence accumulated which showed that this was not so. The most interesting evidence in this respect was found along the modern shore to the west of Derna. At most points on this part of the coast the bases of the fossil dunes were entirely submerged. A few sections were found, neverthe-less, in which their bases could be seen to rise above sea-level. One such section lay at the extreme western end of the Derna chain, 4·5 km. west of the town, and here it was found that the dunes rested, not upon the bedrock, but upon red gravels similar to those which constituted the alluvial fans; the base of the gravels themselves remained submerged. Further to the west, as the end of the dune-chain was approached, the layer by which it was represented in the section became thinner, and a second bed of red gravels appeared above it. Eventually the layer of dune vanished altogether, and the two beds of gravel coalesced with no sign of an intervening unconformity.

Still further to the west, similar relationships were observed in the case of several small fossil dunes lying to the east of the mouth of Wadi bu Msafer. In many places, Middle Palaeolothic implements were found both above and below the dune, those from the two levels showing no apparent typological differences; this fact indicated that the two beds of gravel were in each case as nearly identical in age as in appearance. Sections of still greater interest were provided by the banks of Wadi bu Msafer itself. Having first cut through its own alluvial fan, this wadi, just before reaching the sea, cuts through the eastern tip of the chain of dunes with which the fan is associated. In the resulting sections, which are illustrated in Fig. 7, it could clearly be seen that the whole chain of dunes was actually 'floating' within the gravels of the alluvial fan.

At Chersa, as already mentioned, the headland appears to be composed entirely of fossil dunes. Yet at its north-west corner recent marine erosion has exposed a section 12 m. high in which the dunes were seen to contain no less than three successive layers of gravel, each with a maximum thickness of 1 or 2 m. A fourth layer was represented, in addition, by the gravels resting against the dunes on their landward side.

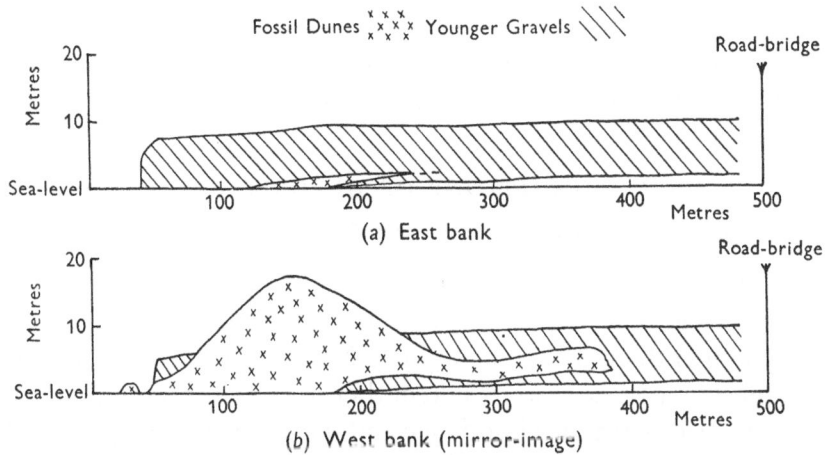

FIG. 7. Relations between Fossil Dunes and Younger Gravels at
lower end of Wadi bu Msafer

Ground levels were taken from the nearest points at which the surfaces of the dunes and of the alluvial deposits appeared to have been unaffected by recent subaerial erosion; these were never more than a few metres from the wadi-banks. In other respects the diagrams represent actual sections in the sides of the wadi. Average width of wadi: 25 m. Vertical exaggeration: × 5.

Between Ras ben Gebara and Ras el Hillal, although large fossil dunes were absent, several sections of alluvial fans were found which contained one or more layers of consolidated dune. In the steeply-inclined fan of Wadi en Nmelaia, $1\frac{1}{2}$ km. west of Ras ben Gebara, one such layer was seen, showing the same inclination as the surface of the fan, and rising to a considerable height above sea-level. Near Marsa el Hillal, the gravels exposed in the sides of Wadi el Glaa at a distance of at least 200 m. from the sea were found to include five or six bands of dune, each one perhaps half a metre thick.

Little further evidence was found to the west of Ras el Hillal; the bases of the fossil dunes seemed always either to be hidden or to rest upon bedrock. An interesting section, however, was seen in a well 2·5 km. to the south-west of Tocra. The well was situated near the foot of the landward slope of the younger fossil dunes, and was found to pass, first, through 1 m. of consolidated dune, and secondly, before reaching the bedrock, through 1·5 m. of terra rossa, apparently water-laid.

The Younger Dunes and Younger Gravels were not, therefore, laid down at completely different times. On the contrary, although the formation of the dunes ceased before that of the fans, and although most of the dunes appear to be older than most of the gravel, some of the dunes at least were laid down after the deposition of the gravel had begun. In the case of the dunes which are now submerged, their formation must have occupied a very considerable period of time, and may well in fact have continued throughout the whole of the time in which the gravels were being laid down.

4. CONDITIONS OF DEPOSITION

DEPOSITS POSSIBLY OLDER THAN THE 6 M. SHORELINE

Because of their extremely uncertain chronology, there is little to be said concerning the conditions of deposition of those dune-deposits which cannot be shown to be younger than the 6 m. shoreline. Nevertheless, some mention should be made of their remarkable scarcity and of their relatively poor topographical expression, as compared with the abundance and impressive development of the undoubtedly 'younger' dune-deposits. These characteristics are all the more remarkable in view of the fact that many of the occurrences, though lying on the landward side of the 6 m. shoreline, may still, after all, belong to the Younger Dunes.

The truncated state of the older marine terraces is not a sufficient explanation; the Younger Dunes themselves often lie adjacent to, or even on top of, the cliff of the 6 m. shoreline. Subaerial erosion is a possible cause, but not very probable, since the existing patches of dune on the higher terraces have often become as hard and as compact as the bedrock itself. It can only be supposed that conditions on the Cyrenaican coast, after the withdrawal of the 6 m. sea, were more favourable to dune-formation than at any other time during the Pleistocene. At present, however, it is impossible to suggest any convincing reason why this should have been so.

In this connexion, it may be pointed out that the relative development of the 'marginal' chain of dunes is even more striking on the eastern shore of the Gulf of Sirte than between Benghazi and Tocra. In the absence of any information about the position and present levels of the 6 m. shoreline, it is not yet possible to say to what extent the dune-belt to the south of Benghazi can be described as 'younger'. Nevertheless, it is strongly suspected that the marginal chains in both areas are of similar ages, and that there may be similar reasons for their relatively large dimensions.

YOUNGER DUNE-DEPOSITS

As in the case of the Younger Gravels, the nature of the Younger Dunes does not in itself imply any special conditions of deposition. The mere existence of the deposits means only that supplies of loose marine sand were available, and that the wind was at times strong enough to transport it. The bedding of the fossil dunes was observed in many places, but these observations led only to the obvious conclusion that the winds which carried the sand had come generally from a northerly direction. As regards the fauna of the deposits, almost all the species found are still living in the territory. The fauna from the fossil dunes of Benghazi does, indeed, include one species of land-snail, *Helicella cespitum*, which at present lives no nearer than northern Tripolitania.[1] Since, however, it is still common on most other parts of the Mediterranean coast, its disappearance from Cyrenaica does not necessarily imply any climatic change.

It is the distribution of the Younger Dunes which is particularly informative. This will now be discussed under two headings: the vertical distribution of the deposits in relation to present sea-level, and their horizontal distribution along the coast.

Vertical distribution

Although it has long been realized that the fossil dunes of the Cyrenaican coast must to some extent have been submerged since their formation, there has been little agreement on the causes of the submergence. The latest discussion of the subject is due to Desio.[2] In his opinion, the varying depths to which the dunes have been submerged indicate that the coast from Apollonia to El Agheila has undergone differential warping in recent or Pleistocene times. Desio also considers that the same movements have affected the eastern part of the Cyrenaican coast as far as the Egyptian frontier, the evidence in this area being provided by the depths of water in the harbours of Tobruk and Bardia and in the inlet of Gazala, each of which is clearly a drowned valley.

In his deductions from the drowned valleys, Desio may well be right. As regards the submerged dunes, there is, indeed, no proof that those in the Gulf of Sirte may not lie at their present great depths partly, or even wholly, as a result of movements of the land. Between Benghazi and Derna, however, the constant level of the 6 m. shoreline shows that here, at any rate, the relative lowering of the sea-level must be ascribed entirely to a movement of the sea itself. For the formation of the Secche di Driana, the sea-level must therefore

[1] Gambetta (1934), p. 272.
[2] Desio (1939), pp. 52-3.

have been at least 17 m. lower than it is at present. There is, of course, no evidence that it was not lower still.

Horizontal distribution

One of the most striking features of the younger fossil dunes is their wide distribution, in comparison with that of the modern coastal dunes. At the present day, indeed, dunes are still in the process of formation at many points around the Gulf of Sirte. In this region, they are often as high as the fossil dunes, which in fact they sometimes bury; as for their extension towards the interior, a 'live' dune was observed by the writer near the coast road 24 km. east of El Agheila, at a distance of 5 km. from the sea. Live dunes continue to be present as far north as Benghazi. A short distance to the north-east of this town, however, they finally vanish; from here to Derna, no dunes of any size are now forming, and the same is probably true of the coast as far as Ras et Tin.

The restricted distribution of the modern coastal dunes cannot be ascribed to local climatic differences. It is obvious, in fact, that it is closely related to the widths of the modern beaches. Wherever dunes are forming at the present day, there appears to be a wide, gently sloping foreshore, upon which large expanses of loose sand are deposited during storms. Between Benghazi and Ras et Tin, on the other hand, the beaches are for the most part very narrow. To some extent, this is due to the presence of cliffs, composed either of the bedrock or more usually of Pleistocene alluvium or consolidated dune. Very largely, however, it is due to the steep initial slope of the surfaces upon which the beaches are laid down; on these surfaces no amount of stormy weather unaided by tides could spread out expanses of sand sufficient to support the existence of large dunes.

It is clear that in most places these surfaces are nothing more than direct, almost unchanged continuations of the 6 m. terrace. Yet it was upon this same terrace that the Younger Fossil Dunes were laid down between Benghazi and Derna. These features, therefore, cannot have depended upon occasional storms for their supplies of sand. It can only be supposed that their formation did not begin until the sea had already fallen far below the 6 m. shoreline, thus exposing an adequate quantity of loose sand. In other words, all the Younger Fossil Dunes on the north coast of the Gebel, even those which are still high and dry, must have been laid down at a time when the sea-level was considerably lower than it is at present. This conclusion, of course, applies equally to the formation of the Younger Gravels, about which, on the evidence of the deposits themselves, it could only be said that it took place when the sea-level was 'no higher' than today.

Even the fossil dunes, however, are not absolutely continuous; from Tocra to Ras et Tin the dune-belt shows many wide gaps, whose significance must now be discussed. To some extent the gaps are due to burial beneath alluvium, as between Tocra and Tolmeita; to an even greater extent, they are due to the rugged nature of the coast, as between Tolmeita and Wadi Giargiarummach, between Derna and Ras et Tin, and at certain points between Ras Aamer and Ras ben Gebara. Nevertheless, there are many stretches of coast which show gentle slopes with no excessive cover of alluvium, and which are still free from fossil dunes.

It is impossible to ascribe such gaps to local deficiencies of sand; the sand was composed almost entirely of organic remains which would be continually supplied by the sea itself. By far the most likely explanation for their existence is suggested by a study of their own distribution. Without exception, every one lies to the east of Ras Aamer; all gaps to the west of this point can be ascribed to one of the two causes already mentioned. It is at Ras Aamer, however, that the coast-line of northern Cyrenaica undergoes a decided change of direction; having faced north-westwards from Benghazi onwards, it here begins to face northwards and, ultimately, towards the north-north-east. Almost certainly, therefore, the peculiar distribution of the Younger Fossil Dunes is an indication that the prevailing winds, at the time of their formation, blew from the north-west. It may be added that this is the direction of the prevailing winds on the north coast of Cyrenaica at the present day.[1]

This conclusion can hardly be invalidated by the fact that fossil dunes are not indeed entirely absent from the coast between Ras Aamer and Ras et Tin; the major occurrences, apart from those at Ras et Tin itself, are only seven in number, and all are fairly restricted in area. It is interesting, nevertheless, to consider what local peculiarities might have allowed the formation of these seven groups of fossil dunes. Two, at Apollonia and Ras el Aslab, can easily be explained; at both localities the coastline again faces approximately towards the north-west for a short distance. The five remaining occurrences, however, are the five headlands between Ras ben Gebara and Derna, and this part of the coastline, as has already been mentioned, faces steadily towards the north-north-east, with scarcely a deviation apart from the headlands themselves.

The most probable explanation for the presence of these particular dunes is suggested by their invariable association with the mouths of wadis. The positions of these mouths being fixed within narrow limits by the positions of the gorges, it must be supposed that the wadis themselves provided conditions in some way favourable to the growth of dunes. In at least three cases the dune-chains as-

[1] *Med. Pilot*, vol. v, pp. 10–11.

sociated with the headlands are known to be partly founded upon alluvial deposits. This may well be true also for the two remaining headlands. If so, there can be little doubt that the favourable conditions were provided simply by the presence of incipient alluvial fans. Since the western sides of the fans would face northwards, or even north-westwards, their surfaces might well have been capable of causing the north-west winds to drop their load, even when the rest of the coastline, being wrongly orientated, would have had no such effect. The resulting dunes would naturally tend to lie to the west of the wadi-mouths, and at an oblique angle to the original coastline; this, in fact, is exactly what has occurred.

Finally, it may be pointed out that the fossil dunes on the eastern shore of the Gulf of Sirte, whatever their age may be, are both higher and more abundant towards the head of the Gulf than elsewhere; in the same area, the development of submarine dunes also reaches its maximum. This again is a stretch of coast which faces towards the north-west.

CONCLUSIONS

The younger fossil dunes were formed at a time when the sea was retreating, thus leaving large areas of loose sand exposed to the air. The retreat, at any rate between Benghazi and Derna, was entirely due to a general fall of sea-level, which must have continued to a depth at least as low as -17 m. From the distribution of the dunes, it seems likely that the prevailing winds were from the north-west.

5. EVIDENCE OF POST-CLASSICAL SUBMERGENCE OF FOSSIL DUNES

At many different localities, artificial excavations have been made in the fossil dunes of the Cyrenaican coast. The majority of these are either stone-quarries or tombs dating from classical times. All such excavations must originally have lain well above sea-level, yet many are known which are now partly or wholly submerged. This condition has hitherto been ascribed to local tectonic movements.

In view of the stability of the 6 m. shoreline, however, this explanation must now be abandoned; it can only be concluded that the sea-level itself has risen since the excavations were made. As Oakley points out, strong indications of a post-classical rise of sea-level have in fact already been found both in Flanders and in Southern England.[1]

From the evidence on the coast between Benghazi and Derna, it may eventually be possible to obtain some exact information, in terms of altitudes and absolute

[1] Oakley (1943), p. 57.

dates, concerning the history of this late eustatic movement. So far, only one piece of evidence has been examined. In the fossil dunes on the west side of the Apollonia peninsula, tombs have been cut, according to Marchetti, which now lie at a depth of 3 m. below sea-level, and which contain Greek pottery.[1] The rise of sea-level since the tombs were excavated can hardly have been less than 4 m.; although the exact age of the tombs does not appear to be known, it is at least certain that the movement must have occurred since the second century B.C., when Apollonia was founded.

[1] Marchetti (1935), pp. xciv–xcvi.

CHAPTER VII

TUFACEOUS DEPOSITS

1. INTRODUCTION

In many of the wadis which were examined on the north coast of the Gebel, continental deposits were altogether lacking. Where any such deposits remained these, in the majority of cases, were of an alluvial nature, and could almost invariably be assigned to the so-called Younger Gravels. Nevertheless, other types of deposits also occurred which consisted partly or wholly of tufa; this material in fact was an almost constant feature of those wadis which contain perennial springs at the present day.

Most of these occurrences were hardly worthy of a detailed examination. The great majority were small and local, associated either with the springs themselves, as in Wadi Bent, 4·5 km. east of Derna, or with waterfalls, as in Wadi el Glaa near Ras el Hillal. Some were clearly modern, and enclosed more or less unaltered plant-remains; by far the largest number could not be dated at all. Four occurrences were found, however, which were relatively extensive, and each of which could be shown, for one reason or another, to be of some antiquity. These in consequence were studied with considerable care. Since each one differed in various ways from the rest, and since all four were widely separated, they will be discussed below under individual headings. A final discussion will then deal with the conclusions which arise when all four deposits are considered in relation to one another.

2. WADI DERNA

TOPOGRAPHY AND HYDROLOGY

Wadi Derna, with a total length of 75 km., is the second longest wadi on the north coast of the Gebel, and the longest of all those which pass through the coastal escarpment before reaching the sea. For much of its course it is a relatively open valley cut in Miocene and Oligocene rocks. 12 km. from its mouth, however, it reaches the top of the hard and massive Eocene limestones, and from this point onwards it becomes steadily deeper, taking the form of a steep-sided gorge with an almost diagrammatically V-shaped cross-section. Having cut through almost the whole of the Eocene succession, the gorge then terminates abruptly at the coastal escarpment. For the last 1300 m. of its course, the wadi is no more than a wide, shallow trench across the alluvial fan upon which Derna is built.

Wadi Derna is remarkable not only for its size, but also for the fact that it contains two perennial springs, both of which are among the most copious of any in Cyrenaica. The higher of the two is situated 12 km. from the sea, and is known as Ain bu Mansur; it owes its existence to the Lower Oligocene spring-forming horizon. The lower spring is Ain Derna, or Ain el Bled, which issues

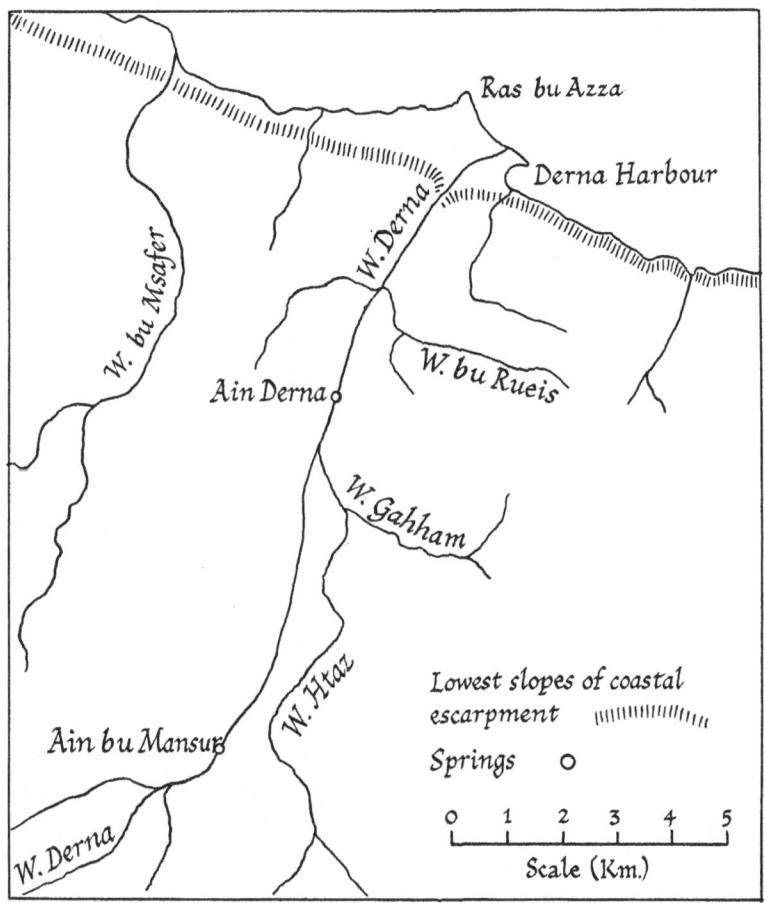

FIG. 8. Map of Derna area, showing wadis and perennial springs

from the Middle Eocene limestone about $5\frac{1}{2}$ km. from the sea. Much of the water from both sources is diverted into irrigation channels, and some of the water from Ain Derna is piped into the town for domestic use. Nevertheless, the remaining water from Ain Derna is still so abundant that it does not finally sink into the wadi-bed, even in summer, until it is within about 4 km. of the sea. The positions of these springs are shown in Fig. 8.

WADI DEPOSITS: GENERAL INTRODUCTION

The terrestrial deposits of Wadi Derna have been the subject of many previous references. The earliest of these seems to be an account by Spratt, who described a deposit of whitish marl and travertine, rising to a height of 100 ft. above the stream-bed on either side of the entrance of the gorge. Within the deposit were abundant stems of reedy plants, and a few fragments of fresh-water shells.[1] The same deposit was later described by Gregory, who considered it to be a delta-fan. In addition, Gregory described a second deposit, consisting of 20 ft. of calcareous sands and gravels interbedded with travertine, and forming a platform between two branches of the Wadi 'about 6 miles from Derna...at the springs of Bon-mansur' (presumably Bu Mansur). The surface of the platform was covered with flint implements.[2] Within the sands, shells were found which were later identified by R. B. Newton as *Hygromia sordulenta* (Morelet).[3]

In 1920, Marinelli made the first reference to the existence of a terrace within the lower reaches of the gorge. The upper surface of this feature was described as lying about 30 m. above the bed of the Wadi.[4] The level of 30 m. was then quoted by Stefanini, who also mentioned Gregory's platform, with its surface 'a few metres above the modern stream-bed'.[5] Finally, Stefanini himself was quoted by Huzayyin, who concluded that Wadi Derna contained two terraces, rising respectively to 30 m. and a few metres above its modern bed.[6]

On investigation, the present writer found that two terraces were indeed present in Wadi Derna; moreover, that their heights above the modern stream-bed were in rough correspondence, at any rate in the lower reaches of the Wadi, with those given by Huzayyin. Certain other observations were made, however, which could not be reconciled with previous descriptions.

The deposits of the higher terrace rested directly against the rock walls of the gorge, and, in the few places where this was exposed, could be seen to extend downwards to its original rock floor. The last remnants of these deposits lay just inside the entrance of the gorge. From here upstream, they were followed for 13 km., for the whole of which distance they were continuously present, though often on only one side of the present wadi-bed; at the highest point examined, moreover, they still showed no signs of disappearance. As stated in the earlier reports, the deposits included large masses of tufa.

The deposits of the lower terrace lay on either side of the wadi-bed, banked up against those of the higher terrace. They were preserved only as isolated frag-

---

[1] Spratt (1865), pp. 378–9.
[2] Gregory (1911), pp. 578–80.
[3] Newton (1911), p. 619.
[4] Marinelli (1920), p. 71.
[5] Stefanini (1930), p. 25.
[6] Huzayyin (1941), p. 80.

ments, often without well-defined upper surfaces; when these surfaces were present, they generally stood 4–5 m. above the wadi-bed. These deposits, however, included no tufa, neither were they found anywhere near Ain bu Mansur. On the contrary, they consisted almost entirely of well-rounded pebbles and boulders, with a certain proportion of terra rossa, while as regards their distribution they were never identified for certain at a distance of more than $2\frac{1}{2}$ km. from the mouth of the Wadi. Although no physical contact remained, it was in fact clear from their levels and lithology that the deposits had originally been continuous with the gravels of the Derna alluvial fan.

In other words, the genuine lower terrace of Wadi Derna was a feature which apparently had never been previously mentioned. Many attempts were made to find Gregory's 20-ft. platform, upon the description of which the idea of a second terrace had originally been based. It was eventually decided that this description must have referred to a great mass of tufa, resembling the prow of a ship, which occupies the angle between the main wadi and a large tributary, Wadi Gahham, at a distance of $6\frac{1}{2}$ km. from the sea. In many respects this mass of tufa corresponded closely to Gregory's description, but there were two discrepancies. First, it lay $5\frac{1}{2}$ km. downstream from Ain bu Mansur. Secondly, its upper surface lay nearer 100 than 20 ft. above the wadi-bed; it formed, in fact, part of the higher terrace.

HIGHER TERRACE: MORPHOLOGY

Before describing the deposits of the higher terrace, it will be convenient to give an account of its variations of level. The upper surface of the deposits, that is to say the terrace itself, was almost always easily recognizable, in spite of the effects of erosion, cultivation and the accumulation of slope-deposits. It was therefore decided to make an accurate survey of its levels over as great a distance as possible, partly in the hope of throwing light on the conditions under which the deposits were laid down, partly in order to make sure that the whole of the surface was of the same age. Levels were taken with a telescopic alidade; distances were obtained mainly from maps and air-photographs, but sometimes by tacheometry. In the lower reaches of the wadi the sea itself was used as a datum; farther upstream one spot-height was available, at 111 m., on the tufa platform by the mouth of Wadi Gahham. During the two expeditions, the terrace-surface was surveyed as far as a point about $8\frac{1}{4}$ km. from the sea; for much of this distance the present wadi-bed was also surveyed.

The resulting profiles are shown in Fig. 9. The last occurrence of the terrace-deposits, from which the lowest point on the profile was taken, lay just inside the entrance of the gorge, 1·5 km. from the sea. It consisted of an isolated and much-quarried bluff projecting from the west side of the wadi; its upper surface, occupied by an Arab cemetery, rose to a height of 24 m. above the wadi-bed and 35 m. above sea-level. From this point almost as far as Wadi Gahham, both the terrace and the wadi-bed showed a fairly steady upward gradient, with an average value of 1 in 90. The vertical distance between the two profiles varied between 25 and 35 m., a variation which could largely be ascribed to the irregularity of the wadi-bed. About 200 m. below Wadi Gahham, however, the gradient of the terrace suddenly increased to an average value of 1 in 18; then, equally suddenly, after another kilometre it once more flattened out.

The survey was only continued for a short distance above this steep rise. It was possible, nevertheless, to obtain a level for the terrace-surface at Ain bu Mansur, since the altitude of the spring itself was known from the maps to be

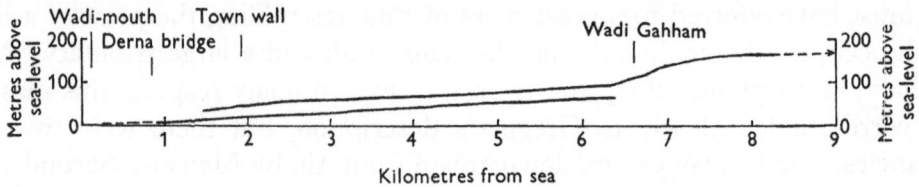

FIG. 9. Profile of higher terrace of Wadi Derna

The profile of the lower part of the modern wadi-bed is also shown; for the final kilometre, this profile is artificially controlled, and is therefore indicated by a broken line. Vertical exaggeration: × 5.

177 m. The terrace at this point was found to lie 15 m. above the spring, its altitude above sea-level being thus established as 192 m. In this way it could be shown that the terrace-surface, in the 3·8 km. between the furthest point of the survey and Ain bu Mansur, rose through a vertical distance of only 25 m., giving an average gradient of no more than 1 in 150. As far as could be judged by eye, a gradient of this order persisted for at least 2 km. beyond Ain bu Mansur. Pl. 4b shows the general aspect of the terrace, looking up the wadi from the furthest point of the survey.

The profile of the wadi-bed, although not surveyed above the mouth of Wadi Gahham, appeared to show a gradient far more regular than that of the terrace. Thus, in the part of the wadi where the gradient of the terrace showed its rapid increase, an increase was also noted in the vertical distance between the two profiles; this finally reached 50 m. or more. Above the drop, however, the distance again decreased. At Ain bu Mansur, as already mentioned, it was only 15 m.; 2 km. further upstream it was no more than 7–8 m.

HIGHER TERRACE: LITHOLOGY OF ASSOCIATED DEPOSITS

For the description of the deposits of the higher terrace, it will be best to proceed in the downstream direction, since it is in this direction that the deposits become increasingly abnormal.

At the highest point examined, 14 km. from the sea, the deposits consisted of a coarse gravel, with well-rounded pebbles and boulders, abundant grains of quartz and glauconite (derived from the Miocene beds), and a certain amount of terra rossa. The first change was noticed about 1 km. above Ain bu Mansur, just below the mouths of two large tributaries. The proportion of pebbles and boulders

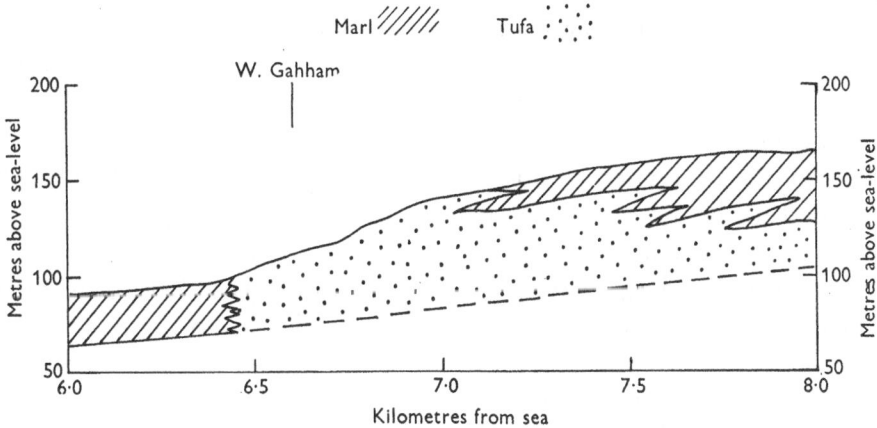

FIG. 10. Deposits of higher terrace of Wadi Derna, near mouth of Wadi Gahham
Geological boundaries purely diagrammatic. Levels of modern wadi-bed not accurately
known above Wadi Gahham. Vertical exaggeration: × 5.

here began to diminish, and the deposit soon became a soft, fine-grained cal-careous marl, yellowish, grey or white in colour. Only a small amount of coarse material remained, mostly in the form of pebbles, which were often arranged in bands, and some broken chips of tufa; it was noticeable that the pebble-bands always increased in size and number below the mouths of tributaries. The marl itself, on subsequent examination under the microscope, was found to be very largely composed of minute grains of calcite, the majority of no particular shape but a few showing well-marked rhombic forms. The bedding of the deposits was remarkably persistent and regular, and was parallel, as far as could be seen, to the surface of the terrace.

About 9 km. from the sea, tufa made its first appearance in large quantities, occurring in bands at different levels within the deposits. At first these bands were mainly confined to the lower levels, but they rapidly increased in number and width until, perhaps $\frac{1}{2}$ km. above Wadi Gahham, the deposit was composed entirely of tufa throughout its thickness. 150 m. below the mouth of Wadi Gahham, however, the tufa suddenly ended, and gave way once more to yellow and grey marls. The relations of the deposits in this part of the wadi are shown diagrammatically in Fig. 10.

This occurrence of tufa deserves to be described in some detail. At its first appearance, while still in the form of thin layers, it was generally brittle and incoherent, and contained few recognizable organic remains other than incrustations around reed-stems; these were present sometimes as broken fragments, sometimes as originally formed, around reeds which had remained in their positions of growth (Pl. 5a). Further downstream, as the layers thickened, the tufa assumed its more typical appearance. To a very large extent, it now consisted of incrustations which could be recognized as having once enclosed the stems and branches of plants, and in some cases leaves. The resulting material was fairly hard, but generally too porous, or even cavernous, to be described as compact. The colour of the tufa when freshly broken was light brown or pink; on all surfaces exposed to the air it had weathered to a deep rusty brown.

Even after the tufa had spread throughout the deposit, it still retained regular and well-marked bedding-planes for some distance. Just before its final disappearance, however, two features developed which caused the bedding-planes to be almost entirely obliterated. First, much of the tufa was now in the form of large blocks, lying at all angles, which had obviously been moved from the positions in which they had originally been formed. Secondly, it also included large masses of petrified moss, which often appeared to be arranged in concentric curved sheets, with convex sides facing downstream and upwards (Pl. 5b). All these features could be studied on both sides of the tufa platform at the mouth of Wadi Gahham; at this locality also the tufa contained many large cavities, often lined with stalactites or botryoidal forms.

The abrupt transition between this chaotic development of tufa and the even-bedded marls on its downstream side was unfortunately obscured by both erosion and slope deposits. Nevertheless, a few clear sections were found, and in these the two materials could be seen to interdigitate in a normal manner.

For several kilometres below this area of transition, the marls remained very similar to those found higher up the wadi. Tufa was absent, except in occasional layers of broken fragments. Pebbles and boulders, often angular or only slightly rounded, occurred locally, their presence being associated, as before, with the mouths of tributaries, many of which now entered as 'hanging' gullies, high up on the sides of the gorge. A particularly large mass of unrounded stones, resembling scree, was seen to form almost the whole of the deposit for a short distance below the mouth of Wadi bu Rueis, 3 km. from the sea. It should also be mentioned that pebbles, well rounded and often cemented with iron-pan, were a fairly constant feature of the lowest visible layers of the deposit, from Wadi Gahham downwards.

Just below the mouth of Wadi bu Rueis, tufa appeared for the second time, at first only within the lower 10 m. of the visible sections, but soon spreading upwards. At the town wall of Derna, 2 km. from the sea, magnificent sections were available, in which the deposits were seen to consist mainly of tufa, with in addition many layers of grey or white marl and some layers of pebbles (Pl. 4 a). The last remaining bluff of the deposits, already mentioned, was composed of much the same materials. The tufa in this part of the wadi was in many ways similar to that near Wadi Gahham. It was, however, softer and less compact, and was grey or white in colour, instead of brown. Also, it consisted mainly of fragments arranged in a 'jumbled' fashion, as if they had been broken up and re-deposited by running water. Bedding was conspicuous throughout and generally regular, though some current-bedding and channelling were present in the last kilometre of the gorge.

In conclusion, some mention must be made of the associated deposits found in some of the tributary wadis. In Wadi bu Rueis, they were preserved for only a short distance upstream, and consisted, from top to bottom, of poorly sorted and poorly rounded pebbles and boulders. Similar gravels were found in another wadi immediately opposite. The most interesting deposits, however, were those found in Wadi Gahham, one of the largest of all the tributaries. For about 100 m. upstream of its mouth, the west bank of the wadi was composed entirely of tufa, continuous with that in Wadi Derna. Further upstream, layers of silt and marl then appeared, and the tufa became confined to rapidly dwindling bands, often consisting of incrustations around reed-stems in their original positions of growth. After 200 m., the deposit had been completely transformed into a grey and white marl, extremely fine-grained and with regular, almost horizontal bedding. The marl had been badly eroded and could not be traced for more than a few hundred metres upstream. Still further up the wadi, a bright red marl was found, perhaps 10–15 m. thick, but its relationship to the grey marl could not be determined.

On the west bank of Wadi Gahham, 250 m. from its mouth, the grey marls were found to contain abundant Palaeolithic implements, together with bones and teeth of vertebrates. Excavations were carried out at this site, and the archaeological remains are described below under the heading 'Sidi el Hajj Creiem' (see Pl. 7).

HIGHER TERRACE: ORGANIC REMAINS OF ASSOCIATED DEPOSITS

Animal

Animal remains were by no means abundant within the deposits of the higher terrace. Those which were found were of two kinds: vertebrate remains, from the Wadi Gahham deposits only, and the shells of terrestrial snails.

The vertebrate remains from Wadi Gahham were examined and reported upon by the late Miss D. M. A. Bate. As will be seen in Appendix A, four forms were identified with certainty: *Homoioceras* sp. (extinct buffalo), *Ammotragus* sp. (Barbary sheep), *Hippotigris* sp. (zebra), and *Testudo* sp. (small tortoise).

Gastropod shells were not generally common, but were found in large numbers at certain localities. Collections were made at several different points; of the specimens obtained, however, it will only be necessary to mention those from the excavations at Wadi Gahham, since these include representatives of every species found elsewhere in the deposits. In this case, part of the collection was made on the site, but many of the smaller specimens were recovered by sieving from blocks of sediment sent back to England for the extraction of archaeological or vertebrate remains. Identifications were carried out by the present writer, for the most part in the Museum of Zoology, Cambridge, and in the British Museum (Natural History). The following ten species were identified:[1]

Amnicola pychnocheilia Bgt.	*Vitrea cristallina* (Müller).
Orcula orientalis (Parr).	*Xerophila chadiana* Pall. *darnensis* nov.
Clausilia klaptoczi Stur.	*X. icmalea* (West).
Rumina decollata (L).	*Albea candidissima* (Drap).
Caecilianella sp.	*Helix melanostoma* Drap.

This gastropod fauna shows two interesting features. In the first place, the only aquatic form represented is *Amnicola pychnocheilia*, whereas the modern fauna of Wadi Derna includes, in addition to this species, two species of Planorbidae.[2] It may also be mentioned that Spratt claimed to have found a cast of *Limnaea* in the tufa at the lower end of the wadi;[3] if the presence of this genus could be confirmed, it would constitute a further difference from the modern fauna, since no living *Limnaea* has yet been recorded from the Gebel.

In the second place, although the majority of the forms are still living in the vicinity, there are three which, according to the latest available information,[4] have not yet been reported from Cyrenaica. These three are the *Caecilianella*,

---

[1] D. 4996–5060. [2] Zavattari (1930), p. 356.
[3] Spratt (1865), p. 379.
[4] Zavattari (1934), supplemented by Gambetta (1934), pp. 241–2.

Vitrea cristallina and *Xerophila chadiana*. The first two, indeed, are minute and fragile; in view of the fact that the same species of *Vitrea* and several species of *Caecilianella* are still found elsewhere on the North African coast, it is probable that both forms are in reality still living in Cyrenaica.

The shell of the snail described as *Xerophila chadiana darnensis* is, however, both conspicuous and relatively robust, and would be much less likely to have been overlooked. It is therefore probable, if not entirely certain, that this snail has since left the territory. It is also very probable that its disappearance took place before the time of the Younger Gravels; although after its first discovery a continual watch was kept for its presence in these latter deposits, no specimens of it were ever found. In the deposits of the higher terrace, on the other hand, it is by far the commonest fossil, and there is no doubt that this is the form described by Newton as *Hygromia sordulenta*. A description of this shell, and the reasons for which the writer disagrees with Newton's identification, will be found in Appendix D. The species *Xerophila chadiana* is one which now lives in Morocco.

Vegetable

Two types of vegetable remains were found in the deposits of the higher terrace. First, the archaeological remains in Wadi Gahham were accompanied by abundant fragments of carbonized wood; these have been identified, with the kind assistance of Dr H. Hamshaw Thomas and Dr C. R. Metcalfe, as belonging to a conifer. Secondly, all the deposits of tufa, as has already been mentioned, consisted very largely of incrustations around various parts of plants. The original tissues of these plants were never preserved, but their forms were often revealed as casts and impressions.

The great majority of the incrustations were around the leaves and stems of reeds. Near the mouth of Wadi Gahham, however, abundant traces were found of plants belonging to other genera. Some leaf-impressions from this locality had already been identified by Tongiorgi as *Laurus canariensis* Webb.[1] During the 1947 and 1948 expeditions large collections of specimens were made at this same locality and were subsequently examined by the present writer. By comparison with specimens in the herbarium of the Cambridge Botany School, it was found that the following four species were represented in the collections:

> *Rubus ?ulmifolius* Schott. (bramble)
> *Arundo* sp. (reed)
> *Laurus canariensis* Webb. (Canary Island laurel)
> *Pinus halepensis* Mill. (Aleppo pine)

[1] Tongiorgi (1935).

Details of the fossil specimens, and of the characteristics by which they were identified, will be given in Appendix E.

Of these four plants, *Rubus ulmifolius*[1] and *Arundo*[2] grow abundantly in Wadi Derna at the present day, the latter genus being represented by the species *Arundo donax* L. There is no definite record of the occurrence of *Pinus halepensis* within the immediate vicinity of Derna, but it is still found in wadis on the coast between Apollonia and Chersa,[3] and the distance between Chersa and Derna is only 20 km. On the other hand, the sole species of *Laurus* now living in any part of Cyrenaica is *L. nobilis* L., which is widespread, though apparently of rather local distribution, throughout the Gebel.[4] *L. canariensis* no longer occurs, indeed, anywhere in the Mediterranean region; it is confined to Madeira, the Canary Isles and the Azores.

The absence of the Aleppo pine from the modern flora of Wadi Derna can doubtless be ascribed to recent deforestation. Thus, of all the plants identified from the tufas, the *Laurus* is the only one which can definitely be said to have left the region as a result of natural causes.

CHRONOLOGY

As regards the deposits of the lower terrace of Wadi Derna, there can be no doubt that these are nothing more than the local representatives of the Younger Gravels. This is shown both by their lithology and by their apparent connexion with the Derna alluvial fan which, as has already been seen, is itself partly contemporary with the fossil dunes. It is at least certain, therefore, that the deposits of the higher terrace are older than the Younger Gravels, the formation of the two deposits having been separated by a phase of down-cutting.

The determination of the age of the former deposits with respect to the marine shorelines presents, however, a problem of considerable difficulty. The lower members of the series of high shorelines are no longer present in the immediate vicinity of the wadi; the last traces of the higher terrace lie, in any case, some 150 m. short of the entrance of the gorge. Direct evidence, therefore, is entirely absent.

After the 1947 expedition, it seemed as though some evidence might be obtained from the present levels of the terrace-surface, since it was then thought that the elevation of this feature was probably due to its having been related to some high sea-level. In the absence of the lower shorelines, direct correlation was of course impossible. Moreover, this absence in itself indicated that a slice of the coast had been removed from the Derna area by later marine erosion.

[1] Pampanini (1930), p. 129. [2] *Ibid.* p. 112.
[3] *Ibid.* pp. 84 and 487. [4] *Ibid.* pp. 209 and 489.

Thus, if this theory were correct, the shoreline concerned might have lain far beyond the present entrance of the gorge, and at an altitude considerably less than that of the lowest remaining fragment of the terrace-surface; this seemed all the more likely since the surface showed no signs of flattening out and the deposits contained no marine horizons.[1] Nevertheless, it was hoped that some kind of solution would emerge when the succession of shorelines became better known.

In 1948, as already mentioned, a thorough examination of the Derna area showed that a shoreline at about 20 m. was preserved to within a distance of 500 m. of the wadi, on its east side. Although the feature was absent at the wadi itself and for many kilometres to the west, it must originally have passed very close to the site of the present entrance of the gorge. Only 150 m. inside the gorge, however, the soft and fragile terrace-deposits were still preserved at levels as low as 12 m. Thus, these deposits must at any rate be younger than the 20 m. shoreline.

The situation was further clarified by the results obtained during the survey of the high shorelines on the whole stretch of coast between Derna and Ras Aamer. As has already been explained in Chapter I, this survey showed that the 20 m. shore line at Derna was almost certainly identical with fragments of shore-line found at nearly similar levels elsewhere in Cyrenaica; also, that these fragments together represented the second lowest of all the individual shorelines on the Cyrenaican coast, the lowest of all being at 6 m. The problem, therefore, was reduced to the question of the relative ages of the terrace-deposits and the 6 m. shoreline.

On the evidence from Wadi Derna alone, no solution to this problem has been found. If at the time when the terrace-surface was formed the sea-level stood lower than 20 m., it must be assumed either that this surface was quite independent of the sea or that contact was only established by the intervention of a large delta. The present levels of the surface, therefore, cannot be regarded as having any further chronological significance. At this point, in fact, only two fragments of evidence can be produced which appear to have any bearing on the question. First, the softness of the terrace-deposits suggests that they are relatively young; secondly, no other deposits have been found to intervene between them and the Younger Gravels. Neither piece of evidence is of great weight, but the two together suggest that the terrace-deposits, though older than the Younger Gravels, may yet be younger than the 6 m. shoreline.

[1] McBurney, Hey and Watson (1948), p. 41.

NATURE OF PROCESS OF DEPOSITION

From the information already given, it is clear that the formation of the deposits of the higher terrace was a process both complicated and unusual. Before discussing the conditions under which it took place, it will therefore be useful to give separate consideration to the nature of this process itself.

The more obvious features of the deposits can be summarized as follows:

(i) Three distinct types of material are represented: gravel, marl and tufa.

(ii) The distribution of these materials is peculiar, inasmuch as the wadi, or at any rate its lower reaches, can be divided into a number of successive sections, in each of which the deposits are largely or entirely composed of one material only, to the virtual exclusion of the other two. The boundaries between the sections are not, indeed, perfectly sharp, since there is always a certain amount of interdigitation, but even so each lateral change of facies is accomplished within a relatively short horizontal distance. Thus the nature of the deposits in the final 15 km. of the wadi can be expressed in the following simple form:

Distances from sea (km.)	Dominant material
$1\frac{1}{2}$– 3	Tufa
3 –$6\frac{1}{2}$	Marl
$6\frac{1}{2}$– 9	Tufa
9 –13	Marl
Over 13	Gravel

(iii) Almost everywhere, the bedding of the deposits is remarkably regular and continuous, with a gentle inclination downstream.

From these facts alone certain conclusions can already be drawn. The nature of the bedding, in the first place, shows that the formation of the deposits was a process which was carried on continuously and more or less simultaneously on all parts of the wadi-floor. Hence, three different types of sedimentation must always have been in progress at one and the same time, each type being almost wholly confined to particular sections of the wadi.

As for the sediments themselves, the gravels are a perfectly normal detrital deposit, composed mainly of fragments of the local rocks, while the tufas are, of course, of chemical origin. There is no doubt that the marls, in places, contain much detrital material. The minute calcite grains, however, of which the marls are always chiefly composed, could scarcely have been formed as a result of any normal weathering process. It is believed in fact that these grains, as well as the tufa, owe their origin to some form of chemical precipitation, a belief which is supported by the presence of the rhombs. Since the thickness of the marls remains fairly constant over long distances, it is believed also that their precipitation took place more or less uniformly on the wadi-floor itself.

Thus the conclusion is reached that virtually all sedimentation in the last 13 km. of the main wadi was of a chemical nature. This also appears to have been the case in the lower reaches of Wadi Gahham, though not in any other tributaries, none of which contains evidence for the deposition of any material other than gravel.

One important matter remains to be discussed. There is one particular respect in which tufa, in its behaviour after deposition, differs profoundly from both gravel and marl: once formed, it is not easily removed again by water, and may thus, without any help from outside causes, bring about considerable changes in the form of a stream-bed. It is surprising to find, therefore, that the Derna tufas, on the evidence of their bedding, must to a great extent have been laid down without causing any irregularities in the profile of the wadi.

There is, however, one single locality, the downstream end of the upper occurrence of tufa, where there are unmistakable signs of an interruption. At this point, as has already been described, the change of lithology is particularly rapid; within a few tens of metres, near the mouth of Wadi Gahham, the deposit is changed from tufa to marl throughout almost the whole of its thickness. In addition the bedding of the tufa, just before its disappearance, is completely disrupted, a condition which was observed nowhere else within the terrace-deposits. A study of the nature of the disruptions leaves little doubt that this was in fact the site of a waterfall, itself composed of tufa, yet maintaining its existence throughout the whole of the time in which the tufa was forming. This is indicated, in particular, by the curved sheets of petrified moss, which resemble the tufa aprons associated with many modern waterfalls, and by the detached blocks of tufa, which could only have been moved by gravity and which therefore denote a sharp change of levels in the wadi-bed. It thus appears, not only that the down-stream limit of tufa-formation remained permanently stable at this point, but that the rate of deposition of the tufa on the upstream side of the limit was higher than that of the marl on the downstream side.

Confirmation is provided by the profile of the terrace-surface, a surface which undoubtedly preserves the form of the wadi-bed as it was at the time when deposition finally ceased. As can be seen from Fig. 9, the profile does not, indeed, show any signs that a waterfall existed at this late stage; nevertheless, the very marked increase of gradient near the mouth of Wadi Gahham indicates clearly enough the existence of rapids. These rapids, moreover, in spite of having degenerated from their former state, had a total fall which was as much as 60 m. It can also be seen, from a comparison between the profile and what is known of the levels of the base of the deposits, that the thicknesses of the tufa above the waterfall and the marl below are respectively about 90 and 30 m.; their average rates of deposition, in other words, were in the ratio of about 3:1.

The form of the profile above the waterfall shows, finally, that the growth of the tufa, although it may have caused no further sharp discontinuities, must have had a powerful effect over the gradients of the wadi for many kilometres further upstream. Between Ain bu Mansur and Wadi Gahham, the rock-floor of the wadi has an average gradient of about 1 in 50. Yet, as has already been seen, the gradient of the terrace-surface between Ain bu Mansur and the top of the former rapids is only about 1 in 150. This reduction of the gradient can only be ascribed to a rise in the base-level, caused by the formation of the tufa itself. It should be added that this process in turn may well have been the sole reason why the marls were able to be deposited in this part of the wadi.

It is impossible to be certain whether the lower occurrence of tufa also terminated with a waterfall. Since its downstream end has been cut short by erosion, the evidence would in any case no longer be visible. Nor can any estimate be made of its effects on deposition further upstream; the base of the deposits is always hidden in the last 6 km. of the wadi, so that its levels cannot be determined. Nevertheless, this occurrence, so far as it is preserved, is so similar to the upper occurrence as to suggest very strongly that both had a similar origin. There is little doubt, moreover, that the rock-floor of the lower reaches of Wadi Derna lies far below the present wadi-bed. If so, its average gradient in this area also may well be greater than that of the terrace-surface.

The conclusions reached can now be summarized. Normal detrital deposits were laid down only in tributaries, and in the main wadi above a point 13 km. from the sea. Below this point, deposition was almost entirely chemical. To a large extent, the products were in the form of a loose marl. In two parts of the wadi, however, they were in the form of tufa. The tufa, like the other deposits, was laid down in extensive sheets on the wadi-bed, but succeeded nevertheless, at any rate in the case of the upper occurrence, in disturbing the form of the stream-profile to a considerable extent. In this case, first, the downstream end of the occurrence occupied a constant position which, as a result of the relatively rapid growth of the tufa, was marked by the presence of a permanent waterfall. Secondly the gradient of the wadi-bed for many kilometres further upstream was severely reduced, and this must have assisted, if not caused, the deposition of fine-grained and incoherent sediments in this part of the wadi. Although positive evidence is lacking, it is strongly suspected that the lower occurrence of tufa was of a similar nature, and that it had similar effects.

CONDITIONS OF DEPOSITION

The first part of this section will be devoted to the deductions which can be made from the organic remains within the deposits of the higher terrace. In the second part, the deposits themselves will be considered, together with the conclusions already reached with regard to the manner in which they were laid down.

Deductions from organic remains

At first sight, the faunal and floral remains of the deposits suggest climatic conditions similar to those of today, the *Laurus canariensis* and the one Moroccan snail indicating, perhaps, a more oceanic tendency. The list of species, however, is scarcely large enough to allow any such sweeping generalizations. If any reliable climatic deductions are to be made, they must in fact be based upon a consideration of individual species or groups of species.

Reasons have already been given, in connexion with the alluvial deposits, for doubting the reliability of snails as climatic indicators. In this case, indeed, their significance might be expected to be greater than in the case of the alluvium, since the number of forms recorded is nearly twice as large. On the other hand, the ranges of the Wadi Derna snails extend in several instances far beyond the limits of the Mediterranean region. Once again, therefore, it is not intended to base any deductions upon the molluscan fauna.

From the vertebrate remains, Miss Bate made no climatic deductions. The presence of the zebra and the tortoise, nevertheless, does at least suggest a climate with temperatures generally as high as those of today.

Among the floral remains, the reeds and brambles, in spite of the absence of reliable specific identifications, can certainly be said to denote a perennial water-supply. Of the two plants whose species are definitely known, the *Laurus* is the less informative, for its species happens to be one whose present range is in any case limited by the mere distribution of land and sea. Nevertheless, although it is hardy in the southern and western counties of England, it is also said to be less hardy than *L. nobilis*, the common Mediterranean bay;[1] its presence may perhaps be taken, therefore, as a further indication of generally warm climatic conditions.

With regard to the Aleppo pine, it is possible to be rather more definite. This is a tree which now lives on most parts of the Mediterranean seaboard, but which is seldom found more than 100 km. inland. It is therefore a species with a relatively narrow range; moreover, in this case there is no reason to doubt that its

[1] Bean (1933), p. 205.

range is determined mainly by its climatic tolerance. Its failure to spread south-wards or eastwards from the Mediterranean undoubtedly signifies that it is in-capable of withstanding true desert conditions. Similarly, its failure to spread northwards must be due to a dislike of temperate conditions; this is confirmed by the observation that the Aleppo pine in England, though hardy when once established, is 'tender in a young state'.[1] Since in France, at least, the January isotherms and lines of equal rainfall run more nearly from north to south than from east to west, it can scarcely be the cold, wet winters which are objectionable. It can only be supposed, therefore, that the species is one which requires a fairly hot dry summer.

Deductions from the nature and supposed mode of formation of the deposits

It has already been concluded (p. 111) that contact between the sea and the surface of the higher terrace must have been either remote or non-existent. If it is also true that the entrance to the gorge was occupied by a 'tufa-fall', contact must now be supposed to have been non-existent. As regards the level of the sea all that can strictly be said, therefore, is that it must have been below 20 m. throughout the formation of the deposits. Although the deposits are preserved down to a level of 12 m., they would not necessarily have been destroyed unless the sea had once more remained stable at one particular level, and this is not known to have occurred again before the time of the 6 m. shoreline.

The origins of the various types of deposit will now be considered in turn, together with the manner in which they are believed to have been laid down.

The gravels, by their very existence, suggest winter conditions more severe than those of today, though certainly much less severe than at the time of the Younger Gravels. From their restriction to tributaries and to the upper parts of the main wadi, it can be deduced that they were laid down at a time when the flow of water in the wadi remained always less than that which is observed during present-day floods.

The latter conclusion is also suggested even more strongly by the regular bedding of the marls. The origin of the marls themselves presents, it must be admitted, a problem which has not yet been solved. Although numerous examples are known of the formation of granular calcite on the bottoms of lakes, the writer has failed to find any published reference to the formation of the same substance in running water. This is not thought to mean that the Wadi Derna marls were not after all laid down under these circumstances; there are certainly no signs of any lake from which the material might have been derived. It does mean, how-ever, that nothing can be said about the conditions which caused the marl to be

[1] Bean (1919), p. 181.

precipitated. It can only be suggested that the process was probably due to the activities of algae, and that deposition may well have been assisted by the presence of vegetation, alive or dead, by which the flow of water was hindered.[1]

A similar difficulty arises in the case of the tufa. Many descriptions of the formation of this material at the present day have been published, and many theories have been proposed to account for its origin. In almost every instance, however, the process is described as taking place either at a spring or on the face of a waterfall. In Wadi Derna, on the other hand, neither occurrence of tufa happens to be in contact with any of the four well-defined spring-forming horizons of the region; both occurrences, moreover, except in the immediate vicinity of the mouth of Wadi Gahham, appear to have been laid down for the most part on gently-sloping surfaces.

Nevertheless, a few instances have been described in which the formation of tufa is taking place in an environment of yet a third kind. In the beds of streams in certain limestone regions, it appears that tufa is now being deposited, not only at waterfalls, but also at places where lesser changes of gradient have given rise to rapids. Examples of this kind of deposition have been described by Branner in Brazil,[2] and, in greater detail, by Emig in the Arbuckle Mountains of Oklahoma, U.S.A.[3] As far as can be made out from Emig's descriptions and illustrations, the tufa formed in these circumstances may be laid down over a considerable length of the stream-bed, and in such a way as to cause within that part of the stream-bed a continual lowering of the gradient; at the same time the downstream end of the mass of tufa forms a true waterfall of gradually increasing height. Under favourable conditions, it appears that the growth of such a deposit, once begun, may continue almost indefinitely.

There is a striking resemblance between this process and that by which the Wadi Derna tufas are thought to have been formed. For this reason it is believed that the upper occurrence of tufa, at any rate, may have owed its origin in the first instance to some local steepening of the rock-floor of the wadi, just above the mouth of Wadi Gahham. The formation of the lower occurrence also may well have been initiated by the existence of a similar steepening, situated perhaps

---

[1] The late Professor F. E. Fritsch examined some samples of the material, and confirmed that it contains only a minute proportion of recognizable organic remains. Among these remains, he identified *Cymbella* spp. and ? *Navicula* spp. (diatoms), also *Cosmarium* (a desmid); no signs were found of any organism by which the granular calcite could have been precipitated.

Samples of a similar material have been collected by the writer from Pleistocene or Recent deposits in the sides of Oued Gabès (Tunisia) and Wadi Scersciara, Tarhuna (Tripolitania). Both deposits formed terraces, and both had clearly been laid down in running water. These facts are mentioned as an indication that the material may in reality be of widespread occurrence.

[2] Branner (1911).   [3] Emig (1917).

just outside the present entrance to the gorge. In neither case, admittedly, can this belief be confirmed by reference to the levels of the rock-floor, but the changes of gradient might in any event have been relatively insignificant.

There are many places, however, where the modern wadi-bed also has a fairly steep gradient. Yet, although some tufa was found to be forming at the present day in the irrigation channels, none appeared to be forming in the wadi-bed itself. More important still, there were no signs that tufa had been deposited at any time since the formation of the higher terrace; in other words, the process had already stopped long before the flow of the water was interfered with by artificial means. This can only mean that the growth of the Pleistocene tufas must have been encouraged, not only by irregularities of the wadi-bed, but also by some further cause or causes no longer in operation.

Although little is known about the formation of tufa in rapids, the nature of these causes is fairly clear. From Emig's observations, it appears that most of the tufa is laid down on surfaces which, though not submerged, are constantly wetted by spray; this also agrees with observations made by the present writer. If it is true, the large-scale deposition of tufa in rapids must require a large-scale production of spray; in other words, the flow of the water must be very turbulent.

At the present day, there are, indeed, certain occasions when the flow of Wadi Derna, as of all other wadis in the Gebel, becomes highly turbulent. Such occasions, however, are rare, whereas the rapids responsible for the tufas must presumably have been in existence for a large part of each year. The water of these rapids, moreover, must have been extremely hard, which implies, since this is an area of steep topography, that it was derived mainly from springs; the water of the present-day floods, on the contrary, comes for the most part directly from the surface of the ground. As for the normal flow of water in the modern wadi, this certainly is perennial and is wholly supplied by springs, but it is everywhere smooth and gentle. For the lowest part of the wadi, this fact is of course of little significance, since so large a proportion of the water is artificially removed. Yet the intake of the first large irrigation channel lies only a short distance above the mouth of Wadi Gahham, and much tufa was formed on the upstream side of this point in Pleistocene times.

Hence, it is believed that the absence of recent tufa can simply be ascribed to the fact that the normal flow of water, though undoubtedly charged with calcium bicarbonate, is too small to throw up spray. Conversely, it is believed that the tufas were laid down at a time when the flow of water was far greater than at present.

Lastly, one curious and unexplained point may be mentioned. As has already been stated, the water-supply of the wadi at the time in question must have come mainly from springs. Yet it is found that the deposits of the higher terrace show no changes of any kind at the sites either of Ain bu Mansur or of Ain Derna, or indeed against any points on the outcrops of their respective spring-forming horizons. This implies that neither spring was in existence at the time when the deposits were laid down, and that the entire water-supply must have come from other springs still higher up the wadi. These could only have been situated on the Langhian (Lower Miocene) spring-forming horizon, which now gives rise to numerous small but perennial springs in the higher parts of the Wadi Derna drainage system.

CONCLUSIONS

The presence of the Aleppo pine suggests that the deposits of the higher terrace were laid down at a time when summers were somewhat warm and dry. The gravels, on the other hand, indicate winters colder than those of today.

The reeds and brambles show that the water-supply of the wadi was perennial, and the presence of the tufa is believed to show that the volume of water, during most of the year, was far greater than it is under normal conditions at the present day. The water itself is thought to have been derived from springs in the higher tributaries of the wadi, rising from the Lower Miocene; those in the lower reaches, now of great importance, were apparently inactive. There are various indications that violent flooding never occurred; hence, two further conclusions are suggested: first, that the supply of surface-water to the wadis may have been regulated by a growth of vegetation more abundant than at present, and secondly that the rainfall may have been more evenly distributed throughout the year.

As regards the contemporary sea-level, it can only be said that it was lower than 20 m.

3. WADI EN NAGA

Wadi en Naga enters the sea about 9 km. to the west-north-west of Derna. It is much shorter than Wadi Derna, and its gorge is somewhat narrower; nevertheless, it is still one of the major wadis on the northern face of the Gebel. Like Wadi Derna, it contains several groups of perennial springs, of which the two lowest issue from the base of the Oligocene and from the Middle Eocene respectively. The first of the two lies about 13 km. from the sea, the distance being measured along the bed of the wadi, and is called Maaten Gerem. The second lies about 10 km. from the sea; its name is not known. No previous references appear to have been made to the existence within this wadi of any Pleistocene deposits.

It has already been mentioned (p. 76) that the lower part of the gorge of Wadi en Naga contains an aggradation terrace composed of the usual red Younger Gravels. When followed upstream these deposits were at first found to rest in the normal manner directly upon the rock-floor of the wadi. 4 km. from the sea, however, a thin layer of grey and yellow marl was found to intervene between the gravel and the bedrock. This layer could be followed continuously for several kilometres further upstream, its thickness seldom exceeding 1 or 2 m. Lithologically, the material appeared to be identical with the marls of Wadi Derna. The resemblance was further enhanced by the frequent presence of bands of tufa, either in the form of broken fragments or of incrustations around reeds in their positions of growth. No animal fossils were observed in the short time available.

About 10 km. from the sea both deposits became discontinuous. Still further upstream, the only deposits present were large banks of coarse, yellowish gravel, rising to heights of several metres on either side of the wadi-bed. As far as could be seen, these were connected with the marls rather than with the Younger Gravels; they contained, at any rate, some lenses of marl, but no traces of terra rossa. Finally, near Maaten Gerem, numerous 'aprons' of tufa could be seen descending into the wadi-bed from the Lower Oligocene spring-forming horizon, where the latter was exposed in the sides of the wadi. Some of these aprons were still in the process of formation; whether any were older than, or contemporary with, the gravel-banks could not be determined.

At no point did the marls themselves form a terrace-surface; where they were most typically developed, indeed, they were always buried beneath the Younger Gravels. Nevertheless, they had clearly been deeply eroded before the Younger Gravels were laid down, and were undoubtedly the products of a separate phase of deposition. It seemed almost certain, in fact, that these deposits, like those of the higher terrace of Wadi Derna, represented a time when the water-supply of the wadi had been unusually plentiful.

4. AIN MARA

The name Ain Mara is applied to a group of copious perennial springs which emerge from the Lower Miocene, about 430 m. above sea-level, on the floor of a small valley 25 km. west of Derna. The valley itself is in fact Wadi en Naga, though this name, strictly speaking, is applied only to its lower reaches. Measured along the bed of the wadi, the distance between the springs and the sea is about 32 km.

The valley at Ain Mara is shallow and open, and its sides, composed of soft

Miocene limestones and marls, have a relatively gentle slope. To this extent it is similar to any other wadi in the Miocene areas of the Gebel. Its section, however, is not V-shaped; on the contrary, at the springs themselves and for about 1 km. downstream, it has a flat floor, 200 m. or more in width. Much of the water from the springs is used for the purpose of irrigating this miniature plain, which, as a result, is highly cultivated. The remainder of the water flows down the length of the plain in a natural bed, whose depth for the most part is only 3–4 m., though it appears to be increasing rapidly at the present day. Eventually, the water sinks into the ground.

It is in the sides of this stream-bed that the deposits underlying the plain can be examined. As Marchetti has already pointed out, the deposits are mainly alluvial, with beds of pebbles and frequent intercalations of tufa.[1] To this lithological description the present writer has little to add, except that he noted many exposures of black silt and clay, the colour of which was later found to be due to the presence of finely divided plant-remains.

It was also found, however, that the same series of deposits, and, in a sense, the plain itself, could be traced for considerably more than 1 km. below the springs. The plain, in fact, did not altogether disappear; it was merely reduced, rather rapidly, to a narrow terrace on either side of the stream-bed, which at this point lay on or near the bedrock floor, some 10 m. below the terrace-surface. This feature was followed downstream as far as a point about 3 km. below the springs, and in this distance showed no further diminution in width. The deposits themselves also underwent a change as the plain dwindled into the terrace. Clay and pebbles disappeared, and the visible sections now showed little else but pale yellow tufa and marl. There was, indeed, a striking resemblance to the deposits of certain parts of the higher terrace of Wadi Derna. This type of lithology persisted throughout that part of the wadi in which the terrace was examined.

The recognizable organic remains obtained from this series of deposits were of three kinds: pollen grains, ostracods, and the shells of molluscs. The molluscan shells were collected from the black silt exposed in the sides of the stream-bed, about 500 m. downstream from the springs; the pollen-grains and ostracods were later recovered from samples of the same material.

Mr P. Tallentire has very kindly examined the pollen at the Botany School, Cambridge. It is by no means abundant, and only five genera could be even tentatively identified: *Plantago*, *Chenopodium*, *Salix*, *Sparganium* and *Pinus*; in addition, the following families were represented: Cruciferae, Rosaceae, Umbelliferae, Compositae, and Graminaceae. All of these genera and families are still living in the Gebel.

[1] Marchetti (1938), p. 111 and fig. 21.

Among the ostracods, Mr Sylvester-Bradley has identified *Ilyocypris biplicata* (Koch) and *Candona* sp., together with two other forms doubtfully referable to *Prionocypris olivacea* (Brady & Norman) and *Cypridopsella* sp.[1] None of these forms appears to have been recorded as living in the Gebel at the present day, but it is not yet possible to be certain that they are genuinely absent.

The molluscs were identified by the present writer, with the exception of the *Pisidia*; these were identified by Mr A. W. Stelfox, for whose assistance the writer is very grateful. Nine species are represented, of which five are aquatic and four terrestrial:[2]

Aquatic:	Terrestrial
Amnicola pychnocheilia Bgt.	*Succinea pfeifferi* Rossm.
Ancylus sp.	*Rumina decollata* (L.)
Planorbis numidicus Bgt.	*Xerophila chadiana* Pall.
Pisidium casertanum Poli	*var. darnensis* Hey
P. personatum Malm.	*Helix melanostoma* Drap.

All of the aquatic forms are known to be still living in the Gebel, with the exception of *Pisidium personatum* and the species of *Ancylus*;[3] of the terrestrial forms, only the *Xerophila* has not been reported.[4] Since the two unreported aquatic forms are both minute, it is quite possible that both in fact are still present. As with the Wadi Derna marls, therefore, the *Xerophila* appears to be the only species in the fauna of the Ain Mara deposits which is fairly certainly absent from that of the modern Gebel.

The age of the deposits was clearly Pleistocene; this was shown by the recent aspect of their fauna, and more particularly by the fact that they yielded a small collection of Middle Palaeolithic implements of types similar to those found in Wadi Gahham. No direct correlation could be established with any of the other Pleistocene deposits of the Gebel; there could be no question even of attempting a correlation with the deposits of the lower reaches of Wadi en Naga, since it was already known that these deposits became discontinuous when traced into the middle reaches of the wadi. Nevertheless, it seemed likely that the Ain Mara beds must at least be older than the Younger Gravels, in view of the apparent absence from the latter of *Xerophila chadiana*.

As regards their conditions of deposition, there could be little doubt that here, as in Wadi Derna, some irregularity in the wadi-floor had given rise to rapids which in turn had induced the formation of tufa. As the tufa developed, it would have lowered the gradient of the wadi on its upstream side, causing the deposi-

[1] D. 5117–20.   [2] D. 5061–5116.
[3] Zavattari (1930), p. 356.   [4] Zavattari (1934), and Gambetta (1934).

tion of silt and mud and perhaps even the formation of a marsh. Here again the water-supply at the time of deposition must have been greater than at the present day. In this case, it should be added, there could be no question about the sources of the water; each of the present springs of Ain Mara emerges from the rock almost exactly at the level of the plain.

5. EL ATRUN

INTRODUCTION

The village of El Atrun (also known as Zahra or Fiorita) lies on the coast about 30 km. east of Apollonia and 12 km. in a direct line to the east-south-east of Ras el Hillal lighthouse. It is not thought that there is any previous description of the Pleistocene deposits at this locality. The present writer has only been able to spend a short time on their investigation; this is unfortunate, since they are of unusual interest.

As has already been mentioned, the coastal escarpment in this area reaches its greatest heights; it is also particularly steep, and in some places plunges straight into the sea. At El Atrun itself the slope of the escarpment slackens off along a line rather less than a kilometre from the shore and about 50 m. above sea-level. On the seaward side of the line there is a strip of country with gentle slopes, forming a relatively well-defined Sahel; along the shore itself a line of high cliffs falls more or less vertically into the sea.

The modern village of El Atrun lies in and around a deep notch in the escarpment, formed by the emergence of a wadi bearing the same name. Strictly speaking, Wadi el Atrun originates only a few hundred metres above the village, at the junction of two wadis named Wadi Zaigh and Wadi es Seghi. Neither is of any great size; Wadi es Seghi, the more easterly, is the longer of the two, but its total length is not more than 7 km. Both, however, are remarkable for containing very copious springs whose waters, rising from the Middle Eocene, are sufficient to maintain a perennial stream in Wadi el Atrun almost as far as its mouth. For the most part these wadis show the usual V-shaped cross-section; an exception is the lowest part of Wadi el Atrun, which crosses the Sahel as a narrow, slot-like gorge.

DESCRIPTION OF DEPOSITS

A heavy in-filling of alluvium was found in the lower reaches of both Wadi Zaigh and Wadi es Seghi. The material in each case was coarse and not especially well rounded; the upper surfaces of the deposits formed single, well-defined terraces lying at considerable heights above the modern stream-beds. In the case of Wadi es Seghi, the deposit, together with its terrace, was found to be preserved for at least 3 km. upstream from the junction of the two wadis. Alluvium was not the only material present; at two or three places the entire deposit was composed for a short distance of reddish-brown tufa. So far as could be seen, each mass of tufa was associated with a sharp drop in the original rock-floor of the wadi, and thus owed its existence to an ancient waterfall. At certain points the tufa contained leaf-impressions; the majority were impressions of reeds, but some could be assigned to *Laurus canariensis*.

At the junction of the wadis, their alluvial deposits merged, their corresponding terrace-surfaces united, and a single set of deposits continued down-

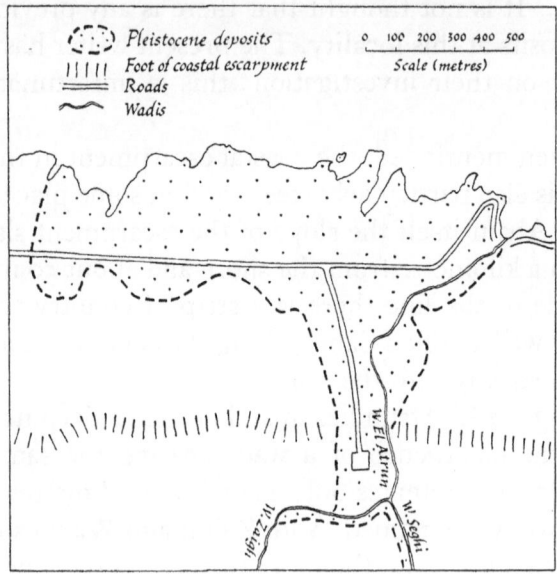

FIG. 11a. Pleistocene deposits near El Atrun

stream into the upper part of Wadi el Atrun. Then as Wadi el Atrun left the escarpment a change took place: the deposits spread out over the Sahel to the west of the wadi, forming what appeared at first to be an alluvial fan; as for the wadi itself, its narrow gorge across the Sahel was obviously relatively modern, and contained scarcely any deposits at all (Fig. 11a).

The 'alluvial fan' extended westwards from the wadi for about $1\frac{1}{2}$ km. For the whole of this distance it occupied at least half the width of the Sahel, on the side nearest the sea. Its surface was under heavy cultivation, and showed little else but a dark red soil. Some idea of the nature of the underlying deposits could be obtained, nevertheless, from continuous sections available both in the west bank of the wadi and in the modern cliffs.

The sections in the wadi-bank showed for the most part about 5 m. of alluvium, similar to that which was present further upstream, resting upon white Cretaceous limestone. The thickness, of course, was of little significance, since the wadi runs along the extreme eastern edge of the 'fan'. The sections were only varied by the presence of some reddish marl, and towards the sea by the reappearance of tufa.

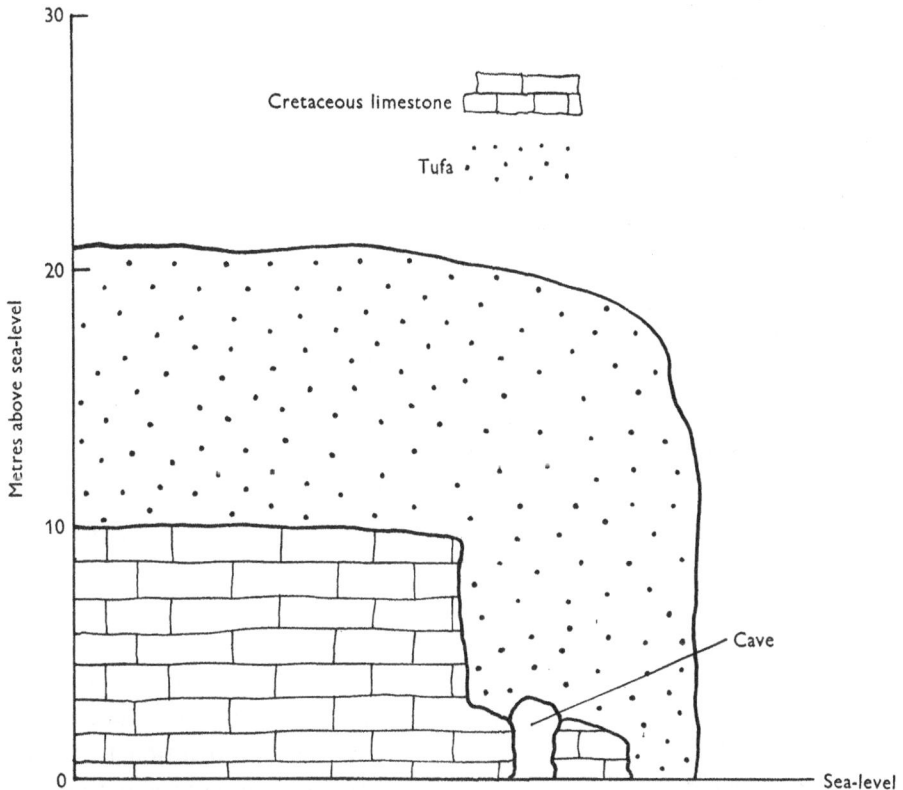

FIG. 11 *b*. Section on west side of cove, 400 m. west of mouth of Wadi el Atrun. Vertical heights obtained with a hand-level; horizontal distances estimated by eye. No vertical exaggeration.

The cliff-sections were more remarkable. As with the wadi-bank, the lower portion of the cliffs was at most points composed of Cretaceous limestone. The upper portion, however, was almost always composed, not of alluvium but of tufa, of a deep rusty red which made a striking contrast with the white of the limestone below. At a few places, indeed, the entire cliff, 20 m. or more in height, consisted of tufa from top to bottom. Most of this material had obviously been laid down around the stems of reeds and other plants; bedding was sometimes well-marked and horizontal, but more often obscure or absent.

125

It was clear, nevertheless, that the tufa formed no more than a narrow band along the seaward margin of the 'alluvial fan'. Apart from the wadi itself, several other deep gullies were seen which had been formed by marine erosion, and in all these it was found that the tufa gave way to beds of pebbles and red marl 100–200 m. behind the general line of the coast. These presumably were the materials underlying most of the cultivated area.

Certain other interesting observations were made in the cliff-sections. Wherever the tufa descended to sea-level, it could be seen to have overwhelmed an ancient cliff, itself cut in the bedrock. At all such places, the bedding of the tufa was chaotic, petrified moss was abundant, and much of the material was in the form of irregular blocks, obviously transported. Only one of these buried cliffs was accessible enough to be examined close at hand. This stood about 400 m. to the west of the wadi-mouth. The foot of the cliff was here found to be about 3 m. above present sea-level, and to form the landward boundary of a narrow rock-cut platform, itself buried beneath the tufa. On the platform rested a thin layer of pebbles and broken marine shells. A rough sketch of this section is here reproduced as Fig. 11 b.

Apart from the plant-impressions in the tufa, the only organic remains found within these continental deposits were land-snails. These occurred in the red marls of the coastal sections, and the following five species were identified: *Rumina decollata, Xerophila chadiana* var. *darnensis, X. icmalea, Albea candidissima* and *Helix melanostoma*.[1]

CHRONOLOGY

Since much of the El Atrun tufa lies at or below present sea-level, it can scarcely be older than the 6 m. shoreline. It is probable, indeed, that this shore-line can be associated with the buried cliffs. As was mentioned in Chapter 1, marine platforms rising to about 3 m. are of common occurrence between Ras el Hillal and Ras ben Gebara, and are believed, in spite of their low altitudes, to be the local representatives of the shoreline which elsewhere lies at or near 6 m.

The association of these deposits with single terrace-surfaces in the wadis behind El Atrun suggests one further conclusion: that they are the equivalents of the Younger Gravels. This view is supported by their containing so much gravelly material. They also include, however, masses of tufa large enough to have been themselves responsible for the aggradation of the wadis. This, therefore, is a case in which geomorphological arguments must be regarded with suspicion. Indeed, the presence of *Xerophila chadiana* is a strong indication that the deposits are actually older than the Younger Gravels. A similar indication is the fact that

---

[1] D. 5121–35.

several banks of consolidated dune were seen resting against the tufa in the cliff-sections, whereas no tufa was found resting upon consolidated dune. If this idea is correct, it must be assumed that the Younger Gravels of Wadi el Atrun and its tributaries have been entirely removed by later erosion. Such a process could well be ascribed to the narrowness of the wadis, further restricted by the presence of earlier gravels; this in fact is exactly what has occurred in most parts of Wadi Derna.

CONDITIONS OF DEPOSITION

The deposits of Wadi Zaigh and Wadi es Seghi suggest conditions similar to those which are thought to have led to the formation of the older deposits of Wadi Derna. The gravels suggest fairly cold winters; the tufa suggests a flow of water far greater than that of today, and its flora indicates that summers at least were warm.

Some further explanation is needed, however, for the curious features shown by the deposits on the Sahel to the west of the wadi-mouth. It must be admitted that no final explanation for these features can yet be offered. For this reason it is intended to give merely a bare statement of what seems to be the most reasonable hypothesis, with little attempt at justification.

One conclusion at least seems fairly certain: that the buried cliffs, though now short and isolated, once formed part of a continuous line of cliffs of marine origin, lying for the most part only a short distance in front of the modern shore. Immediately before the deposits were laid down, this feature would still have existed intact, though the sea must already have retreated from its foot to allow tufa to form below present sea-level. Also, Wadi el Atrun would already have cut itself a gorge across the Sahel. This, of course, could hardly have been the same as the present gorge, since it is unlikely that the modern wadi would have resumed down-cutting on exactly the same site as before. There is in fact an inlet about 450 m. west of the present wadi-mouth which has every appearance of being a buried gorge whose infilling of marl and tufa is now being removed by marine erosion.

It is believed that this original gorge was in some way blocked, perhaps by an early development of tufa. It then overflowed, and its waters, far more abundant than at present, wandered freely over the surface of the Sahel. At certain points these waters would reach the edge of the cliffs, and would form waterfalls, perhaps small but numerous. Tufa would immediately begin to form on and around the waterfalls; most of it would form below the edge of the cliffs, but some would accumulate on the edge itself. In this way a dam would be created at the top of each waterfall, and the water itself would be diverted, sooner or later, to some other part of the cliffs. The process would then start again. In time, every part of the old line of cliffs for $1\frac{1}{2}$ km. to the west of the present wadi-mouth would have been occupied at some time by a waterfall, and the whole of it would eventually be buried beneath a great bank of tufa. The tufa, in turn, would establish a local base-level, permitting the deposition of gravels and marls on the Sahel, and presumably controlling deposition in Wadi el Atrun.

Finally, when the flow of water dwindled to its present volume, the formation of tufa would cease, the wadi would confine itself to a permanent bed along the eastern edge of the deposits on the Sahel, and down-cutting would begin once more. At the same time, marine erosion would proceed to destroy much of the coastal tufa, together with the greater part of the old line of cliffs.

6. TUFACEOUS DEPOSITS: GENERAL CONCLUSIONS

Each of the four deposits described is thought to represent a period in which a particular spring, or group of springs, reached a degree of activity which has never subsequently been equalled. The supposed dates of the deposits, as deduced from internal evidence, can be summarized as follows:

Locality	Date
Wadi Derna	Older than Younger Gravels; younger than 15–25 m. shoreline; ? younger than 6 m. shoreline
Wadi en Naga	Older than Younger Gravels
Ain Mara	? Older than Younger Gravels
El Atrun	? Older than Younger Gravels; younger than 6 m. shoreline

In any individual case, there are two possible reasons for the unusual activity of the springs. On the one hand, some accident might have caused a temporary concentration of ground-water at that particular locality; the accident itself might have been tectonic, or, since the Gebel is mainly composed of limestone, it might have involved the sudden diversion of underground water-courses. On the other hand, the Gebel might have undergone a period of exceptionally high rainfall. As for the dates of these periods of unusual activity, none are known precisely, and in theory all might be different.

These conclusions are far too indefinite to be of much interest; yet no further deductions can be made from the study of the individual deposits. When all four are considered as a group, however, certain alternatives stand out as being by far the most probable.

At the present day, the springs of Ain Mara and El Atrun are among the most active of any in the eastern Gebel. A relatively high level of activity is still shown by the Langhian springs believed to be responsible for the Wadi Derna tufas, and there are many copious springs in the upper reaches of Wadi en Naga, including those of Ain Mara itself. If during some period of high rainfall any one of these springs or groups of springs was producing enough water to cause the formation of tufa, it is likely that the other three would be doing the same. Moreover, there is no obvious reason why the products should not stand a roughly equal chance of preservation in all four cases. This implies that there are only two real possibilities. Either the deposits are all of climatic origin and all contemporaneous with each other, or else they are all accidental, in which case all may still perhaps be of quite different ages.

Neither theory can be proved or disproved, but there can be no doubt which is the more probable. Since 'accidents' of the kinds mentioned above would be unlikely to affect more than one spring at a time, and would in any case be rare, the second theory would imply that the formation of the four deposits was spread out over a long period of time. On this theory also all four deposits would have to be later than the last period in which the rainfall in Cyrenaica was high enough to bring about the formation of tufa. Yet it is hard to believe that no such period of high rainfall has occurred in recent geological times; on the contrary, such periods are likely to have occurred repeatedly throughout the Pleistocene. It is concluded, therefore, that the first theory is by far the more probable of the two. This conclusion conflicts with none of the existing evidence, whether geological or archaeological. It agrees also with the fact that all the deposits are weathered and cemented to roughly the same extent; with the resemblance between the older deposits of Wadi Derna and Wadi en Naga; and finally, with the resemblance between the floras of Wadi Derna and Wadi es Seghi.

It is believed, in fact, that all four deposits were laid down at one and the same time, between the formation of the Younger Gravels and the formation of the 6 m. shoreline, and that the rainfall during this time was very much heavier than today. The climatic conditions would also have included all those which have already been deduced for the older deposits of Wadi Derna. These conclusions admittedly involve a contradiction, for the Aleppo pine, whose remains are found in Wadi Derna, is nowadays associated with dry climatic conditions. It can only be suggested that the rainfall, though heavy for most of the year, may have ceased altogether for a few months each summer.

CHAPTER VIII

CHRONOLOGY

I. NATURE OF REMAINING PROBLEMS

The internal chronology of the continental Pleistocene deposits of the Cyrenaican coast has already been discussed sufficiently. According to the conclusions reached, by far the greater part of these deposits was laid down during two distinct periods, both subsequent to the formation of the 6 m. shoreline, and hence, almost certainly, subsequent to the Last Interglacial.[1] The conditions which are believed to have prevailed during these two periods, and the events which occurred, can be summarized as follows:

(i) Summers hot, though winters probably fairly cold. Heavy annual rainfall, resulting in the deposition of tufa in some, at least, of those wadis which contain perennial springs at the present day. Sea-level below 6 m., but exact position unknown.

(ii) Winters very cold; rainfall probably moderate and seasonal, as today. Severity of winter climate produced large quantities of rock-waste, subsequently deposited as alluvial fans and terraces, and to some extent as mud-flows. Sea-level falling, to some unknown depth below − 17 m.; the loose sand thus exposed was then picked up by the wind and formed into dunes, some of which now lie below, or intercalated with, the coastal alluvial deposits, while others lie at considerable distances offshore.

It is now necessary to consider whether the dates of these two periods can be more exactly expressed in terms of any other Pleistocene time-scale, preferably again one which is regarded as 'standard'. As has already been stressed, the deposits themselves contain no remains, whether fossil or archaeological, upon which correlations could safely be based. Nor is the evidence of the low sea-levels of any assistance. There are, as will appear below, many other parts of the world where there is evidence of a general recession of the sea subsequent to the formation of the lowest Pleistocene shorelines. It appears, however, that no geological event has yet been recognized as having occurred between the time of the 6 m. high sea-level and the time when the sea-level, during its subsequent fall, had reached a depth of − 17 m.

The one remaining hope lies in comparisons between the Cyrenaican climatic succession and those obtained elsewhere. Since the present discussion is con-

---

[1] See, once again, the reservation made in the footnote to p. 65.

cerned only with that part of the Cyrenaican succession which is later than the 6 m. shoreline, and since the latter feature is believed to be relatively recent, this is a case in which such comparisons can be expected to have some significance.

2. LATE PLEISTOCENE CLIMATIC SUCCESSIONS ELSEWHERE IN THE MEDITERRANEAN

In the entire Mediterranean region, the number of deposits from which any Pleistocene climatic succession has been obtained is unfortunately rather small, and of these only a few have been shown with any certainty to belong wholly to the time of the last glaciation. At present, there seem to be only four Mediterranean localities outside Cyrenaica where such deposits have been found containing clear evidence for more than one climatic phase. These are as follows: the French and Italian Rivieras, the plain of the Lower Versilia (north-west of Pisa), the Pontine Marshes (south-east of Rome) and Grotta Romanelli (near Castro, in southern Apulia). The chronological and climatic evidence in all four cases has been summarized and discussed by Zeuner.[1]

The caves and rock-shelters of the Riviera, studied by many different workers, have provided a composite climatic succession which is assigned to the Würm glaciation on the grounds that one of the caves, Grotte du Prince, contains evidence of a sea-level at 22·7 m.; this level is thought to date from the Last Interglacial. The succession itself shows evidence for two cold periods, separated by mild periods, and, according to Zeuner, a final damp period represented by the formation of stalagmite among certain accumulations of breccia.[2]

The Pleistocene deposits of the coastal plain of the Lower Versilia lie for the most part below sea-level. Much information as to their nature has been supplied, nevertheless, by A. C. Blanc, from an examination of sands brought to the surface by pumping, and from a study of samples and records from boreholes.[3] The lowest level reached was −95 m. The succession revealed consists essentially of three layers of marine deposits alternating with three of terrestrial deposits, the lowest known layer being marine and the highest terrestrial. All the marine layers contain temperate faunas, and the two highest contain in addition seeds of the vine. In the two highest terrestrial layers, cold floras have been found; regarding the climate of the lowest terrestrial layer no definite information is available. The lowest marine layer is believed to represent the lowest level reached by the sea during the Würm glaciation, and the remainder of the

---

[1] Zeuner, *op. cit.* pp. 179–96. Mount Carmel, also discussed by Zeuner (pp. 199–200), has been omitted from the list since its exact chronology is still in doubt.

[2] Zeuner, *op. cit.* pp. 179–82.   [3] Blanc, A. C. (1937), pp. 635–41.

series would thus provide a record of the climatic changes during the subsequent rise of sea-level, sometimes known as the 'Flandrian transgression'. By analogy with the other two, the lowest terrestrial layer is thought to represent, if not a cold phase, at least a time when glaciers were advancing elsewhere.

In the Pontine Marshes, the Pleistocene succession, again studied by A. C. Blanc, begins with deposits of Calabrian and Sicilian Age.[1] The succession includes a marine layer which contains *Strombus bubonius* and which rises to a maximum level of +10 m.; the deposits above this layer, all of terrestrial origin, are therefore considered to belong to the time of the Würm glaciation. The organic remains found in the latter deposits indicate two climatic phases: the first, cold and oceanic, the second, cold and continental. In this case, however, there is some evidence that deposition was not continuous.

Grotta Romanelli is a cave of marine origin. The investigation of its deposits has been carried out largely by G. A. Blanc.[2] In this cave, the lowest deposit of all is a beach-conglomerate, containing few identifiable shells but lying about 7·5 m. above present sea-level; this, once again, is thought to show that the bulk of the later deposits can be assigned to the time of the Würm glaciation. The climatic interpretation of these later deposits has given rise to some controversy. According to Blanc the climatic succession, from bottom to top, is as follows: (i) warm (faunal evidence); (ii) damp (stalagmite); (iii) warm and dry (faunal and lithological evidence); (iv) damp (stalagmite); (v) cold (faunal and lithological evidence).[3] Zeuner disagrees with this to the extent of regarding the fauna of the highest layer (the 'Terra Bruna') as cool and continental, rather than Arctic.[4] Another objection has been raised by Vaufrey in relation to Blanc's first warm stage. The layer by which this is represented is composed mainly of angular rock-waste, and this material, in Vaufrey's opinion, suggests a climate with great thermal changes, indicative of the onset of a glaciation.[5] In spite of these objections, however, it is still certain that the cave contains evidence for a first damp period, an intermediate warm and dry period, and a second damp period followed by a phase of continental and relatively cold climatic conditions. Before the first damp period, the climate was either warm, in which case it can be regarded as continuing the conditions of the Last Interglacial, or, on the other hand, it was in some way 'unstable', and can be regarded as the first intimation of the damp period itself.

There are many obvious ways in which all these deductions can be criticized. For example, there appears to be no real proof that the various marine deposits

---

[1] Blanc, A. C. *op. cit.* pp. 628–35.   [2] Blanc, G. A. (1921) and (1928).
[3] Blanc, G. A. (1921), pp. 88–102, and (1928), pp. 367–409.
[4] Zeuner (1945), pp. 195–6.   [5] Vaufrey (1929), p. 148.

and shorelines, to which reference is made in each instance, have not been disturbed by earth-movements. Moreover, there is no guarantee that any of the successions described are complete. If, however, the chronological arguments in each case are correct, it appears that the Würm glaciation was represented at the two most northerly localities by not less than three periods which were cold or damp or both, these periods being separated by intervals which were warmer and in some cases drier. At the two most southerly localities, on the other hand, even on Vaufrey's interpretation of Grotta Romanelli, it cannot be said that there is evidence for more than two separate cold or damp periods, with evidence for an intervening warm and dry period in the case of Grotta Romanelli itself.

3. CORRELATIONS WITH SOUTHERN ITALY AND SICILY

Of the four localities mentioned, the one nearest to Cyrenaica is Grotta Romanelli, which lies about 850 km. to the north-north-west of the most northerly point on the Cyrenaican coast. This distance is relatively great, and the difference of latitude is about 7°; nevertheless, any large-scale climatic changes affecting the one area might reasonably be expected to affect the other. Hence, if it is assumed that the succession of Grotta Romanelli really does provide a complete record of local climatic changes throughout the time of the Last Glaciation, any resemblance which may exist between this succession and that deduced for the Cyrenaican coast can hardly be regarded as mere coincidence.

There is, in fact, a strong resemblance. In both cases there is first a damp period and secondly a cold continental period. In Cyrenaica, admittedly, no evidence has been found for an intervening warm and dry phase, but there are many indications that the damp phase and the cold phase were separated by an interval in which much erosion took place; judging by present-day conditions, a warm and dry phase in Cyrenaica would in any case be likely to cause erosion of the land, rather than deposition upon it. Thus there appears to be good justification for establishing a correlation between the two successions.

Without making any further assumptions, it is possible from these conclusions to reach one further chronological conclusion of some interest. This possibility arises from the results of Vaufrey's work on the caves of Sicily. Having examined the reports of numerous previous excavations, and having carried out certain additional excavations himself, Vaufrey found that it was often possible to distinguish in the deposits of these caves two successive vertebrate faunas, but never more than two. He found also a strong resemblance between the corresponding faunas of different caves. In particular, the lower faunas included various extinct forms, of which the most remarkable belonged to three dwarf races of *Elephas antiquus*; the upper faunas, on the other hand, consisted entirely of forms which were still living, though in most cases no longer present in Sicily.[1]

It can reasonably be assumed that the disappearance of the extinct forms from each individual site does in fact represent their final disappearance from the fauna of Sicily. Presumably, also, this event was caused by some considerable climatic change. On internal evidence, it was not possible to deduce either the nature of this change or its date. Vaufrey noted, however, that there is a strong resemblance between the early faunas of the Sicilian caves and those of the lower layers in Grotta Romanelli, *Elephas antiquus* being represented at the latter locality by individuals of the normal size. He also noted that there is a level in Grotta Romanelli, as in the Sicilian caves, above which the extinct forms characteristic of the early faunas are no longer found; this level is the base of the Terra Bruna. These facts, in Vaufrey's opinion, indicate that both faunas of the Sicilian caves can be assigned to the time of the Würm glaciation; also, that the disappearance of the extinct forms from Sicily took place shortly before the Terra Bruna of Grotta Romanelli began to be laid down—in other words, at the beginning of the last period of severe cold in southernmost Italy.[2] If this is so, and if the correlations with Cyrenaica are correct, it can then be said that the tufaceous deposits and the Younger Gravels of the Cyrenaican coast are respectively contemporary with the lower and upper faunas of the caves of Sicily.

4. CORRELATIONS WITH MORE NORTHERLY REGIONS

Further correlations can only be made if it is assumed that a complete Würmian succession is present, not only at Grotta Romanelli, but also in the Pontine Marshes and the Lower Versilia, and on the Riviera. If this assumption is made, it then follows that the two cold or damp periods in the south must be contemporary with two of the three similar periods in the north. A suggested correlation between the northern and the southern localities has been put forward by Zeuner. In his opinion, it is the third of the northern cold or damp periods which is missing in the south. This view is based largely upon the comparative unimportance of the third period, both on the Riviera and in the Lower Versilia. It is Zeuner's opinion also that the three cold or damp periods in the northern Mediterranean are contemporary with the three generally accepted phases of the Würm glaciation in the Alps.[3] If these views are accepted, the tufaceous deposits of Cyrenaica may then be assigned to the first phase of the Würm glaciation, and the Younger Gravels to the second.

[1] Vaufrey (1929), pp. 45–8.   [2] Vaufrey, *op. cit.* pp. 145–50.
[3] See Zeuner (1945), pp. 179–202, and correlation table, p. 203, fig. 64.

CHAPTER IX

SUMMARY AND FINAL DISCUSSION

I. SUMMARY OF RESULTS

(i) Most of the Pleistocene continental deposits noted in the coastal regions of Cyrenaica could be assigned to one of three lithological categories: alluvial deposits, consolidated dune-deposits, and deposits composed mainly or entirely of tufa.

(ii) The greater part of the alluvial deposits occurred as fans outside the mouths of the gorges which emerge from the lower escarpment; in certain cases the same material was found within the gorge itself, its upper surface forming a single terrace. Many of these features could be seen to overlie the 6 m. shore-line, and all, in fact, are believed to be contemporaneous. The material of which they are composed is referred to as the 'Younger Gravels', to distinguish it from those few patches of material of the same general nature which may yet be of greater antiquity. The only organic remains found within the Younger Gravels were land-snails, of species still living in the region. In spite of this, the deposits are thought to have been laid down at a time when winters were very cold, the rainfall perhaps being moderate and seasonal.

(iii) Of the deposits of consolidated dune which were examined, all were composed of material of marine origin, and most were in the form of genuine 'fossil dunes', adjacent to the modern shore. Between Derna and Benghazi, the vast majority of these latter features could be shown to be younger than the 6 m. shoreline; where this was so, they are referred to as 'Younger Fossil Dunes'. Many of the Younger Fossil Dunes both rest upon, and are covered by, deposits assigned to the Younger Gravels. From their distribution it is deduced that all were formed at a time when the sea-level stood considerably lower than at present; there are, indeed, numerous features on the adjacent sea-floor which are believed to be submerged fossil dunes, and one of these denotes a sea-level no higher than -17 m. Their distribution also indicates north-westerly prevailing winds.

(iv) Four extensive deposits were examined which were composed to a great extent of tufa, together with a certain amount of gravel and much fine-grained, highly calcareous marl, the latter probably of chemical origin. Each of these deposits lay either in, or in association with, one of the wadis of the eastern Gebel. On somewhat indirect arguments, all four have been attributed to a single period of high annual rainfall, intermediate in age between the 6 m. shoreline and the Younger Gravels. Three of the deposits have yielded faunas of snails, and one, in Wadi Derna, has yielded remains both of vertebrates and of plants. The vertebrate remains include an extinct buffalo; the most interesting feature of the flora is a species of *Laurus* no longer found in the Mediterranean. Both flora and fauna indicate a warm climate.

(v) The bulk of the continental deposits are thus believed to belong to not more than two distinct periods. In the first of these, to which the tufas are assigned, the climate was generally warm and wet. In the second period, represented by the Younger Gravels and the Younger Fossil Dunes, the climate was cold and drier; the sea-level, at the same time, was falling far below its present position.

(vi) In view of their supposed relationship to the 6 m. shoreline, the deposits assigned to these two periods are all thought to be younger than the end of the Last Interglacial. A correlation has been suggested with the deposits of Grotta Romanelli, in southern Italy; also, with the deposits of numerous caves in Sicily. It is pointed out, finally, that a very tentative correlation may be made, on certain assumptions, with the glacial succession of the Alps; according to this scheme, the two climatic periods referred to above would be the respective equivalents of the first and second phases of the Würm glaciation.

2. FINAL DISCUSSION

As with the shorelines, certain implications of the results obtained will now be considered. Once again, there is nothing more to be said concerning local implications. The results, however, have a bearing upon at least two general problems: worldwide changes of sea-level, and the climatic changes which occurred during glacial periods in regions remote from the ice. Both of these points require further discussion.

CHANGES OF SEA-LEVEL

It has already been mentioned that a worldwide fall of sea-level is believed to have occurred during the Würm glaciation. Evidence for this belief, derived from many different parts of the world, is quoted by A. C. Blanc.[1] Blanc himself has provided further evidence by his own investigations in the Lower Versilia, and has, in addition, found indications that the lowest level reached was below −90 m.

Much of this evidence can be questioned, however, on the grounds that there is little real proof of the tectonic stability of the regions from which it was obtained. Such objections can hardly be raised, on the other hand, in connexion with the submerged dunes of Cyrenaica. In this case, there seems to be convincing evidence that the regression did occur, that its maximum depth was

[1] Blanc, A. C. (1937), pp. 625-7.

greater than −17 m., and that a part of the subsequent transgression has taken place since classical times.

CHANGES OF CLIMATE

The published evidence, as has already been seen, shows clearly enough that the coast of Italy and the Mediterranean coast of France were subjected during the Würm glaciation to repeated periods of cold or damp climatic conditions, alternating with periods in which the climate was milder. There are also some indications that, whereas in the northern Mediterranean all three phases of this glaciation had a strong effect upon the climate, in regions south of Rome one of these phases, probably the third, left no traces.

For the North African coast, on the other hand, the information available on Late Pleistocene climates is meagre, and, such as it is, implies no very drastic changes. In Sandford's view, it was 'at the close of the last glacial epoch' that desert conditions, moving gradually northwards, ultimately reached Lower Egypt.[1] In the Western Desert of Egypt, the climate, according to Murray, was generally arid, with, however, one rainy interlude in 'Middle Palaeolithic times';[2] there is also some evidence for a similar interlude on the Tripolitanian coast (see below, pp. 223–4 and 270). Finally, for the coast of French North Africa, Wulsin, after an examination of all the published evidence, could only say that the climate was wet and warm at the time of the '18 m. strand-line' (believed to date from the Last Interglacial), and that it has since become drier and cooler.[3]

According to the results of the present investigation, it now appears that in the Gebel Akhdar at least, which admittedly is the most elevated area on the eastern half of the North African coastline, the climate was influenced profoundly by the Würm glaciation. The effects, indeed, seem to have been much the same as in southern Italy, and scarcely less intense.

There are also some small indications that this glaciation may have caused in Cyrenaica biological changes corresponding to those which it caused in southern Italy and Sicily. In the latter regions, the last cold phase of Würmian times resulted in the final disappearance of a number of animals. In Cyrenaica this phase, which is thought to be represented by the Younger Gravels, may well have been responsible for the disappearance of the Canary Island laurel and possibly of the once abundant snail *Xerophila chadiana*. It may be mentioned that the Canary Island laurel has actually been found near Palermo in deposits whose fauna includes the dwarf races of *Elephas antiquus*, and whose flora also includes *Persea indica*, another plant which now lives only in Madeira and the Canaries.[4]

---

[1] Sandford (1936), p. 76.   [2] Murray (1951).
[3] Wulsin (1941), pp. 144–5.   [4] De Stefani (1946), quoted in Gignoux (1950), p. 670.

PART II

PREHISTORY

CHAPTER X

THE ARCHAEOLOGY OF
THE EARLIER VALLEY-DEPOSITS OF
THE EASTERN GEBEL AKHDAR

The purpose of this and the succeeding chapters is first to describe observations on the prehistoric lithicultural traditions of Cyrenaica collected during our two expeditions in 1947 and 1948, secondly to review the Stone Age finds from neighbouring territories, and finally to discuss such general conclusions as may be drawn from the material as a whole. Chapters x–xiii present data regarding the equipment and subsistence of the hunter-gatherer cultures of the Pleistocene and early post-Pleistocene epochs in Cyrenaica itself. Evidence regarding the closing phases of the Stone Age in this district, when the earliest methods of food-production may be presumed to have made their first appearance, is meagre.[1] In Marmarica to the east and Tripolitania to the west, however, relatively rich finds of well-concentrated surface occupation sites go some way towards filling the gap. The two final chapters are accordingly devoted to presenting and discussing this latter material and attempting some measure of synthesis with recent finds and theories in other parts of North Africa.

In the present stage of investigations the most convenient method will be to subdivide the material partly on a geographical, rather than a strictly chronological, basis. The first group of finds to be dealt with will accordingly be those from the remarkable series of tufaceous deposits in the Derna region, described from a geological point of view in Chapter VII of the preceding section. It will be recalled that it was concluded, on geological grounds, that all the deposits in question could be attributed to a single depositional phase following immediately on the 6 m. sea-level.

All the climatic evidence pointed to conditions with a rainfall far higher than at present, resulting in a perennial flow in valleys now dry for the greater part of the year. There was no indication that temperatures differed very greatly from the present, and they must in any case have been in sharp contrast to those of the much colder phase which followed, characterized by extensive frost action.

---

[1] Briefly summarized in McBurney (1947), pp. 82–3 and fig. 16, earlier finds consisted almost entirely of surface material. A single stratified site became available after the present work was already in the press; the salient results have been included in footnotes.

1. SIDI EL HAJJ CREIEM

This is the most considerable archaeological discovery so far made in this series of deposits. It occurred at a locality on the left bank of Wadi Gahham, less than 250 m. from its confluence with Wadi Derna, immediately below the marabout from which the locality is named, some 6½ km. from the coast. The remains take the form of a well-preserved settlement scatter *in situ* towards the base of deposits associated with the high terrace of the main Wadi Derna. Full-scale excavations were carried out here during both seasons. The resulting collections, though perhaps small by the standards of some cave sites, are nevertheless believed to provide the most complete picture of an industrial assemblage of this stage so far obtained from the North African littoral.

CHARACTER OF THE SITE AND NATURE OF THE EXCAVATION

The site was located during the first season in the course of a search for artificial and organic remains in the exposures of the high terrace in the lower and middle reaches of Wadi Derna. Wadi Gahham is one of the largest tributaries of Wadi Derna, and the deposits near its mouth can be shown to be continuous stratigraphically with those of the high terrace in the main valley.

When discovered, the remains consisted of flint artifacts and fossil bones projecting from the vertical face of a formation of marl and fine silt, grading into tufa a few metres downstream, and exposed by recent torrential erosion (Pl. 7).

Subsequent search of Italian literature showed that this exposure had apparently been noted earlier, though as far as we have been able to ascertain no collections were made from it.

The deposit was horizontally bedded and the implementiferous zone confined to a thickness of not more than about 50 cm., so that it was clear from the outset that excavation would require the prior removal of a very large mass of overburden increasing rapidly with distance from the modern face. This was in fact done, and during 1947 and 1948 a total of some 40 sq. m. of the ancient settlement were laid bare.

Once the implementiferous horizon was uncovered investigation was carried out by a process of splitting off large lumps of the deposit, which were then crumbled apart by tapping. Owing to the extremely tough nature of the marl and silt matrix it was found quite impracticable to work in the normal manner with trowels and hand-picks, as both bones and flints were frequently far too fragile to be detected by feel. On the other hand, our method took advantage of the even bedding of the formation, which had a tendency to break off in small lumps exposing portions of specimens which could then be packed in their original matrix after slight trimming. The final preparation of the organic remains was carried out in most cases after shipment, either at Cambridge or at the British Museum (Natural History). Particularly large or fragile pieces such as the large horn core shown on Fig. 39 were encased in a solid block of plaster of Paris,[1] and subsequently consolidated (if necessary in the course of extraction) by the use of dilute polyvinyl acetate, following a method developed by the staff of the Geological Department of the British Museum.[2]

Mention has already been made of the geology, the invertebrate fauna, and the fossil flora of the deposit; it will now be useful to combine the conclusions they suggest regarding the appearance and character of the site at the time of its prehistoric occupation. Briefly, the picture is that of a long narrow pool, not more than 50 or 60 m. wide and perhaps four or five times as long, occupying the bottom of a deep rocky ravine. The pool was doubtless supplied with water by some perennial spring, no longer active, higher up the wadi, and owed its existence to the fact that the outlet of the wadi was dammed a short distance below the site by the rapid formation of tufa within Wadi Derna itself. As in many parts of Wadi Derna at the present day, dense vegetation, including brambles and reeds, would have grown in and around the pool; it may be mentioned, however, that no traces were found of the oleander, now one of the most common plants of the wadi. The Aleppo pine and a species of laurel occurred in the vicinity, and presumably grew high up on the sides of the ravine; the pine, at any rate, still grows abundantly on the sides of many Cyrenaican wadis, though not indeed in the immediate vicinity of Derna.

The ancient camp site itself, used at most for a few seasons and perhaps only for a few days or weeks, was situated close to the rocky wall of the right bank not more than a metre or two from the position in which the majority of the specimens now occur. This can certainly be deduced from the immense quantity of small chipping waste, and debris of broken food bones. In one case two flakes, struck one after the other, were found in contact. As far as the information given by the surviving portion goes the scatter of rubbish was not more than about 20 m. long and of unknown width, but perhaps not more than about 10 m. The only remains found outside this limit were two large horn-cores of a buffalo which lay with some teeth and skull of Barbary sheep closely packed together some 20 m. upstream.

---

[1] We should like to thank Dr F. Barber (of the B.M.A. Hospital, Derna) for most kindly undertaking this work, with entirely successful results.

[2] The success of the whole operation and consequent identifications were due in no small measure to the unfailing help and advice of the members of this department.

THE STONE INDUSTRY

General characteristics and raw material

The artifacts collected amount to about 1500 in all. While this is not a large figure compared with the more prolific cave sites, their interest is somewhat enhanced by the fact that they provide an unmixed picture of the stone-working activities of a single community at one time.

The raw material used is for the most part remarkably homogeneous and consists of a very fine grade of grey to greyish-brown flint. The surface of the nodules seen on the cortical fragments is almost invariably that of unrolled and unweathered native lumps worked soon after removal from their parent matrix.[1] That some of the nodules were collected by the ancient artificers at or immediately *below* the contemporary sea-level is shown by the presence of intact limey tubules of marine worms adhering to them.

At two localities within a few miles of Derna, beds of fine quality flint (in the Lower Eocene limestones) lie exposed on the surface of the 6 m. marine terrace, and are strewn with thousands of waste flakes which bear eloquent witness to the importance they held for the Palaeolithic and later inhabitants of the area. It was no doubt to some such locality that the occupiers of the Hajj Creiem settlement came for their flint. This careful selection of raw material by the Hajj Creiem people is in interesting contrast to the extremely haphazard choice of raw material frequently shown by artisans of Middle Palaeolithic type industries elsewhere. Then as now the plateau a mile or less away must have been strewn with the products of earlier industries in every stage of patination and decomposition, yet less than 1 per cent of the cortical flakes betray the use of this source.

This peculiarity is no doubt to be connected with the very exceptional level of craftsmanship about to be described.

Primary flaking

Owing to the character of the site it is possible to reconstruct the process of tool manufacture in some detail. The primary production of flakes was mainly from flat circular cores, of the classical 'Mousterian' type. The largest of these actually found measures 8·7 cm. in diameter by 3·0 cm. thick, but the majority (as is not unusual in cores of this type) are considerably smaller, running down to a mere 3·2 cm. in diameter by 0·8 cm. thick.

It is sometimes doubted whether flakes of the minute size yielded by cores of such proportions can really have served any useful purpose, and whether it was

---

[1] This is shown by the absence of incipient cracks or abrasion (characteristic of flint fragments transported by water) or other signs of weathering.

not rather the cores themselves that were the primary object of manufacture. While this is of course a possibility and no final decision can be reached in any particular instance, it should not be forgotten that carefully trimmed flake tools less than 3 cm. in greatest dimension are by no means unknown in the Middle Palaeolithic both in Europe and elsewhere (see for instance Aterian examples described on p. 228). Moreover, the mean size of the scars on a given core represent at most the size of flakes produced at the closing stage of the flaking process rather than the mean size yielded throughout the course of their exploitation.

In the case of Hajj Creiem further evidence in favour of regarding the flakes as the primary products can be adduced from several sources. Thus several of the cores of the smallest size are far too irregular, occurring as they do in an industry of outstanding regularity and finish, to be regarded as members of a definite tool class. Another indication pointing in the same direction is provided by the striking-platforms of the flakes in general. In Middle Palaeolithic industries of large size it is not unusual to find that the fine faceted preparation of the striking-platform, while well represented on the larger and medium-sized flakes, is only rarely observed on those of the smallest class, many of which show completely haphazard preparations and can indeed frequently be identified from other characteristics as trimming waste from the manufacture of points, reshaping of cores, and the like. At Hajj Creiem, however, a very high proportion of flakes between 1·5 cm. and 3 cm. show unmistakable fine-faceted preparation of the platform.

Admitting, then, that most of the cores in question are by-products, it remains to reconstruct as far as possible the whole process of flaking, from the 'quartering' of the original nodule onwards. Generally speaking, there are two ways in which a core of discoid pattern can be started—either by selecting a more or less flat or 'tabular' nodule, or by striking a massive flake on an anvil. In the latter case a large flake of this kind will normally (though not always) have a large cortical area on the dorsal surface and this frequently remains as a patch of cortex on the base of a discoid core. The next stage consists in making the peripheral platforms, and finally the convergent removal of the initial flakes intended for use. Where the core is started from a large flake in this way the first flakes struck from it will in general show traces of the convex surface of the original bulbar face on their dorsal side. At Hajj Creiem a considerable proportion do in fact show this feature, including some of the large pieces subsequently trimmed into points and side-scrapers. In addition many cores in various states of exploitation show portions of the bulbar surface of the original flake (see Fig. 14, nos. 2 and 4).

Although the evidence suggests that the above was by far the commonest method used at Hajj Creiem, slight traces of two other distinct techniques can also be detected. The first is suggested by the presence of a few large rectangular flakes of the type now often known as 'flake-blades'. These demand more elongated cores more nearly on the pattern of the flat, rectangular blade-cores not infrequently associated with the European Levalloisian. Rare but quite unmistakable indications of another method of greater cultural significance can also be found at Hajj Creiem, namely, true 'tortoise-cores' of Levalloisian type.

The term 'tortoise-core' is often used in a loose sense to describe any rounded core from which flakes are struck convergently across a flat or slightly convex surface from a series of faceted platforms. By this loose use of the term an interesting cultural distinction which has long been recognized in Europe is obscured, namely, the difference between the true tortoise-core in the strict sense (as found in innumerable examples in the Lower and Middle Levalloisian from the fluviatile and solifluction deposits of North-west Europe) and the much more generalized disc or discoid technique characteristic of the European cave Mousterian. The difference between these two devices as seen on typical specimens is perfectly clear-cut. The former is designed for the production of a *single* circular, oval, or rectangular flake struck from a domed or slightly convex surface formed by a large number of small preparatory strokes each yielding merely a *waste* flake. In general, the whole core has to be prepared afresh before a second main flake can be struck. The number of preparatory strokes recognizable on both flakes and cores is generally of the order of eight or nine and not seldom as high as nineteen or twenty.

On the 'Mousterian disc' on the other hand the production of useful flakes is a continuous process, each flake forming part of the preparation of a successor, and the number of preparatory scars shown on the dorsal side of each flake is seldom more than four or five and more usually two or three. This simple quantitative difference forms a convenient yardstick to measure the extent to which any given flake industry leans towards one or the other method of flaking. Another striking distinction between Mousterian and Levalloisian at the eponymous sites and neighbouring areas in Northern France and Britain is that of size—most European Levalloisian sites yield a mean size of flake at least twice that of normal cave Mousterian sites, while the difference between specifically tortoise-core and disc-struck flakes in the two traditions is greater still. In the report on the excavations at Mount Carmel in Palestine, Professor D. A. E. Garrod drew attention for the first time to the peculiar character of the Middle Palaeolithic from this and other Levantine cave deposits. The name Levalloiso-Mousterian was chosen to describe this, since analogy with the Levalloisian was shown by the presence of true tortoise-cores, and with the Mousterian by the small dimensions of the implements, abundance and general character of the trimmed tools (and, it may be added, by the considerable abundance also of perfectly normal Mousterian-type discoid cores).

Although the qualitative evidence in support of this idea at Mount Carmel is convincing, the quantitative information contained in the report is insufficient to enable an exact comparison, and the following analysis of the Hajj Creiem material will comprise a direct comparison with the two eponymous traditions of Western Europe. The comparative figures quoted were obtained in the course of the investigation of a somewhat different problem and are here published for the first time. It has seemed advisable to exclude observations on artifacts less than 2 cm. in maximum dimension since these are not generally preserved in available collections from the classic European sites.

These results, then, are given in Table I. On the score of absolute dimensions the figures for length and width of 'raw' flakes (i.e. flakes in the state in which they left the core without subsequent or 'secondary' trimming) are seen to show that Hajj Creiem falls well within the range of normal Mousterian cave sites. In absolute and relative thickness, however, a statistically significant contrast is demonstrated.[1] Of the twenty-four Mousterian industries I have examined from this point of view in Europe, I know of none giving evidence of as low a ratio of thickness to area as Hajj Creiem. Many indeed (and by no means the most primitive) show a relative thickness in the order of 25 per cent greater. This means in practice (apart from other advantages such as a lower angle of incidence along the edge—i.e. greater sharpness) that the Cyrenaican hunters with their improved technique were getting appreciably more flakes of a given size from a given cubic unit than were their contemporaries in Europe, who were nevertheless often obliged to go much greater distances for their raw material. In other words the industrial tradition at Hajj Creiem resulted in an appreciable saving of effort.

As regards the use of the tortoise-core technique the presence of flakes apparently requiring this type of core has already been noted, and it is of interest to see how far this conclusion can be demonstrated statistically, both as corroboration of the proposed archaeological classification and as a possible test in doubtful instances elsewhere. From the results given in Table I it seems clear that the flakes in question are numerically too infrequent to affect appreciably the *mean number* of preparatory scars; the difference, in so far as it can be detected statistically, resides in the long 'tail' of the frequency distribution extending to much higher maximum values than any that I have observed in exhaustive analysis of twenty-four different Mousterian cave sites in western and central Europe.

---

[1] It may be objected that this result is partly due to the quality of the raw material at Hajj Creiem. While this is of course possible, it is difficult to believe that the effect was such as to explain so wide a difference.

TABLE I. *Statistical summary of Hajj Creiem Levalloiso-Mousterian site, near Derna.* (Dimensions in mm.)

Hajj Creiem		Maximum observed mean in group of twenty-four Mousterian sites	Minimum observed mean in group of twenty-four Mousterian sites	European Levalloisian (Baker's Hole)
Raw flakes				
(1) Length	Mean **41·57 + 5·17** S.D. 13·28 Number 702 C. of V. 32·2%	**69·32** (La Gare de Couze, rock shelter, S.W. France)	**35·43** (San Francesco, rock shelter, N. Italy)	**123·90**
(2) Width	Mean **28·08 ± 3·87** S.D. 9·04 Number 708 C. of V. 32·2%	**47·04** (Le Moustier, rock shelter, S.W. France)	**22·58** (San Francesco)	**53·39**
(3) Thickness	Mean **6·88 ± 0·13** S.D. 3·58 Number 717 C. of V. 52·1%	**17·82** (Merveilles, rock shelter, S.W. France)	**7·65** (Torre di Talao, rock shelter, S. Italy)	**21·02**
(4) Mean number of primary scars per flake	**4·081 ± ·075** S.D. 1·981 Number 705 C. of V. 48·2%	**4·07** (Les Festons, rock shelter, S.W. France)	**2·11** (Spy, cave, Belgium)	**7·39**
Ratio $\frac{(3) \times 100}{(1)}$	**16·55%**	**17·63%** (La Gare de Couze)	**26·13%** (Bay Bonnet, rock shelter, Belgium)	**16·96%**
Ratio $\frac{(3) \times 100}{(2)}$	**24·50%**	**25·95%** (Hastière, cave, Belgium)	**34·83%** (Combe Capelle, Upper Shelter, S.W. France)	**28·14%**
Ratio of raw flakes to flakes with secondary working	**78·66%**	**13·00%** (Chapelle-aux-Saints, rock shelter, S.W. France)	**99·88%** (Belcayre, rock shelter, S.W. France)	**98 + %**
(5) Maximum number of primary scars per flake observed	**18**	**9** (Le Moustier, layer 'G')	**6** (Schulerloch, rock shelter, S. Germany)	**18**

TABLE I (*continued*)

Hajj Creiem		Maximum observed mean in group of twenty-four Mousterian sites	Minimum observed mean in group of twenty-four Mousterian sites	European Levalloisian (Baker's Hole)
Trimmed tools				
(1) Length	Mean 58·83 ± 1·11 S.D. 19·075 Number 141 C. of V. 32·6%	85·06 (La Ferrassie (main rock shelter, S.W. France)	41·96 (San Francesco)	—
(2) Width	Mean 36·96 ± 0·63 S.D. 19·075 Number 141 C. of V. 51·8%	51·10 (Merveilles)	23·83 (San Francesco)	—
(3) Thickness	Mean 8·010 ± 0·292 S.D. 3·619 Number 153 C. of V. 44·2%	8·44 (San Francesco)	20·10 (Merveilles)	—
Ratio $\frac{(3) \times 100}{(1)}$	13·622	16·38 (Les Festons)	27·88 (Merveilles)	—
Ratio $\frac{(3) \times 100}{(2)}$	21·693	29·69 (Petit Abri, La Ferrassie)	41·03 (Krapina, N. Yugoslavia)	—
Cores				
(1) Length	Mean 50·165 ± 2·49 S.D. 13·208 Number 28 C. of V. 25·6%	(2) Width	Mean 40·459 ± 3·106 S.D. 16·462 Number 28 C. of V. 40·8%	
(3) Thickness	Mean 13·86 ± 0·98 S.D. 5·38 Number 30 C. of V. 38·7%		Mean number of flakes per core 29	

Specimens with secondary working

Trimmed implements at Hajj Creiem fall into the two main classes of points and racloirs common to Middle Palaeolithic industries throughout their distribution. In general, they are identical to the variants found in Palestine though one or two special tools seem to be missing. There are, for instance, no burins. As shown in Table I the absolute dimensions of length and breadth fall well within the range of corresponding measurements in European sites, while their generalized shape as indicated by the length-thickness and width-thickness ratios

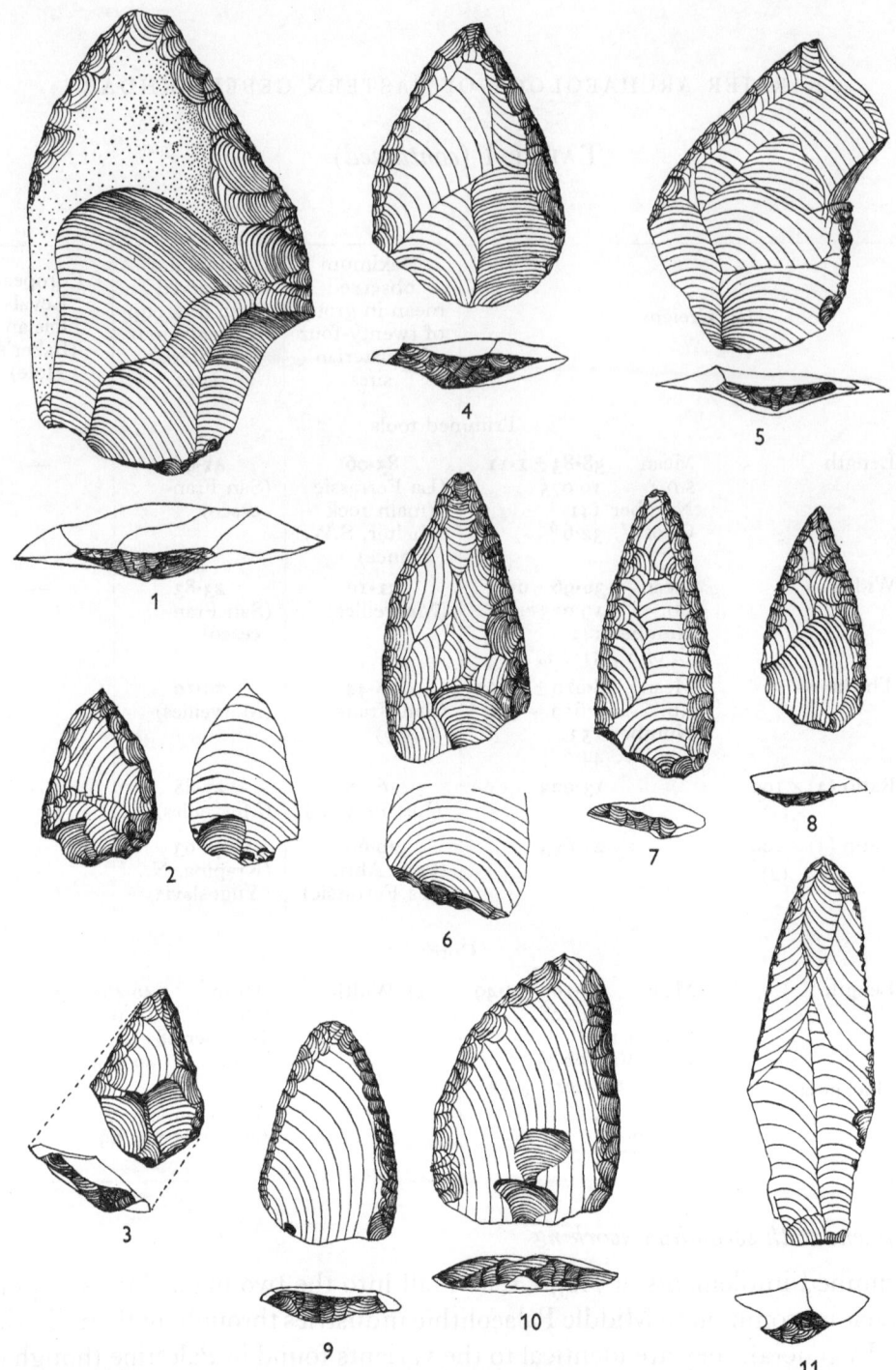

FIG. 12. Hajj Creiem: points

No. 1, the largest specimen found; No. 2, bulb removed by flat flaking; No. 5, unfinished specimen in two pieces, broken in course of manufacture; Nos. 9 and 10, with rounded tips, made on initial flakes struck from disc cores started from very large flakes showing original bulbar surface; No. 11 made on flake-blade. All × ½.

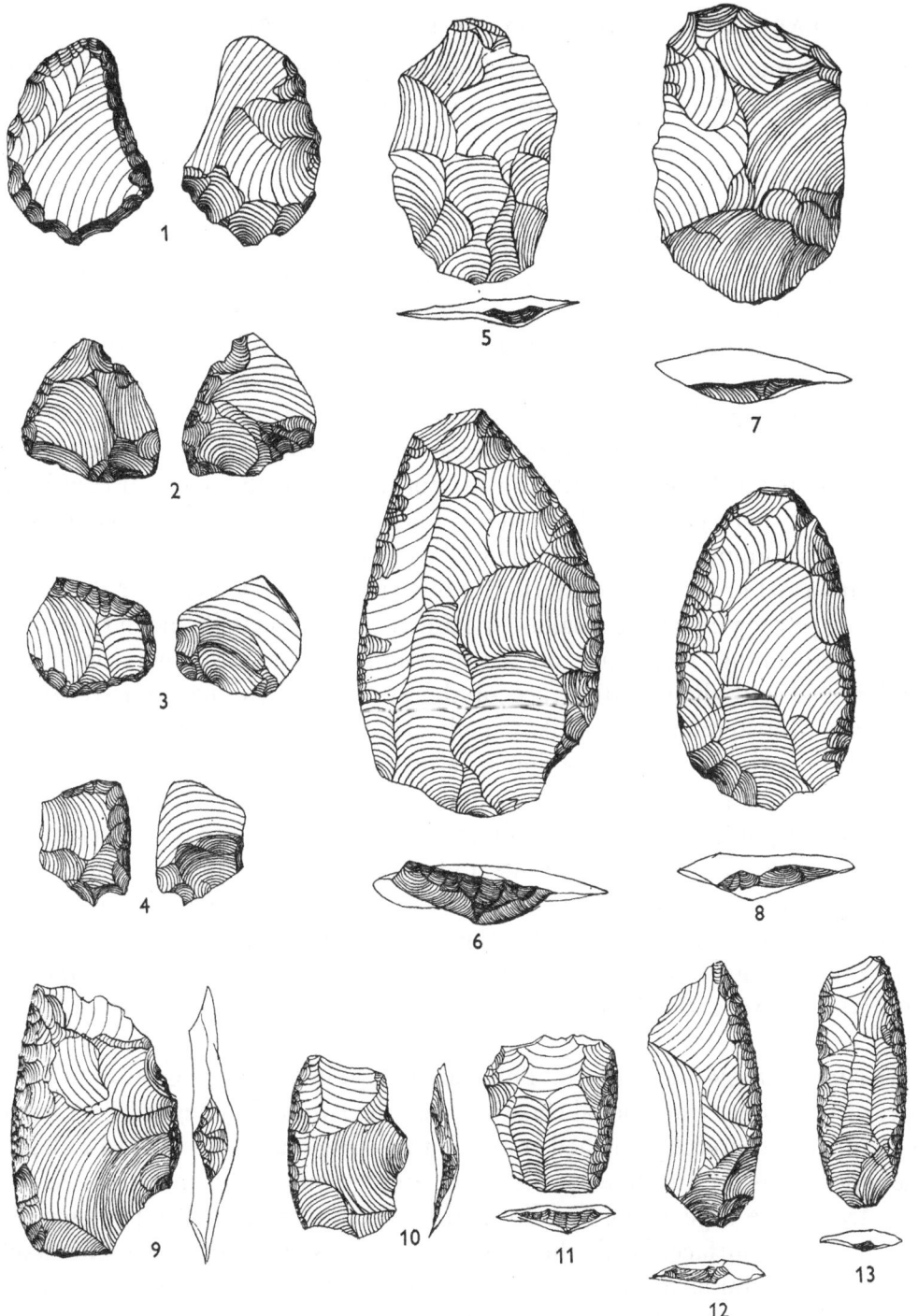

FIG. 13. Hajj Creiem: scrapers and flakes

Nos. 1–4, scrapers with bulbar face trimming; Nos. 6, 8 and 13, double side-scrapers; Nos. 9–12 single side-scrapers; Nos. 5 and 7, Levallois flakes from true tortoise-cores. All × ½.

shows the same peculiarity noticed among the flakes to an enhanced degree. This must in the main be due to the nature of the flakes of which they are made, although exaggerated to some extent by fairly frequent removal of prominent bulbs of percussion. It is difficult to imagine that this latter feature was not connected in some way with a method of hafting.

Typological details of the two classes are as follows:

(i) *Points* (48). These show a smooth gradation from narrow and elongated to wide and stumpy shapes. The great majority retain a finely faceted striking-platform at the base, and are made from relatively thin disc-struck flakes. Removal of the bulb is not uncommon (8 pieces) and occasionally secondary thinning was carried out on the dorsal surface as well—a feature also noted at Mount Carmel. The percentage of pieces with bulbar flaking compares closely with that noted at Layer D of Et Tabun,[1] for instance—about 14 per cent in the latter and 17 per cent at Hajj Creiem. There are, however, no specimens in which the trimming is *confined* to the bulbar face. The dimensions of the points are of the usual order for Mousterioid industries—$12 \cdot 7 \times 7 \cdot 9$ cm. for the largest piece and $4 \cdot 85 \times 3 \cdot 25$ for the smallest specimen certainly classifiable in this group.

The retouching technique on both points and side-scrapers is closely standardized and shows little trace of the step-fractures characteristic of the European Mousterian, the squills usually passing right through the relatively thin flakes used. In this connexion it is interesting to note that the trimming squills themselves never show the signs of heavy wear on the outer margin of their striking-platforms commonly observable in the European Mousterian sites, and suggesting that there both points and racloirs were frequently resharpened after being blunted by use. An interesting broken specimen at Hajj Creiem in which it was possible to reassemble the two fragments was apparently a point broken in the course of trimming, and shows that the characteristic shape of a point was in process of being produced directly from the raw flake (Fig. 12, no. 5).

(ii) *Side-scrapers—double* (16). These are placed in a separate group since they appear to provide a transitional class between the two main categories. A few of the points have carefully finished rounded tips—Fig. 12, no. 9, for example. In other specimens the two lines of retouch converge but do not meet, and leave a sharp transverse edge at the tip—formed by the intersection of a primary scar with the bulbar surface. While these are technically side-scrapers it seems likely from their general shape that they might well have served some of the same purposes as points, particularly in view of the round tipped pieces just mentioned (Fig. 12, no. 10 and Fig. 13, nos. 6, 8). On other examples, however, the

[1] Garrod and Bate (1937), pp. 73 *et seq.*

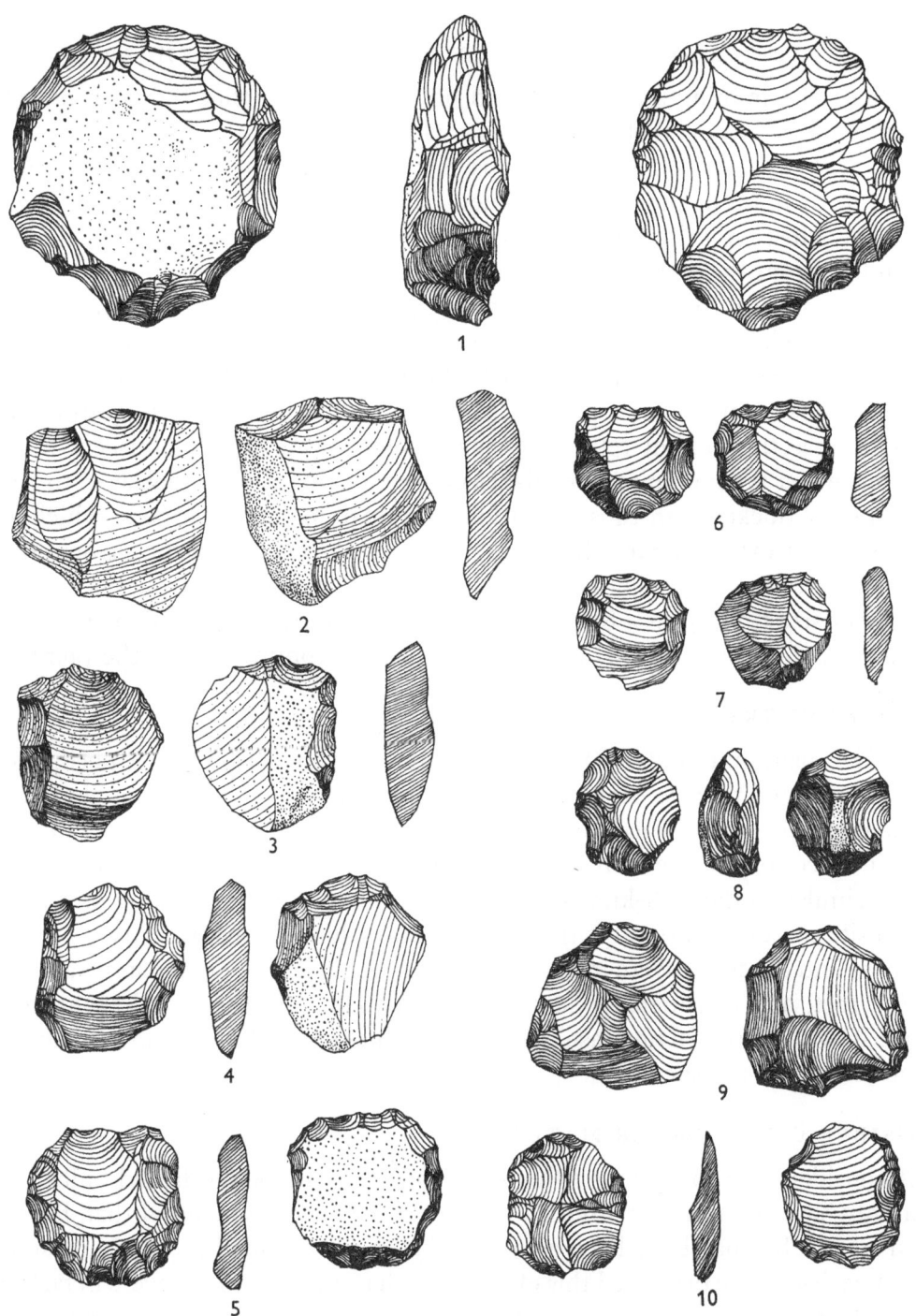

FIG. 14. Hajj Creiem: cores

No. 1, the largest core discovered; Nos. 2, 4, 7 and 9, disc-cores showing trace of bulbar surface of flake from which they were started; Nos. 3, 6 and 10, disc-cores showing dorsal surface of original flake; No. 5, normal disc; No. 6, biconical core (the smallest core obtained). All $\times \frac{1}{2}$.

two lines of trimming are so separated on a broad flake that they can only be interpreted as two parallel scraper edges (Fig. 13, no. 13).

(iii) *Side-scrapers—single* (61). This is the largest single class of trimmed artifacts; in it are included all specimens showing a single well-defined line of retouch. Most typically this consists of a convex line of trimming 6–7 cm. long extending from the right-hand end of the striking-platform up the right-hand margin of an oval or rectangular flake (45 pieces). Much less commonly it is found on the left-hand margin of the flake (8 pieces). In a small group (corresponding to the laterally-struck type at Mount Carmel)[1] the platform and trimmed edge are discontinuous and parallel (9 pieces). Occasionally a suggestion of a short intersecting line of retouch forms a point at one end reminiscent of the shorter, wider type of point. The flakes chosen for this type of tool seem to have been rather thicker than those used for the remaining side-scrapers, which are frequently of extraordinary thinness. A few were certainly struck from small tortoise-cores but the majority came from normal discs. The group as a whole ranges in size from 9.3×4.8 to 3.5×2.5 cm. Thinning on the bulbar face, and occasionally on the dorsal face as well, is about as common as on the points.

BONE INDUSTRY

Of the many hundreds of fragments of bones which littered the occupation site all appeared to have been broken in the course of food extraction, and none could certainly be identified as owing their form to intentional shaping. The only certain traces of the use of bones were zones of crushing on fragments of limb bones similar to the well-known 'compresseurs' of Western Europe on Mousterian sites.[2] It was noticeable that only slivers of long bones were used for this purpose and the traces were in general only just sufficient to demonstrate activity of the sort, and considerably less prominent than, say, those at La Quina in South-west France. In one case (Pl. 6a) two zones of crushing could be detected at either end of the specimen.

POSSIBLE USE OF OTHER MATERIALS

A considerable proportion of the flakes show well-marked diffused bulbs that can only have been made with a hammer of more yielding material than flint or hard rock. This may have been a wooden baton. A rounded piece of limestone might perhaps have produced this effect, though no suitable pieces were in fact recovered, and the nature of the percutors in general remains an unsolved problem.

[1] The term used to describe these—'side-blow'—is here avoided owing to possible confusion with the quite different tool of Neolithic Egyptian context.
[2] Henri Martin (1909), pl. XVIII.

ECONOMIC CONTEXT AND OUTSIDE AFFINITIES OF THE HAJJ CREIEM CULTURE

Whatever may be the case in cave sites intermittently occupied by man and various other creatures, at Hajj Creiem the conditions of discovery were such as to ensure that all the remains obtained belonged to animals killed and brought to the site by its human inhabitants. Moreover, the way in which the bones were treated strongly suggests that all the creatures were hunted primarily for their value as food. The list below gives the numbers of specimens attributable to the principal food animals and may be taken as providing a fairly complete glimpse of the economic basis of life in the Wadi Derna during or shortly before the first phase of the last glacial maximum in Europe.

Gazelle	(1) (1%)
Carnivore	(2) (1%)
Large Bovine	(21) (14%)
Zebra	(30) (20%)
Barbary sheep	(73) (49%)
Small land tortoise	(21) (14%)

It may be noted that in contrast with later North African Cultures no use seems to have been made of marine shell-fish, snails, or birds. These and other sources of food-supply may conceivably have been exploited at other places and seasons of the year, but here at any rate the mainstay of life was large powerful animals that must have required considerable skill in stalking, and slaughtering. It is possible that trapping played a part,[1] but it is difficult to believe that the whole of this varied assemblage can have been obtained other than by hunting. We have positive evidence from Mount Carmel that stout wooden spears were in use at this time[2] though it is not easy to imagine which of the small and delicate-looking stone tools can have been used in shaping them.

As for the natural environment of the culture as a whole some light is thrown by the species of the plants and animals identified. While the rainfall was, as stated earlier, appreciably higher than at present (p. 129), it is unlikely to have differed so widely as to alter the general pattern of the environment. The belt of evergreen scrub in the immediate vicinity of the coast may have been wider and more continuous than at present, but then as now the greater part of the plateau to the south must have been occupied by a still wider zone of dry steppe shading into true desert. This southern zone was, no doubt, the main habitat of the

---

[1] Evidence for trapping young mammoths has been detected at the Weimar-Ehringsdorf Mousterian station in Germany—see Soergel (1922).

[2] McCowan and Keith (1939), pp. 74–5 and pl. XXVIII.

Barbary sheep and antelope, and to a lesser extent, of the zebra, while creatures such as the extinct buffalo must have been obliged to stay relatively close to the permanent water in the network of coastal valleys in the Gebel Akhdar. It was here, no doubt, when they came down to water from the surrounding plateaux that they were attacked by Levalloiso-Mousterians. Indeed, the very nature of the steep cleft-like valleys must have favoured hunting of this sort.

It might be thought that such sharply varying conditions as those just described would lead to a high degree of local cultural divergence. In Europe, however, where an at least equal contrast can be observed between the environment of the cave-bear-hunters of the high Alps and the mammoth- and horse-hunters of the surrounding lowlands, an astonishing uniformity of material culture is the keynote of the later Middle Palaeolithic industries. It now looks as if a comparable degree of cultural continuity existed across the wide regions of desert and steppe separating the Cyrenaican Gebel from the hills of Judea.[1]

2. MATERIAL FROM OTHER LOCALITIES EQUATED ON GEOLOGICAL GROUNDS WITH HAJJ CREIEM

In Chapter VII it was stated that archaeological material from certain other deposits in the Derna region was not inconsistent with the theory of their strict contemporaneity with the Wadi Derna high terrace suggested by the geological evidence. This archaeological material may now be examined in the light of the detailed description just given of Hajj Creiem. All the collections were small and were made in the course of brief geological reconnaissances. No traces of well-defined habitation sites were observed and the specimens occurred widely disseminated throughout the deposits in question.

WADI EN NAGA

Over a distance of about 6 km., the following artifacts were collected from the older (tufaceous) deposits of Wadi en Naga:

Cores (2)

One large biconical core showed the method of production probably used for the few triangular flakes observed at Hajj Creiem (of which one served as a point).

---

[1] Other finds that can probably be classed in the same Middle Palaeolithic continuum have been reported from further west (see pp. 224 *et seq.*). The most recent finds indeed, at the time of going to press, suggest that there was a community in the physical type of the population as well. A mandible fragment from a sounding in a newly discovered cave in the Apollonia region, associated with a Levalloiso-Mousterian industry, shows a number of the peculiar features which serve to distinguish the human type of Tabun from that of the Neanderthaloids of other regions (McBurney, Trevor and Wells (1953).

One fragment of a typical large flat discoid core with a cortex base, broken in the attempt to remove a final flake.

Primary flakes (14)

The most interesting is a large flake-blade, similar to Fig. 12, no. 11, at Hajj Creiem, but larger, with a finely faceted platform. There is also a small flake showing an earlier bulbar surface on the dorsal side in the manner of those described at Hajj Creiem, and one resharpening flake from a point or side-scraper. The remainder are mostly small with plain or simply prepared platforms.

Retouched specimens (1)

A singularly well-finished double side-scraper with even trimming in the Hajj Creiem manner, and a finely faceted striking platform.

The most important points of resemblance with Hajj Creiem are offered by the retouched piece and the fragment of a flat disc, especially the former.

AIN MARA

A short search in the deposits of Ain Mara yielded a number of artifacts of some interest.

Details are as follows:

Cores (3)

(i) A fragment representing about one quarter of a large flat disc core with cortical base, originally perhaps 12–15 cm. in diameter by 2·5–3 cm. thick. The striking-platforms run evenly round the whole of the periphery preserved and are of the typical faceted variety seen on the majority of Levalloiso-Mousterian flakes (Fig. 15, no. 6).

(ii) A very small rather irregular discoid core, 3·7 × 3·8 × 1·5 cm., similar to several at Hajj Creiem except that it is worked over both faces. This feature is a common one on all large series of cores of this pattern and no cultural significance is to be attached to its absence from the relatively small collection at the latter site (Fig. 15, no. 4).

(iii) Fragment of a very small (?) tortoise-core originally about 2–2·5 cm. thick and 5 cm. in diameter.

Primary flakes (17)

The striking-platforms may be classified as follows—five finely faceted, five plain or only slightly prepared, and seven indeterminable. There is one clear example of a miniature tortoise-core flake 4·7 × 3·8 × 1·1 cm., with a finely

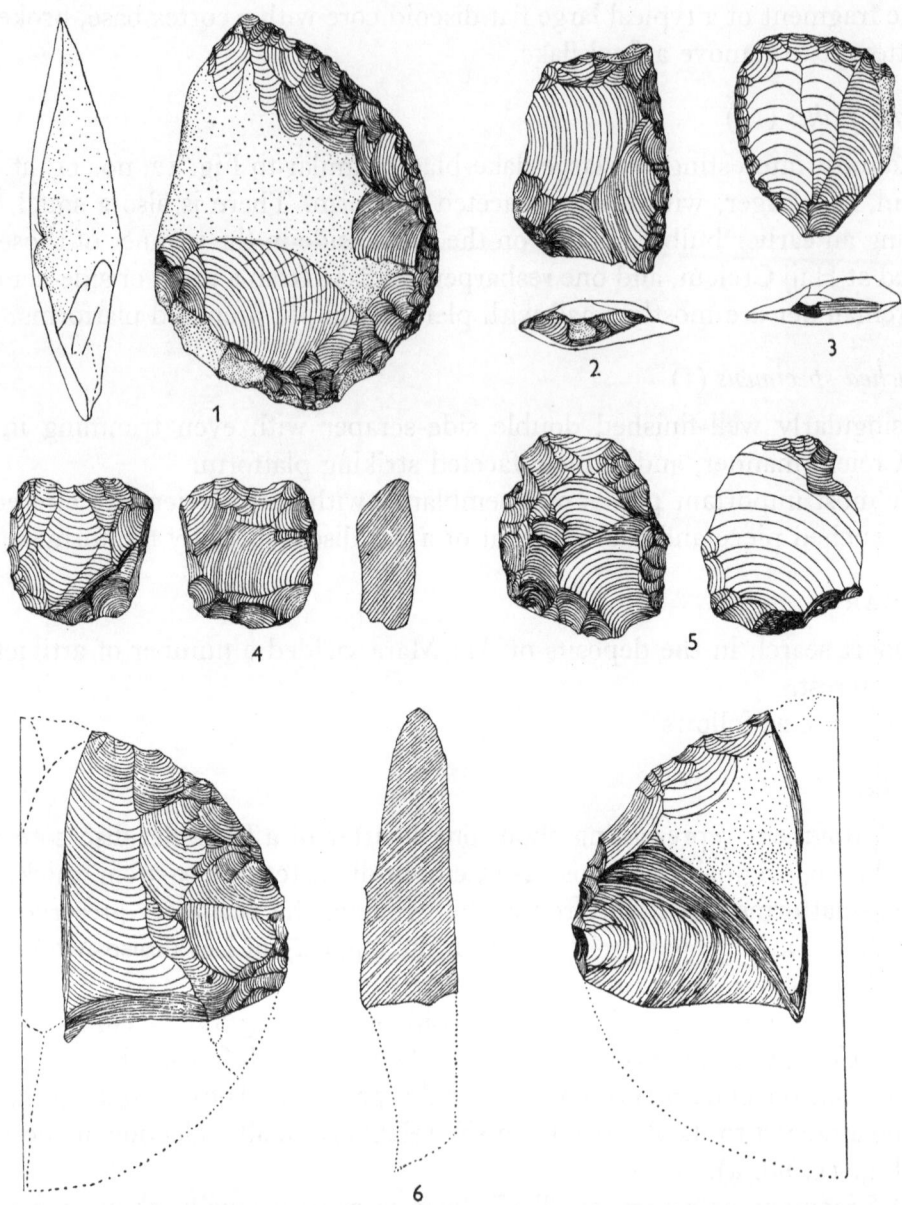

FIG. 15. Ain Mara: implements collected *in situ* from silts correlated with Hajj Creiem

No. 1, side-scraper on large vertical flake, slightly damaged; No. 2, small side-scraper on flake with faceted platform; No. 3, end-scraper on flake; No. 4, small discoid core exploited on both faces; No. 5, ? steep-scraper; No. 6, plunging flake from large disc, with suggested reconstruction of core. All × ½.

faceted platform, and one small flake-blade with the same. The remainder are small flakes mostly with little dorsal preparation.

Retouched specimens (7)

(i) A very small side-scraper with step-fracture trimming and bulb removed by steep flakes, possibly an unsuccessful attempt at dorsal thinning.

(ii) A normal side-scraper of a type common at Hajj Creiem, more or less rectangular in outline, with the retouch running up the right-hand margin of the flake from the right-hand end of the striking-platform. A short second line of retouch makes a rough point at the distal end (Fig. 15, no. 2).

(iii) A small steep scraper with the bulb removed by steeply flaked platforms for dorsal thinning scars (Fig. 15, no. 5).

(iv) A small end-scraper on a cortical flake-blade.

(v) A finely retouched end-scraper on a triangular flake.

(vi) A large well-made side-scraper on a thick cortical flake with a cortical platform. The retouch shows a high proportion of step-flaking (Fig. 15, no. 1).

(vii) A small damaged implement with the bulb removed by flat flaking.

This highly typical sample of Levalloiso-Mousterian work provides one or two points of detailed information not found at Hajj Creiem. The probable fragment of a true small tortoise-core is of interest since no actual cores of this type were collected at the latter site (though the technique could be deduced from the flakes). The very small tortoise-core flake is also a rather more typical example than any of these dimensions at Hajj Creiem. The two carefully made end-scrapers are of a type well-known in the Levalloiso-Mousterian of Palestine, though unknown at either of the two other Cyrenaican sites. The stepped trimming technique of the large side-scraper is no doubt the result of the proportions of the flake chosen.

EL ATRUN

Three large flakes were collected from the tufa deposits of this locality. There is nothing sufficiently characteristic about them to enable their attribution to the cultural stage just described, though they would be quite consistent with this idea.

CHAPTER XI

THE CULTURAL SUCCESSION OF THE SAHEL

The previous chapter contained an account of the earliest well-defined cultural stage so far identified in Cyrenaica: we now have to describe the archaeological record of the coastal region proper, in which there is direct evidence of recent changes of sea-level with respect to the land. This archaeological record starts, though in a very fragmentary fashion, at a much earlier period than that just discussed.

I. THE IMPLEMENTIFEROUS BEACH CONGLOMERATES OF THE 15–25 M. SHORELINE AT RAS AAMER

With the headland of Ras Aamer the Cyrenaican coast attains its most northerly latitude. The region in the immediate vicinity of the headland is one in which the topographical remains of the 15–25 m. shoreline are particularly well preserved, in the form of wide terraces sloping to seaward and delimited on the landward side by clearly recognizable wave-cut cliffs. Just west of Wadi Zemalia the terrace corresponding to this shoreline attains a width of over $\frac{1}{2}$ km., with a maximum height of 17 m. at its landward edge. Patches of an extremely hard conglomerate of well-rounded pebbles, sand, and marine shells can be observed on this terrace at widely separated points. One or two of these contain a few mechanically fractured flints of a probably artificial character, while one case in particular, situated close to the foot of the landward cliff itself, seemed to be of a rather more conclusive nature.

The remains in question may be summed up as just sufficient to establish the fact of human occupation in the territory at the time of the formation of the conglomerates, without providing any real clue to the cultural stage represented. The only finished implement is a small side-scraper shown in Fig. 16, no. 4, a specimen of a type liable to be found at almost any stage of the Lower or Middle Palaeolithic. Two small struck-flakes found a few inches away in the same matrix are doubtless artificial and made at the same time.

2. ARTIFACTS IN SITU IN THE 6 M. BEACH CONGLOMERATES

The next securely dated finds are found embedded in the conglomerates of the 6 m. strandline. Although by no means numerous they do at least afford a general indication of the technique of stone-working as practised at that time.

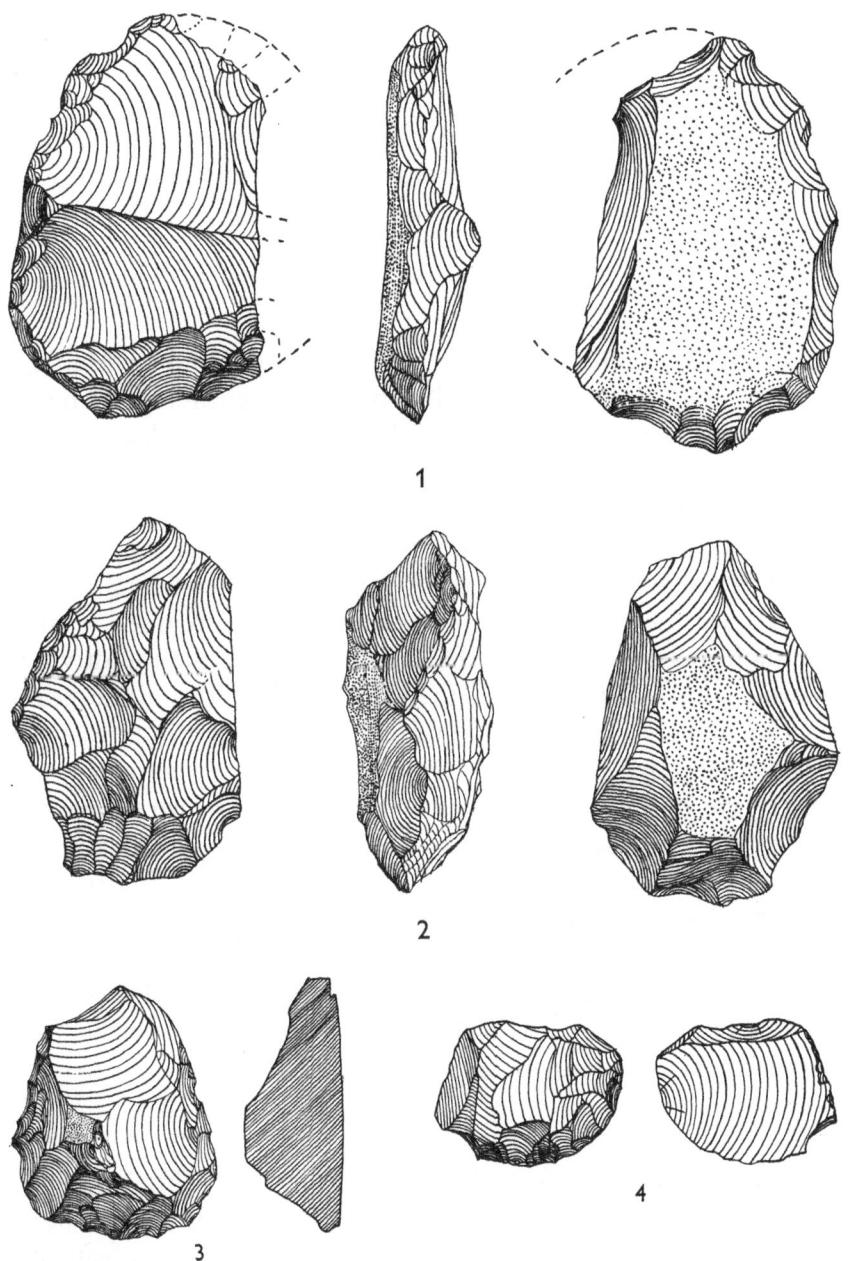

FIG. 16. Wadi Haula, el Atrun and Ras Aamer; specimens *in situ* in beach conglomerates No. 1, fragment of flat disc, Wadi Haula, 5 m. beach; No. 2, thick discoid core, Wadi Haula, 5 m. beach; No. 3, steep scraper, el Atrun, 5 m. beach; No. 4, scraper, Ras Aamer, 18 m. beach. All × ½.

RAS AAMER

In 1947 a single large core was obtained from the 6 m. beach at Ras Aamer. The specimen measures some $11 \times 9 \cdot 7 \times 4 \cdot 9$ cm. and is flaked all over both faces. While too nearly globular in form to be classed as a true Middle Palaeolithic disc core, it is, nevertheless, thoroughly characteristic of the local Late Levalloisian tradition (to be described later in this chapter) and may well be a rough attempt at a tortoise core.[1]

WADI HAULA

$4\frac{1}{2}$ km. west of Apollonia, below the alluvial fan of the Wadi Haula, a second exposure of the 6 m. beach was found to be implementiferous. The specimens obtained comprise two well-characterized cores and eight flakes. All were enclosed in 2–5 cm. of the characteristic matrix of the beach (as opposed to that of the overlying consolidated alluvium). Since, however, the total thickness of the beach-deposit at this point was small, it is just possible that they were resting on or near the original surface of the beach and became accidentally incorporated in it. In this case they would belong to the period of regression immediately succeeding its formation. On the other hand, owing to the rapidity with which these and other similar deposits seem to have consolidated, and the clear evidence that the artifacts were incorporated at a time when the beach was still soft, it seems very unlikely that they were manufactured more than a very short time, if at all, after the beach was laid down. The two cores are shown in Fig. 16, nos. 1 and 2. No. 1 is the greater part of a large, characteristically flat, Mousterian-type disc with the usual faceted platforms surrounding the base of the two main scars. No. 2 is a somewhat cruder attempt at a disc, but it is interesting to note the small size of the flakes which the makers thought it worth while to produce—below 5 cm. in length. Among the flakes is a small specimen of some interest since it appears to be a typical resharpening secondary flake, from a point or side-scraper. The remaining flakes show no features of diagnostic significance.

EL ATRUN

A single specimen from the 6 m. beach in this locality—Fig. 16, no. 3—which lies some 38 km. east of Wadi Haula, appears to be a steep scraper not unlike one from Ain Mara described above, and shows the same curious feature of dorsal thinning.

[1] Illustrated in McBurney, Hey and Watson (1948), p. 38, fig. 1.

3. THE INDUSTRIES OF THE YOUNGER GRAVELS

It will be recalled that the valley silts examined in the last chapter were believed, on geological grounds, to be intermediate in age between the 6 m. sea, and the final depositional stage of the Sahel succession. Except at El Atrun, no traces of this intermediate age have been recognized on the Sahel itself, and the last (and much the most clearly defined) cultural horizon comes from the ubiquitous 'Younger Gravels', which are believed to date from the last period of low sea-level, and to be associated with a sharp drop in mean temperature. Good vertical sections of these deposits were examined near the mouths of many wadis and along the seashore. Such sections, often as much as 5–8 m. high, provided ideal conditions for the collection of archaeological specimens *in situ*, and although it was scarcely to be expected that actual working floors should be identifiable, the freshness and quantity of the artifacts was often such as to provide a very clear picture of the lithicultural habits of the time. Some 200 specimens were obtained in this way, and an attempt to classify and interpret these, together with brief notes on their conditions of discovery, is given below.

LOCALITY I—DERNA WEST

The exposures grouped under this heading are situated some 5–9 km. west of Derna, and comprise vertical faces cut by recent erosion in alluvial deposits in the lower reaches of Wadi en Naga, Wadi bu Msafer, and along the seashore for about 2 km. to the east of the mouth of the latter (Pl. 3 *b*). The deposits in question are all believed to be contemporaneous with one another, and the collection made here is the largest from any single locality, eighty-seven specimens.

In preservation the artifacts range from the virtually unpatinated to those with an even cream patina upwards of 5 mm. thick. Many in the latter state are, however, entirely free from mechanical abrasion and may well owe their condition simply to slight differences in the chemical environment. A surprisingly small proportion show signs of marked wear, considering the size of many of the elements of the deposit, and the conclusion is clear that the great majority can only have travelled a very short distance from the point where they were dropped by ancient man.

Cores (18)

These can be grouped under the following headings:

(i) *Large irregular polyhedric cores* (5). These seem to form a definite class that may well account for many of the less classifiable flakes. In the present selection they vary in size from $14 \cdot 5 \times 13 \cdot 8 \times 9 \cdot 1$ cm. to $8 \cdot 1 \times 7 \cdot 4 \times 6 \cdot 2$ cm. The scars suggest that they were used for making flakes in the order of 7×4 cm., with rather obtuse periphery, and about a third to a half showing some cortex.

(ii) *Discoid cores, large* (6). These seem to represent the dominant method of flake production. They run about 15 cm. in diameter by 4–5 cm. thick and are regularly made with a cortical base and complete ring of roughly prepared platforms. Fig. 17, no. 3, shows the most carefully executed specimen in the collection, and this may just possibly be an unfinished tortoise-core, though it is more likely to have belonged to the present group, and to have been abandoned on account of the development of awkward hinge fractures near the centre. Fig. 17, no. 1, appears to be a large nodule on which the process has just been started.

(iii) *Discoid cores, medium and small* (6). Smaller and neater versions of the above are also well represented and include several quite comparable to Hajj Creiem (Fig. 14) often with carefully faceted striking-platforms, though not as small as the smallest noted there. As a rule they would seem to have been about 6 cm. in diameter, when abandoned.

(iv) *Tortoise-cores* (1). Only one probable example of this type was obtained (Fig. 17, no. 2). It is about the dimensions to be expected from the flakes mentioned above, and appears to have been abandoned owing to the development of pronounced hinge fractures in the course of re-preparation after a main flake had been removed.

(v) *Flake-blade cores*. Apart from one small and rather atypical example, no specimens of parallel-flaked cores for flake-blades were actually collected here, though from the presence of the flakes mentioned below the use of some process of the kind can confidently be inferred. Cores displaying the process in detail are, however, well known from contemporary deposits elsewhere on the Sahel.

Flakes (69)

A fairly high proportion—about 30 per cent—can be attributed to the initial stages of core preparation, about 25 per cent appear to have been struck more or less in the discoidal fashion and not more than one or two show signs of being made on true tortoise-cores. The largest certain disc-struck flake measures $18·2 \times 11·6$ cm. Judging by the scars of the core from which this flake was made, it can scarcely have been made on a core less than about 30 cm. in diameter. One probable tortoise-core flake is of comparable dimensions, and the whole assemblage seems to be on an exceptionally large scale. A group of large flake-blades appears to demand some sort of double-ended core with parallel flaking, and there is a fairly large proportion of crude simply-prepared flakes that cannot be included in any of the above categories with certainty.

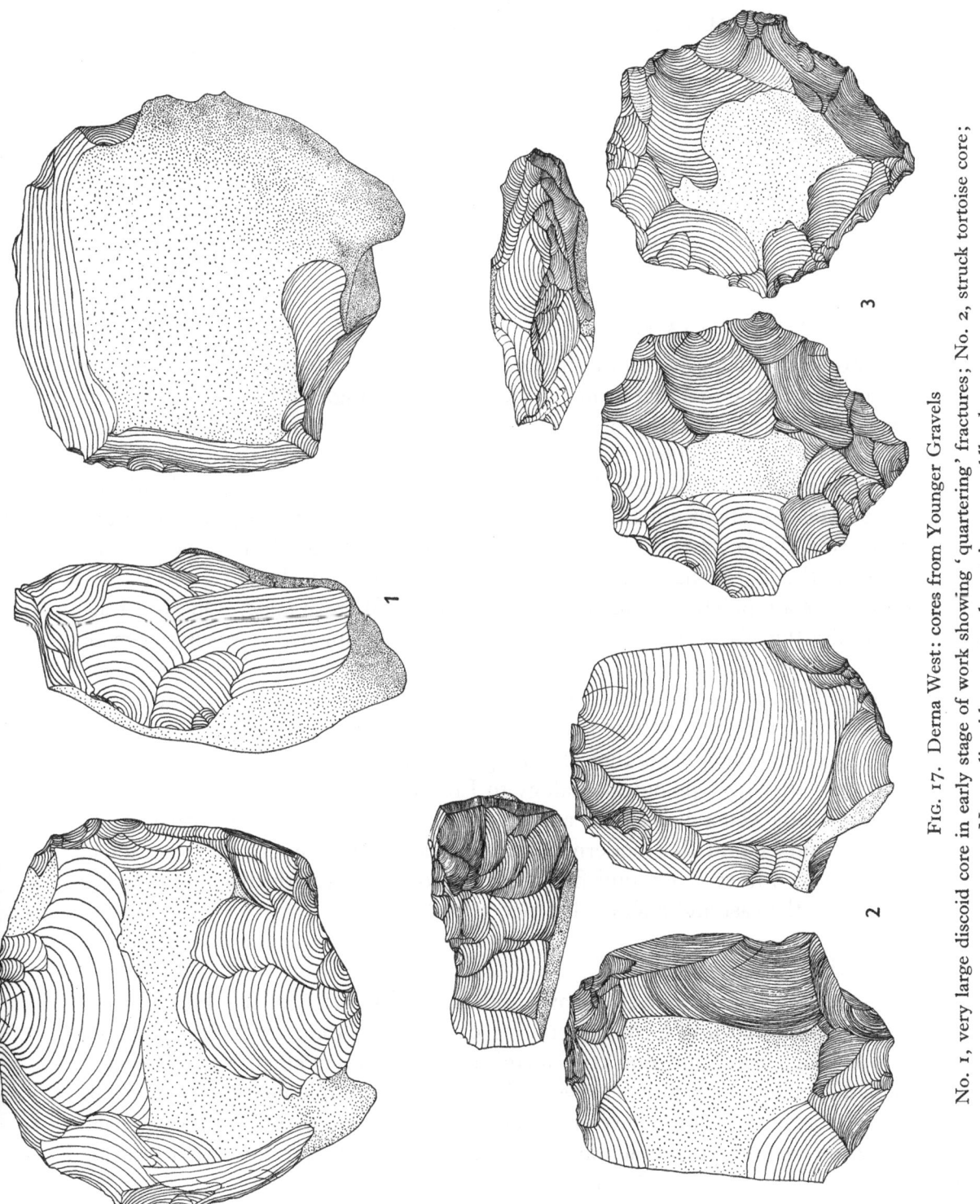

FIG. 17. Derna West: cores from Younger Gravels

No. 1, very large discoid core in early stage of work showing 'quartering' fractures; No. 2, struck tortoise core; No. 3, discoid or unstruck tortoise core. All × ⅓.

1

2

3

Secondary work

Only one unmistakable example of secondary work was found—a small and very carefully finished side-scraper much in the manner of Hajj Creiem. It is made on a subrectangular cortical flake, and the flat retouch is along the nearly straight margin of the flake starting from the right-hand end of the platform. A small asperity has been removed from the left-hand margin by the same technique.

LOCALITY II—DERNA EAST

Two sets of deposits, both situated on the eastern borders of the modern town, yielded the material grouped under this heading. The first is revealed in exposures 2–4 m. in height in a small wadi about 1 km. from the sea, and consists of more or less consolidated alluvial deposits continuous with the main sheet of alluvium which extends right down to the seashore. Eleven artifacts—all flakes—were obtained, about equally divided from each formation. In technique and patination all except two of the specimens correspond closely to those just described, though perhaps showing rather more mechanical damage. The two exceptions are practically unpatinated and in a mint fresh state. One is a fragment of a thin flake blade, and the other a carefully finished side-scraper made on a very thin disc-struck flake with a curved flat-retouched working edge extending from a closely faceted striking-platform. If despite their remarkable appearance of freshness these two are, as they appear to be, contemporary with the rest of the assemblage, then the sample as a whole, small though it is, would provide a striking typological analogy to Locality I.

In the same general area rather more prolific finds were made in a vertical face of alluvium behind the modern beach just east of the mouth of the wadi in question, and in scree or mud-flow deposits continuous with this alluvium still further to the east and a short distance inland.

Cores (4)

These may be classed as follows: one small discoid, one medium discoid possibly trimmed into a rough scraping or chopping tool, one flake converted into a rough core, and one true tortoise-core of rather small size. The last-named specimen was found loose at the foot of the implementiferous deposit, but must almost certainly have been derived from it. Despite very heavy damage by recent salt or gypsum action which has removed the platform there can be little doubt about its attribution.

Flakes (27)

The condition of the flakes is almost without exception one of mint freshness, entirely free from accidental fractures. Except for a higher proportion of pieces showing cortex, segregation into the same categories fails to show any significant difference from Locality I.

LOCALITY III—WADI HAULA

This site has already been mentioned in connexion with artifacts collected from the 6 m. beach conglomerate. Immediately overlying these very hard conglomerates are deposits believed to be a fossil 'mud-flow', in which a number of artifacts were found. Apart from one or two which have suffered heavily from gypsum action, the majority are in an almost perfect state of preservation with little signs of mechanical damage. From the occurrence of minute trimming waste among the flakes and the relatively restricted area in which they occur, it seems likely that the site is close to an old chipping floor.

Cores (4)

A very large flake-blade core $20 \times 15 \times 10$ cm. reveals very clearly a method of production suitable for the large elongated flakes noted at Locality I, and numerous examples on the same general pattern will be noted at Locality IV. Roughly speaking, the principle is somewhat similar to that of the 'double-ended cores' found in most Upper Palaeolithic sites, that is to say, the flaking is carried out from the opposite ends of the long axis of a tabular nodule. Similar specimens are occasionally found in the cave Mousterian of Western Europe, though it is worth noting that the system is entirely different from that normally employed for blade-making in the European Late Levallois. There is one very typical polyhedric core, and two others—one whole and one fragmentary—which represent a fairly characteristic small-scale disc technique.

Flakes (10)

As far as the series goes these show no significant difference from Locality I. Two characteristic examples of disc flaking from closely faceted platforms are worthy of note.

LOCALITY IV—TOCRA, WADI SLEIB

The fairly numerous artifacts found at this locality were all taken from the recently eroded sides of Wadi Sleib[1] where it cuts through a sheet of partly con-

---

[1] A considerable distance downstream of the find-area of the Neolithic specimens described in McBurney (1947), p. 82.

solidated alluvium between the escarpment and the sea. Unlike the localities just described the specimens here are almost all heavily abraded and the flaking is often much obscured by subsidiary fractures of natural origin. Under these conditions it is not surprising that no traces of secondary work should have been recognized.

Cores (17)

(i) Eight fairly characteristic discoid cores, seven with cortical bases, range from 11·4 × 10·4 × 4·4 cm. to 6·6 × 4·6 × 1·6 cm.

(ii) Two rough discoids, the largest measuring 7·4 × 5·7 cm.

(iii) Five flake-blade cores all on the same basic principle as the Wadi Haula specimen, average about 12 × 10 × 7 cm.

(iv) One elongated nuclear piece in an exceptionally battered state may just conceivably have been a rough bifacial tool more or less on the pattern of a hand-axe.

(v) One tabular nodule with a few intentional flakes removed.

Flakes (50)

As far as size and other general characteristics go, it seems likely that if this assemblage could be examined in its original state it would prove to contain essentially the same features as those described at Locality I. As it is, in the broken and damaged condition in which it now occurs, only the most generalized classification can be attempted. No traces of true tortoise-core flaking can be identified, but the proportion of definitely disc-struck flakes seems to have been about the same. Although the proportion of flakes showing traces of cortex is distinctly higher, and so also is the proportion of flake-blades, neither feature need have much cultural significance. At the same time it is interesting to note that the last-mentioned peculiarity is also reflected by the cores.

Although the possibility of some degree of specialization in the Tocra area cannot be ruled out the main point of interest that emerges is that basic flaking techniques in current use were the same as elsewhere, if differing somewhat in relative importance.

STATISTICAL SUMMARY, LOCALITIES I–IV

The quantitative characteristics of the above series are presented in Table II. The contrast that they offer to the Hajj Creiem assemblage, despite a community of basic technical notions, is striking and may be summarized as follows:

(i) Much greater size of both flakes and cores.

(ii) Much lower proportion of flakes to cores.

(iii) Much lower proportion of retouched specimens to raw flakes.

Such retouched specimens, on the other hand, as have been found in the

Younger Gravels or contemporary deposits are, surprisingly enough, remarkable for their close similarity to Hajj Creiem. Putting these observations together, it is difficult to eliminate the possibility that we have here merely the quarry-site and home-site expression respectively of one and the same industrial tradition. Admittedly we have as yet no contemporary of the later Sahel series from an inland site corresponding in geographical position to Hajj Creiem. The occupation of the latter site must be separated from the period of manufacture of the Sahel material by many millennia—ample time in fact for the observed cultural developments to have taken place. A point in favour of this explanation is provided by the surface quarry-site west of Derna described below, since here we have positive evidence of the exploitation of beds of flint exposed by marine erosion along the Sahel. The number and extent of such exposures available before the deposition of the Younger Gravels had reached its present state may well have been appreciably greater.

TABLE II. *Statistical summary of samples of industry found* in situ *in the Younger Gravels at various localities of the Sahel*

(Dimensions in mm.; number of observations in brackets)

	Derna West	Derna East	Wadi Haula (Apollonia)	Lower Wadi Sleib (Tocra)
	Flakes			
(1) Length	74·67 (58)	67·04 (33)	59·69 (10)	67·92 (50)
(2) Width	49·14 (58)	46·95 (33)	39·62 (10)	41·86 (50)
(3) Thickness	17·47 (58)	14·85 (33)	15·80 (10)	15·64 (50)
Ratio $\dfrac{(3) \times 100}{(1)}$	23·39 %	22·15 %	26·47 %	23·03 %
Ratio $\dfrac{(3) \times 100}{(2)}$	35·54 %	31·63 %	25·09 %	37·36 %
Ratio of raw flakes to flakes with secondary working	98 %	97 %	90 + %	98 + %
Mean number of primary scars per flake	3·7 (62)	4·3 (33)	3·8 (10)	2·4 (45)
Maximum number of primary scars observed on one flake	8 (62)	9 (33)	7 (10)	7 (45)
	Cores			
(1) Length	104·14 (17)	76·84 (4)	147·3 (3)	87·94 (16)
(2) Width	90·69 (17)	61·60 (4)	88·3 (3)	84·14 (16)
(3) Thickness	51·41 (17)	27·25 (4)	54·30 (4)	34·50 (16)
Mean number of flakes per core	3·4	8·25	2·5	3·13

IMPLEMENTS FROM THE DERNA BRICKWORKS

A small series of implements and flakes were obtained *in situ* in a deposit of uncemented terra rossa, overlying the main Younger Gravels at a locality 10 km. west of Derna and within a few hundred metres of the sea. The general appearance of this series is that of a homogeneous sample from a single industry, but owing to the soft nature of the deposit and the relative proximity of the specimens to the surface, the inclusion of occasional specimens of a later date is not impossible. Assuming the specimens to be properly associated, however, they would repre- sent a small industry in the Middle Palaeolithic tradition with finely retouched side-scrapers and flakes made on small discoid cores with carefully faceted platforms. In date this would presumably belong to the period immediately following that of the underlying deposits of the Younger Gravels.

SURFACE FINDS OF A MIDDLE PALAEOLITHIC CHARACTER

During the first season a careful record was kept of scattered artifacts found loose on the Sahel, in the hopes of associating these with the various marine terraces. As this method did not appear to yield adequate results it was dis- continued during the second season and only two finds need be mentioned at this point.

The first was the discovery of a very typical Aterian point probably associated with a fragment of thick blade with flat retouch on the bulbar face and a broken flake blade with a carefully faceted platform (Fig. 18, nos. 5–7). All three speci- mens have the same shiny orange patina, and were found lying close together on a recently washed-out surface of terra rossa on the 17 m. platform at Ras Aamer. The specimen (Fig. 18, no. 7) is sufficiently characteristic to confirm the presence of the culture in the area, and its position suggests that here as elsewhere this culture was practised after the retreat of the sea from the second lowest (eustatic) raised shoreline. The second find, of rather greater interest, was that of the quarry site mentioned earlier in connexion with Derna West. This consists of an area some 45 m. wide by 90 m. long, and 45–90 m. from the sea, situated on the 6 m. marine terrace some 20 km. west of Derna. The whole of the area in question for some distance around is littered with countless thousands of large struck flakes of fine quality chert. Most of these are lying on the partially eroded surface of a sheet of terra rossa that seems to have covered the area until a few years ago, but outcrops of the underlying Lower Eocene limestone are also plentiful, and enormous nodules of flint can still be seen in place in them. Several of these nodules have had numerous large flakes struck off them without being removed from their parent matrix.

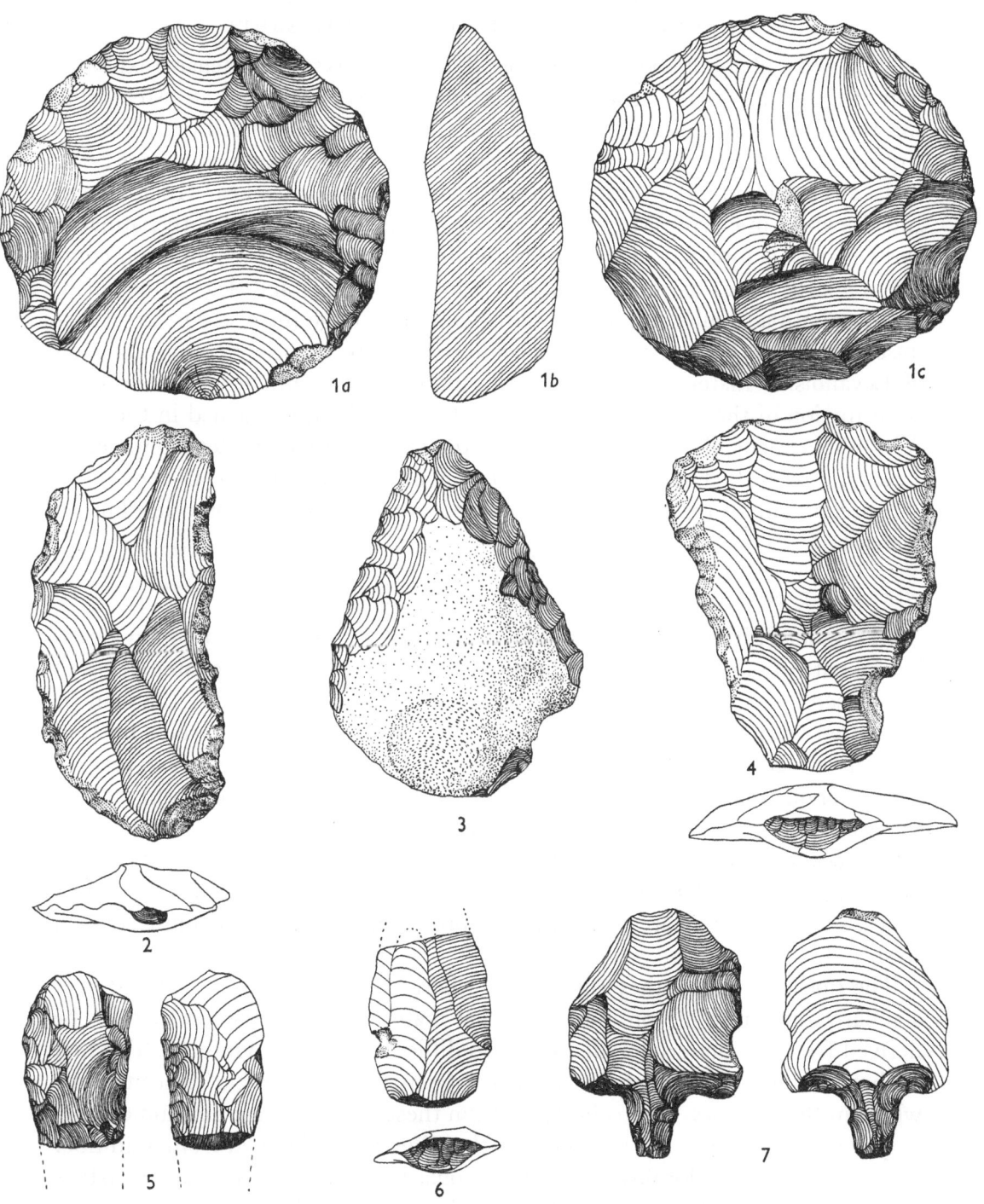

FIG. 18. Surface finds, near Derna and near Ras Aamer

Nos. 1–4, from quarry site on 6 m. wave-cut Terrace west of Derna, heavily patinated with occasional accidental fractures (shown stippled); Nos. 5–7, specimens with yellow porcellaneous patina from eroded loamy surface of 18 m. wave-cut terrace near Ras Aamer. All × ½.

An analysis of about 100 artifacts from this site points to the extensive exploitation of the chert during at least two periods, the Neolithic or Late Mesolithic and the Middle Palaeolithic. Evidence for the former is provided by a series of large coarse unifacial and bifacial pick-like implements, presumably quarrying tools, and strikingly like those from the plateau quarry sites near Kharga oasis and elsewhere in Egypt. Most of the Derna specimens are only slightly patinated and entirely unabraided. Evidence for a Middle Palaeolithic exploitation comprises a large series of very typical disc cores of all sizes and degrees of finish, together with a smaller number of remarkably fine large tortoise-cores and tortoise-core flakes. Two of the former could scarcely be bettered in any collection of Levalloisian material with which the writer is acquainted, and bring clear confirmation of the full development which this technique reached in the area. All the specimens of earlier morphology are heavily patinated often up to nearly a centimetre in thickness, and a few show signs of abrasion (Fig. 18, nos. 1–4).

A second site of the same kind was noted in a similar situation a few km. to the east.

SURFACE FINDS OF A LOWER PALAEOLITHIC CHARACTER

At the outset it was one of the objectives of the investigation to obtain a geological date in terms of sea-levels for the Lower as well as the two later stages of the Palaeolithic. In this we were disappointed, and the only finds that can be attributed with any reasonable confidence to this stage are the surface finds of hand-axes made by N. L. Scott in the Tocra area in 1943.[1] These, which seemed natural enough at the time, are now seen to raise a difficult problem in view of the archaeological and geological evidence just brought forward, for all seven were discovered *on the surface of the Younger Gravels*, which are now known to be younger than the 6 m. shoreline. While several are more or less heavily rolled, at least two of the clearest and most evolved are in an almost mint state of freshness.

The question is how they come to lie where they were found, scattered over a square mile or so of continuous alluvial material? If we assume them to be derived from the underlying deposits two difficulties are encountered. The first is that offered by the negative evidence from the comparable deposits elsewhere in the territory. The collections from these, it is true, are not numerically very considerable, but the fact remains that they were obtained from a number of widely-separated localities which all consistently present the same picture of an evolved Middle Palaeolithic flake industry from which even recognizable hand-axe chipping waste seems to be absent. Moreover, the absence of similar

---

[1] Described in McBurney (1947), p. 78 and fig. 15.

traces from a large collection of post-6 m. sea-level chipping debris such as the surface site west of Derna is also of some importance. A fairly well-defined class of bifacial tool was in fact encountered at the latter site, namely a leaf-shaped point reminiscent in size and outline of a very rough Solutrean 'laurel leaf'. Such pieces are not uncommon in the Neolithic of Egypt, and are also known to some extent in the Aterian and Neolithic of the Maghreb. Somewhat similar pieces, but of rougher workmanship still, have been claimed to occur in a Middle Palaeolithic context in the cave site of Hagfet et Tera described in the next chapter. On grounds of patina and technique the writer is inclined to attribute the Derna specimens to the Neolithic rather than any earlier industrial stage, but however that may be, they differ so entirely as a class from the hand-axes of Tocra that there can be no serious question of the latter being aberrant examples of this type.

Further confirmation of this negative evidence is supplied by the much fuller collections of the preceding Hajj Creiem stage, where there is not the slightest hint of bifacial work of any kind.

In conclusion, the salient facts regarding the chronology of the hand-axe stage in neighbouring territories may be reviewed. It will be remembered that in Palestine the long succession of hand-axe cultures at Mount Carmel comes to an abrupt close during the penultimate period of increased rainfall, and is followed by an unbroken series of pure flake industries in the Levalloiso-Mousterian style until the appearance of the Upper Palaeolithic. At Kharga Oasis in Upper Egypt, the transition from Acheulean to Levalloisian was apparently gradual, and was again associated with the early phases of a marked pluvial increase. In the Nile Valley itself the two great industrial cycles are clearly segregated into two well-defined and perfectly distinct sedimentary phases of long duration. Although it has been suggested that some slight traces of the earlier tradition may have survived into the early part of the later geological phase, survival as late as the last period of marine regression would conflict sharply with any system of correlation between Nile depositional terraces and sea-levels yet suggested.[1]

To the west of Cyrenaica the chronological data regarding the transition from Lower to Middle Palaeolithic traditions is considerably more tenuous. A comprehensive scheme has, it is true, been proposed by Neuville and Ruhlmann for the Atlantic coast of Morocco,[2] but in its later stages this scheme is admittedly only tentative. It appears in any case to be open to some criticism on geological grounds.[3] As far as it goes, however, the suggestion is that the first traces of

[1] See for instance Caton-Thompson (1946a).
[2] Neuville and Ruhlmann (1941), and Ruhlmann (1945).   [3] Bourcart (1943).

Levalloisian activity occur in the period of marine regression between a '28–30 m.' and a '12–15 m.' phase of transgression. The final hand-axes would have been made immediately subsequent to the '12–15 m.' level and would have been followed by a succession of pure flake industries up to the appearance of blade industries of Oranian type. A few small bifacial tools occur in the Aterian, which culture undoubtedly belongs to the latter part of this period, but these are quite different in form and technique from the Cyrenaican specimens under discussion, which are typical of evolved Acheulean.

Elsewhere along the Mediterranean coast of French North Africa considerable evidence is available to support the idea that the hand-axe tradition had long been extinct by the time of the 6 m. sea-level, to say nothing of the subsequent phase of regression. In this connexion Professor Arambourg is kind enough to allow us to mention his unpublished observation of a widespread alluvial deposit in Tunisia corresponding closely in appearance to our Sahel Alluvium, and similarly associated with a recent marine regression, which contains exclusively what he describes as a crude variant of Levalloisian.[1]

From the foregoing considerations it would seem that to suppose that the Tocra hand-axes have been recently eroded from the deposit on which they were found would involve a cultural anomaly of a very unexpected kind. Until more positive evidence is forthcoming, therefore, it is probably safest to assume that however the Tocra hand-axes found their way to this position (and human agency cannot be ruled out in this connexion, since all the specimens are large enough to have provided serviceable quantities of raw material at a later period) it was not through being made and dropped at the spot where they were found, during or after the period of formation of the Younger Gravels. The cultural stage associated with the Younger Gravels is most probably what it appears from the *in situ* remains to be, namely the chipping-floor expression of a Levalloisian or Levalloiso-Mousterian tradition of quite normal character, without any appreciable bifacial element. As far as the Acheulean is concerned, then, we are left with the conclusion that although it must certainly have been practised at some time in the territory, we have as yet no direct evidence of its date and can only suppose on general grounds that it was nearer in time to our 15–25 m. stage than to the period in question.

---

[1] Perhaps really a quarry variant of a more finished type of industry, in the manner suggested above for the Cyrenaican Sahel.

4. SUMMARY OF THE ARCHAEOLOGICAL SUCCESSION ON THE SAHEL

The archaeological results of our survey of the Sahel and neighbouring areas may be tabulated as follows:

15–25 m. stage. First traces of human occupation. *Post-quem* limit of Ras Aamer Aterian find.

6 m. stage. Evolved Middle Palaeolithic of Wadi Haula beach deposit. *Post-quem* date of evolved Levalloisian of Derna West surface quarry site.

Post-6 m. regression. (*a*) Climate wet and warm. Levalloiso-Mousterian culture in wadis of Derna area.

Erosional phase in wadis.

Post-6 m. regression. (*b*) Climate far colder than present, with extensive freezing in winter. Levalloisian practised along the Sahel. *Post-quem* limit of blade industries.

In reviewing this scheme as a whole, attention may be concentrated on one or two particular points. As regards the Levalloiso-Mousterian it will be recalled that at Hajj Creiem this culture occurs at the *base* of the high terrace, i.e. at the beginning of the period of higher rainfall which that formation indicates. The same industry at Skhul Cave in Palestine occurs at a similar stage in the local climatic sequence where, it is interesting to recall, it is also accompanied by evidence of the immigration of African mammalian species into the area.[1] Further, in Palestine this phase is followed by one of climatic change, possibly including a fall in temperature, during which basically the same human culture survived, but which seems to have proved fatal to much of the earlier fauna. In Cyrenaica we have the same phenomenon of the survival of this cultural tradition through the initial phase of a climatic change to much colder conditions. As for the change in fauna and flora in Cyrenaica, we know only that it took place at some time *after* the wet phase in question though, as mentioned above, it seems perhaps most reasonable to associate it with the violent climatic events which immediately followed. Here, however, the analogy between the two sequences seems to end, for whereas the possible drop in temperature in Palestine was accompanied by an *increase* in rainfall, in Cyrenaica there seems to have been a decrease (or at most an insignificant increase) in rainfall with the corresponding, but much more marked, fall in temperature. On the other hand this discrepancy is mitigated by recent discoveries in northern Syria, where climatic evidence associated with the later Levalloiso-Mousterian seems to be strikingly similar to that of the younger gravels in Cyrenaica.[2]

---

[1] Garrod and Bate (1937), p. 201.

[2] Wright, H. E. (1951) showed that at Ksar Akil in Syria deposits apparently identical to our Younger Gravels (under the same topographical conditions) can be associated with the end of the Levalloiso-Mousterian.

CHAPTER XII

THE CAVE OF HAGFET ET TERA

The cave site about to be described is of great interest from several points of view. It provided the first closed find of prehistoric artifacts to be made in the territory, and the first site with any claim to offer a direct stratigraphical sequence from the flake industries of the Middle Palaeolithic to the blade industries of the following cultural epoch. The honour of this pioneering work falls to C. T. Petrocchi, who discovered the cave in 1937, and carried out a series of excavations in the following years which resulted in the enormous collection of prehistoric implements now housed in the Museum at Tripoli. No final publication of this work has yet appeared, and the stratigraphical and other details given below derive in part from the preliminary publication which appeared in 1941, together with verbal details modifying some of his earlier conclusions kindly communicated by Signor Petrocchi when the writer visited him in Tripoli in 1948, and finally, the results of our own soundings undertaken at the end of the 1948 season to try to resolve some of the remaining uncertainties.

Some of these soundings proved relatively rich in specimens of all kinds and most of the artifacts illustrated here, together with all the material reported on by Miss Bate in Appendix A, derive from this source. It will be seen that some of the archaeological and palaeontological conclusions now offered differ appreciably from those contained in Signor Petrocchi's preliminary report.[1]

I. GENERAL CHARACTER OF THE SITE

The Hagfet et Tera or 'Cave of the Birds' is the largest of several cavities formed in the Miocene limestone escarpment that marks the eastern limit of the coastal plain some 10 km. inland from Benghazi. The cave itself faces south-west and the entrance is prolonged into a level terrace roughly triangular in shape with its base at the mouth of the cave, and measuring about 40 m. long by 20 m. wide (Fig. 19). There are signs that at one time part at least of this terrace was roofed with an extension of the vault of the cave, and it still forms a sheltered bay from which a commanding view can be obtained for many miles north and south along the Sahel. The level rocky floor of the terrace was until recently covered by

---

[1] It is necessary to add that we understand that Signor Petrocchi himself has changed his opinions in a number of respects since the issue of his original report (Petrocchi, 1940).

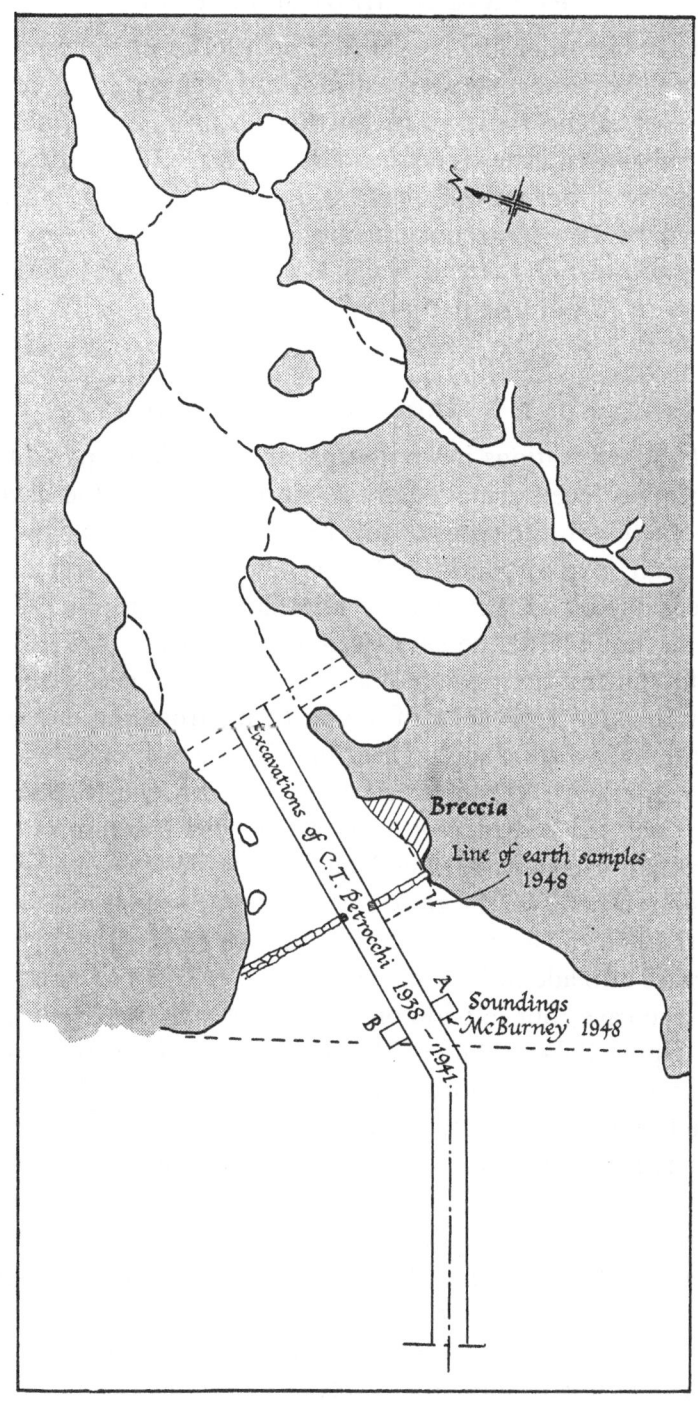

Fig. 19. Hagfet et Tera (Cave of the Birds)
Ground plan after C. T. Petrocchi (1941). Scale 1/400.

cultural deposit in the order of 2 m. thick. The interior of the cave continues as a high, wide tunnel in a north-easterly direction for some 40 m. before bending slightly to the east to end some 10 m. further on in a large cavity with a permanent spring or natural cistern in the floor, frequented by cattle at the present day. An ancient wall, perhaps of classical date, still bars the entrance, with a doorway flanked by two large upright stones. Except at its extreme south-eastern end the foundations of this wall have been dug right through the prehistoric deposits to rest on the natural rock floor.

2. C. T. PETROCCHI'S EXCAVATION

At the time of the excavations it would appear that a large part of the ancient deposits had already been removed from the entrance of the cave. This is a common phenomenon in the caves of this region and is perhaps to be attributed to the exploitation of phosphatic earths as agricultural fertilizer some time during the Greek or Roman occupation;[1] alternatively their removal may simply be due to a desire to enlarge the cavities as living spaces.

Petrocchi's first operation was to cut a trench (Fig. 19) along the main axis of the cave, extending for a distance of about 12 m. from the entrance into the interior and prolonged with a slight bend for about 24 m. on to the terrace. At or near the north-eastern end of this trench he was able to establish, in the interior of the cave, the presence of an intact deposit containing two thin cultural layers of an unmistakably Middle Palaeolithic character. The older of these, 'Layer G', rested directly on the rock bottom and was separated from the latter by over a metre of sterile earthy deposits and a layer of stalagmite—'Layers E and F'. No traces of blade industries whatever were found in the formations in the interior of the cave, but outside on the terrace two extremely rich cultural horizons of an Upper Palaeolithic or Mesolithic character were identified. They occur in an obviously intact deposit 2 m. thick and extending for about 23 m. to the edge of the terrace. North-eastwards, towards the cave entrance, the upper of these horizons runs out at the surface, and only the lower horizon with its capping of moderately implementiferous intermediate formations remains. South-westwards, towards the edge of the terrace, the converse occurs and it is the earlier horizon which seems to fade out at the base and disappear, leaving only the later members of the sequence.

Just inside the entrance and adhering to the south-east face of the cave was found a small patch of extremely hard breccia. Several specimens of a Middle

---

[1] In the other cave site excavated there was some positive evidence to suggest that removal of a large part of the cave-earth took place at this period (see pp. 197–8).

Palaeolithic character illustrated in Petrocchi's report are stated to have been obtained here.

Briefly, the main problem raised by these observations may be said to be the correlation of the two principal intact deposits—that in the far interior of the cave and that on the terrace. It may be noticed in the vertical section published by Petrocchi that there was discontinuity between these two at two points, namely, at the 'unexplored zone' to the north-east and at the 'disturbed zone' round the foundations of the post-prehistoric wall. About the intervening region between these two gaps the first excavator seems also to have felt some uncertainty, since no symbol is given for the nature of the deposits other than the indication of a thin layer of stalagmite resting on the rock floor.

Despite these apparent uncertainties however, Petrocchi felt justified in correlating the *upper* archaeological layer in the interior of the cave with the *lower* horizon on the terrace, and did not hesitate to treat the fauna and archaeological contents from the two as a single unit.

This was somewhat unfortunate from two points of view. On the palaeontological side the faunal context of the Middle Palaeolithic epoch in the Maghreb is known to have differed in a number of respects from that of the later blade cultures. Thus, apart from their intrinsic interest if the two faunas had been kept distinct the degree of resemblance between them would have provided a most useful check of the proposed correlation. On the archaeological side it led him to propose the existence of 'a transitional culture passing gradually from Mousterian to Upper Palaeolithic forms'. Since the existence of a transitional stage of this kind is precisely one of the most difficult and interesting questions of North African prehistory, this also must be regarded as a premature conclusion.

PALAEONTOLOGY AND ARCHAEOLOGY

Turning for the moment from the problems of stratigraphy, the following summary of the contents of the three fossiliferous horizons reported by Petrocchi in his publication may be given.

Layer G, resting on the rock floor in the interior of the cave, yielded only a very small series of artifacts comprising a few flakes and one or two retouched points and side-scrapers. The accompanying fauna is stated to have contained *Capra* sp., *Equus caballus*, *Hystrix cristata*, *Ovis* sp. (?). These and other determinations given in Petrocchi's report have been subject to some criticism since their publication. They are quoted here verbatim as they occur in Petrocchi's report. (See, however, Appendix A, p. 276.)

Layer D (combined collections from the interior of the cave and the terrace) contained archaeological specimens which are rightly considered under two headings, those of Middle Palaeolithic morphology and those of later type. In the former group are a number of typical points, side-scrapers and discs. These could easily be paralleled at any Levalloiso-Mousterian site, or indeed for that matter at any evolved Mousterian site in Europe, but a nearer cultural diagnosis is not possible on so small a series. On the other hand, the bifacial specimens suggest a trait more nearly in the Aterian tradition, though these are said to show signs of double patina and may belong to an earlier phase. In the latter group are a large number of backed bladelets, end-scrapers on true blades, and what is almost certainly a burin spall. No other definite types are illustrated or set out in the show-cases at Tripoli. The general appearance of these is indistinguishable from those of the final layer B–C described below. The fauna from this source is given as: *Bos primigenius*, *Bos* sp., *Rhinoceros* (?)*merckii*, *Equus caballus*, *Capra* sp., antelope sp. (?)*Kobus*, *Equus asinus hydruntinus*.

In addition, there are one or two remarkable bifacial points and one specimen resembling a small cordiform hand-axe.

Layer B–C is stated to have contained a very abundant industry based on small blades with very numerous backed-blades of small to microlithic proportions and a few end-scrapers on blades. It is stressed that both microburins and angle-burins are entirely absent, though Signor Petrocchi tells me that he has since identified a few of the former in his collections. The fauna reported from this layer is: *Bos primigenius* (a large variety), *Bos* (?*primigenius*), *Equus caballus* antelope sp., *Cervus* sp.

PROBLEMS RAISED BY THE FIRST EXCAVATION: SUMMARY

It will be seen from the foregoing that the problems raised by the first investigation of this by no means simple site are of considerable import to the archaeological succession of Cyrenaica as a whole.

Stratigraphically the central problem is simply this—does this site, or does it not, provide us with an unbroken sequence leading from flake to blade traditions?

The purely archaeological evidence raises two main questions. The first concerns the precise nature of the 'transitional industry' postulated by Petrocchi. Well-attested cases of this kind are extremely rare and none are as yet entirely free from controversy. The presence of such an industry in Cyrenaica would profoundly affect not only our conception of the cultural history of north-eastern Libya, but would have an important bearing on that of neighbouring territories as well.

The second archaeological question raised by the suggested interpretation of this site is that of the affinities of the final blade culture. It will be observed that this is exceptional for North Africa, and indeed the southern Mediterranean area in general in several respects, notably the reported complete absence of microburins, burins, or true microliths other than very small backed blades.

On the other hand, the mention of Solutrean-like bifacial artifacts is of great interest, for the analogy afforded with the known, transitional industries of South and East Africa grouped under the heading Still Bay. Of the fairly numerous North African industries claimed at one time or another to possess these negative characteristics, nearly all have subsequently been shown to contain definite signs of the traits in question, even if only to a small extent.

Vaufrey now reports that the absence of microliths from both the Abri Clariond and Beni Segoual was illusory.[1]

3. RE-EXAMINATION OF THE SITE IN 1948

In the hopes of shedding some additional light on the foregoing it was decided to undertake a short re-examination before leaving the territory at the end of the second season. The objectives of this were threefold. First, to see if any traces of the section inside the cave remained and whether the connexion between the two terra rossa deposits could be inspected or tested in any way. Secondly, to recover statistical samples of cultural material from the two layers on the terrace for precise comparative purposes and to see whether any traces of a Middle Palaeolithic Culture contact or Transitional Culture could be detected in the earlier of these. Thirdly, to see whether anything further could be said regarding the stratigraphical relationship of the breccia near the entrance to the deposits on the terrace.[2]

SOUNDINGS INSIDE THE CAVE

These were made at two points. The first, at 5 m. north-east of the wall across the entrance and 1 m. from the rock on the west side of the entrance, revealed only a shallow deposit of obviously disturbed loose brown earth overlying the rock floor. A brief examination of the rest of the entrance suggested that the whole of the original deposits in this part of the cave had been removed.

A second sounding at a point about 13 m. from the entrance wall and 1 m. from the east rock face proved equally unproductive. Over 2 m. of loose brown and red earth were found to contain fragments of modern glass near the bottom. At this stage, owing to pressure of time, further attempts to relocate such deposits as might remain intact in the interior of the cave were abandoned in favour of the more obviously useful work to be carried out on the terrace and the breccia at the entrance.

[1] Vaufrey (1950). [2] C. T. Petrocchi (1940), fig. 13.

RE-INVESTIGATION OF THE TERRACE DEPOSITS

Stratigraphy

The greater part of the longitudinal trench cut by Petrocchi (labelled A–D on his plan) was still open at the time of our visit and the stratigraphy shown on his section could still be followed to a large extent on the south-east face. The north-west face was more seriously obscured by fallen deposits and sounding B, 6 m. south-west of the gateway, showed that the deposits are now all disturbed down to a depth of at least 1 m.

Sounding A in the south-west face was a metre square and continued down to rock bottom. The sequence of deposits as checked by us was as follows:

Description	Depth (cm.)	Corresponding layers in Petrocchi's scheme
Dark brown earth	50	A, B and C
Reddish buff earth	60	D
Lower dark earth	100	D

All the deposits were firm and showed no traces of disturbance,[1] being evenly and horizontally laid down except where interrupted by a few large stones. The upper part of D was the only layer to show appreciable signs of cementing, and Petrocchi's distinction between layers B and C could not therefore be established.

The above appears to agree reasonably well with the scheme proposed by our predecessor, which at this point on the opposite side of the trench appears to be (making approximate measurements from his published section):

Description	Depth (cm.)	Reference letter
Cemented brown earth	20	A, B
Uncemented brown earth	40	C
Red earth	85	D

The main difference is that layer D was considerably thicker on our side of the trench and found to be a reddish buff rather than true terra rossa in the technical sense.

Archaeology

The layers were dug horizontally about 15 cm. at a time, passed through a riddle with 4 mm. mesh and the contents segregated in accordance with the stratigraphy observed by us.

---

[1] Other than one small rodent burrow apparently recently dug in from the trench face near the base. The disturbance of specimens was not appreciable.

A. *Layers A, B and C.* This was by far the richest in artifacts, which occurred at the rate of about 1550 to the cubic metre. The raw material consists for the most part of coarse-grained jasper with smaller quantities of flint, chert, and other fine-grained materials.

(i) *Cores* (24). Those made in the coarser raw material are about equally divided between flake-cores and blade-cores. In both cases the normal length of scar is in the order of 1–2 cm. A few scars run up to 4 or 5 cm. but are rarely larger. The original nodules seem to have measured not much over 10–15 cm. in greatest diameter and the cores themselves average about 7–8 cm. The much less common cores in flint and chert run considerably smaller than the rest and are seldom over 3–4 cm. in length in their worked-out state. The flake-cores in the coarser-grained raw material seem generally to have been started from a slightly flattened nodule, the flakes being struck off more or less alternately all round the periphery so as to produce at last a roughly biconical residual shape. The blade cores in both types of materials are either of the usual single cone shape, or are made in the narrow plane of a more or less tabular fragment, producing a result resembling a rough polyhedric burin. Not infrequently the two types of flaking are combined and a suitable projecting portion of a flake core may have had a few true blades struck off it. A few pieces of this latter type may perhaps have served as some kind of steep scraper, but on the whole the cores show few traces of adaptation for use as tools.

(ii) *Flakes* (about 500). This relatively high figure includes all complete and broken flakes over 1·5 cm. in length. Taken together with the retouched pieces about to be described it implies a ratio of about twenty-four flakes to the core. A very high proportion of these are simply small, more or less elongated flakes. Well-made blades, some of considerable size, occur rarely either whole or in fragments, but the general standard of blade manufacture seems to have been low and good examples contribute rather less than 5 per cent to the total output.

(iii) *Retouched tools*

(a) *Backed blades: fléchettes* (47). The manufacture of these, together with other types of backed bladelets seems to have formed one of the main objectives of stone work at this site. This is a common feature of the mesolithic type blade industries of North-west Africa, though it is usual to include all the specimens in the single class of 'backed bladelets' or 'lamelles à dos'. At Hagfet et Tera, however, there is evidence to suggest the presence of two distinct tool types. The one under discussion approximates in shape to the 'fléchette' class suggested by Bouyssonie for certain European Upper Palaeolithic industries, but is generally considerably smaller. Technically it is simply a crescent of which one extremity is brought to the usual acute point while the other is carefully rounded. Specimens of this sort are by no means uncommon in North-west African sites of the Oranian tradition but I know of no case where they form anything like so numerous or so sharply standardized a group as at Hagfet et Tera. Judging by the density of their occurrence in the intact deposits, their total number in the station before the first excavation can be estimated at not far short of 50,000 for one occupational layer. This certainly suggests that they must have been an important object of daily use, and the possibility of spear- or arrow-barbs offers perhaps the most plausible explanation. One of the few occasions where true crescents have actually been found in their original haft[1] shows that they were in fact sometimes used in this way. The 'fléchettes' would form a fairly logical development of this function; alternatively, of course, they may simply have been arrow-heads, and various arguments on this score can be adduced as well. The range of size of this type is from 17 to 30 mm. with the vast majority at almost exactly 20 mm. (Fig. 20, nos. 1–31).

(b) *Backed blades: miscellaneous* (58). These may be considered under various headings which may or may not have functional significance, but are convenient for purposes of comparison. Thick, narrow specimens with a fairly open triangular cross-section form about one-third of this class. The backing may either be in a straight line producing a very acute point at one or both extremities with a slightly curved cutting edge, or else the backing may be curved leaving a straight cutting edge, and so forming a typical knife-blade outline. This narrow type seems to have averaged about 3–4 cm. long. A second category made on much wider, thinner flakes or blades may occasionally, judging by the broken pieces, have been as much as 6 or 7 cm. long, but a more usual size was in the order of 4–5 cm. Here again the backing might be either rectilinear, leaving a convex or irregular edge, or curved at the tip in a shape appropriate for use without a mounting. A high proportion of all the above types show traces of rough and prolonged usage, but careful search reveals no signs whatever of lustration due to grass or straw cutting.

A small number of blades show oblique, transverse blunting at the tip reminiscent of the well-known type of the European Mesolithic; they are, however, too rare to give definite evidence of a type and may be simply unfinished specimens of one of the above categories. The same applies to occasional pieces showing a scalene triangular outline, none of which are really comparable to the 'pointes scalènes' of the later stage of the Capsian of North-west Africa (Fig. 20, nos. 32–6 and 39–41).

(c) *End-scrapers* (2). As in the Oranian, this tool is rarer than in most blade industries, and sometimes of extremely reduced proportions. Rather better

1 Reisner (1911), pl. 62, no. 5.

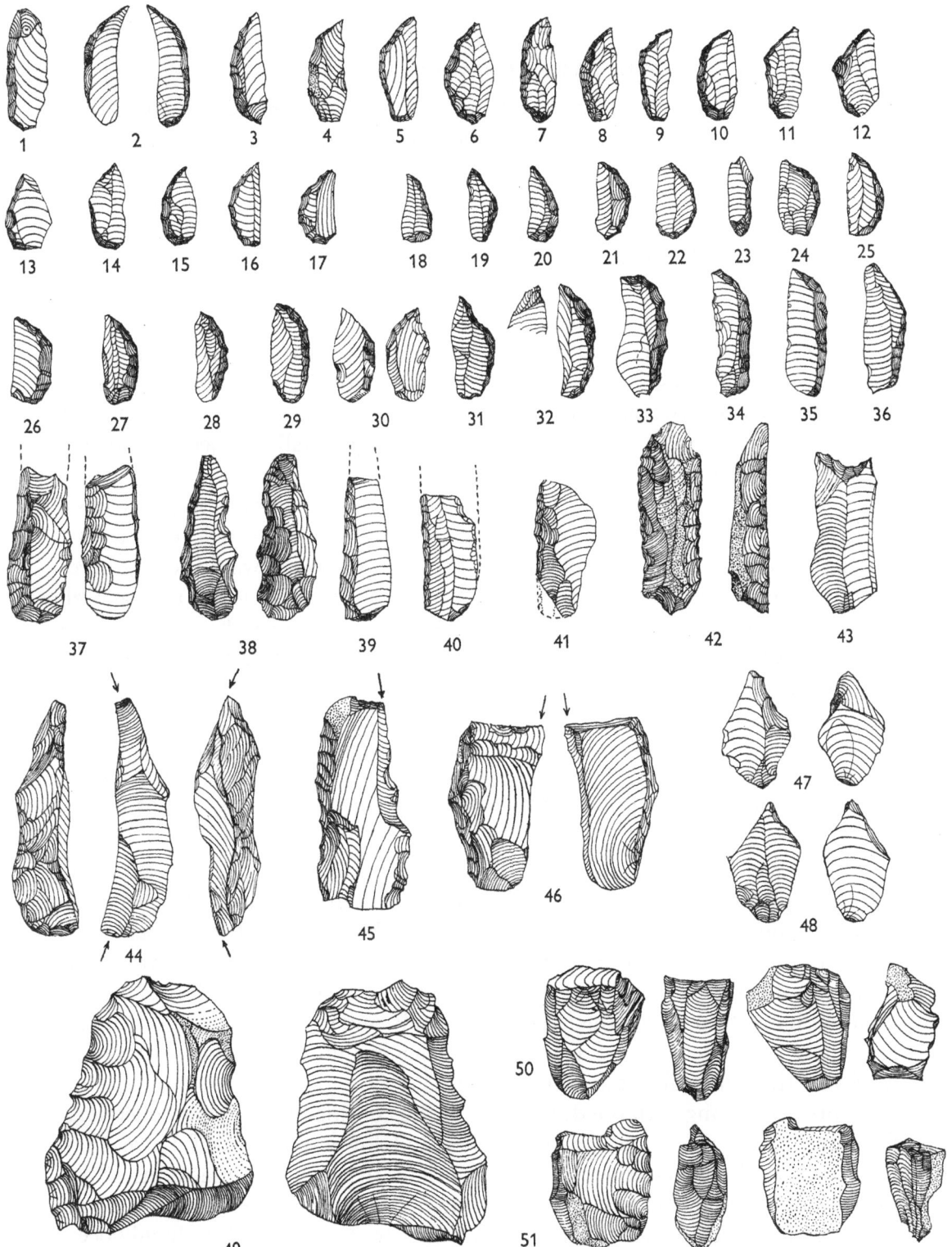

FIG. 20. Hagfet et Tera: representative series from Layers A–C in the terrace deposit

Nos. 1, 33–36 and 39–41, miscellaneous backed bladelets of small to normal size; Nos. 2–32, backed bladelets of the 'fléchette' or barb type; No. 42, rough limace; No. 43 double borer or hollow end-scraper; Nos. 44–46, angle-burins; Nos. 47–48, microburins; No. 48 flake core; Nos. 49–51, bladelet cores. All × ⅔.

specimens were obtained from the disturbed deposits in Sounding B, and others are illustrated by Petrocchi. Two possible hollow end-scrapers may really be a variety of double awl (Fig. 20, no. 43).

(*d*) *Rough round scrapers* (4). The number given represents only an approximation, since possible scraping edges of varying degrees of regularity can be observed on several rough flakes and fragments. There is some evidence for a small variety of thumbnail scrapers. The general status of tools of this sort is again reminiscent of the Oranian.

(*e*) *Rough awls* (4). The evidence for the intentional manufacture of fine points for boring or piercing is comparable to most North African sites, and more noticeable than at any other site examined in eastern Libya before the Neolithic. The majority are made on flakes offering an accidentally projecting portion suitable for the purpose.

(*f*) *Burins* (3). Despite the opinion of the previous excavator (quoted on p. 181) three perfectly typical angle burins were found in this level (Fig. 20, nos. 44–6), and the unmistakable spalls of several others. It would seem, therefore, that although this tool played only a relatively minor role its use was well understood. In addition to the above well-defined examples there are distinct signs of a rough polyhedric form, suggested no doubt by the technique of blade production.

(*g*) *Squamous flakes* (2). As in most blade industries a proportion of wide flakes and blade fragments show intensive flat squilling at one or both ends. The specimens are identical to those noted elsewhere and in particular to the 'lames écaillés' of the French authors. The term used here is that suggested by Professor Garrod at Mount Carmel.

(*h*) *Microburins* (2). As noted earlier, Petrocchi has now discovered a few rare examples of this type in his collections and their presence is further confirmed by our own discovery. The writer feels, however, that it is possible to attach too great cultural importance to this type when it occurs in very small numbers; it has been found by experiment that the backing of small blades is on occasion liable to produce this type accidentally and it seems possible that the same may have been true of a number of ancient industries. (For further remarks on the same subject see Chapter XIII.)

Before proceeding with the description of the considerably sparser collections from the two lower layers, i.e. the upper and lower portions of Petrocchi's Layer D, it will be convenient to discuss briefly the cultural connexions of the material obtained by Petrocchi from Layers A, B and C and re-examined by us as above. The resemblance of certain tool classes to the Oranian has already been mentioned, and this is made the more striking if the percentage of the different classes as listed below be compared with the corresponding figures for the main Maghrebian facies given on p. 215:

Backed bladelets	87
Scrapers	4·5
Awls	3
Microburins	1·5
Burins	2·5

The close resemblance to the Oranian is seen particularly in the absence of geometric microliths, and the relative rarity of microburins, burins, and scrapers of all kinds, as compared with the Capsian group of industries. Taken as a whole, the resemblance of this assemblage to the Oranian is striking, and although strictly speaking geographically intermediate occurrences are required for absolute proof, the writer would suggest that the extension of this culture to Cyrenaica need not be regarded as by any means improbable. The geographical gap appears indeed formidable at first sight—some 1300 km.—but given the remarkable uniformity of this culture over the 1500 km. of its known distribution from Morocco to Tunisia, and also the established fact that another and earlier culture—the Aterian—quite certainly did cross the area, this objection loses a good deal of its weight.[1]

B. *Layer D (upper portion)—reddish-brown earth*

There is a slight decrease in the density of artifacts particularly in the lower part of this deposit, the estimated rate being about 1400 to the cubic metre. The tool classes are the same as those in the overlying layer but apparently significant variations in relative frequency can be detected.

(i) *Cores* (36). The same forms and about the same proportions of the different types of raw material as above.

(ii) *Flakes* (673). Combining these with the retouched tools, a rate of about 22 per core can be deduced.

(iii) *Retouched tools*

(a) '*Fléchettes*' (11). The majority were found in the upper part of the layer and none at all in the lower 20 cm. Their frequency for the layer as a whole is about the same as that at the Oranian station of Beni-Segoual in Algeria.[2]

(b) *Backed blades—various classes* (54). No observable difference of type between these and the corresponding pieces in Layers A–C.

---

[1] Some writers have not hesitated to assume continuity between quite localized variants of the blade tradition such as the Capsian of South Tunisia with Trans-Saharan regions over 3000 miles away. The temerity of such theories when unsupported by intermediate sites, is perhaps only fully apparent to those with some first-hand knowledge of desert travel.

[2] Arambourg (1934).

(c) *Burins* (1–?3). Only one certain and extremely small, made with transverse retouch in a small fragment of patinated blade. The remaining two may be due to accidental fractures.

(d) *Awls* (1–?3). Only one certain, though the remainder are probable.

(e) *Small rough scrapers* (5). As in the later layer, these are a poorly characterized and somewhat crudely made type of tool. Most are about the size of true thumbnail scrapers, though the retouch is not carried the whole way round the periphery.

(f) *Squamous flakes* (1). Normal specimen.

(g) *Bifaces* (1). Small oval bifacial implements with a very flat cross-section, and measuring 5–8 cm. long × 2–5 cm. wide, form one of the commonest classes of pre-historic finds to be collected loose on the Sahel. They are generally worked by free flaking rather than pressure flaking and their date and use remains very uncertain.[1] Our example shows slight traces of lustration which are not usual on other specimens in this layer and it is possible that it was picked up outside the cave by the ancient inhabitants rather than made on the spot. On the other hand, two irregular nuclear pieces from the upper layer look suspiciously like unfinished specimens in an early stage of manufacture, and these at any rate show no signs of derivation. It may be remembered that small rough bifaces of this type occur occasionally in some late Maghrebian industries, notably the Neolithic-of-Capsian-Tradition.

(h) *Microburins* (2). Fairly typical.

C. *Layer D (lower portion)—dark reddish-brown earth*

The density of artifacts falls off sharply in this lower half of the sounding, but the high proportion of cores and the dark colour of the earth suggest that occupation may not really have been less intense and the rarity of artifacts may be a purely local phenomenon, peculiar to this part of the site.

(i) *Cores* (20). No traces whatever can be detected of Middle Palaeolithic techniques and the technique of blade manufacture is, if anything, rather higher than in the later layers.

(ii) *Flakes* (185). The proportions of raw material noted above appear to have been about the same in this layer as well.

(iii) *Retouched tools*

(a) *Burins* (4–?2). All are rough, very much larger than any in the later levels, and made of chalcedony. In view of their technique, however, it is hard to deny them tool status in this category.

(b) *Large scrapers* (3). Made on cores or thick flakes, these are much like specimens recorded from Oranian sites.

[1] Cf. McBurney (1947), p. 79.

Economic evidence

It will be noted from the introduction to Appendix A that Miss Bate, as a result in part of her examination of the remains brought back by us, is not in full agreement with the original faunal list published by Petrocchi from this site. As in Hagfet et Dabba, the other cave excavated in the territory, there are no traces whatever of domesticated species. The numbers of specimens attributable to the food animals identified from the two layers are as follows:

Layers	A, B and C	D (both zones)	Total	
Ox or African buffalo	12	2	14	(13 %)
Zebra	7	6	13	(12 %)
Gazelle	50	16	66	(61 %)
Barbary sheep	8	7	15	(14 %)

Although the sample is relatively small, coming as it does from a sounding, the difference between it and Hagfet ed Dabba (see p. 217) is such that it is difficult not to believe that it represents a profound distinction in habitat, and that the hunters of Hagfet et Tera operated in a significantly more desert and steppe-like environment than that of their possible contemporaries further inland, whose industrial habits will be described in the next chapter. This factor must certainly be taken into account in comparing the two widely differing types of stone-tool equipment.

CONCLUSIONS

The conclusions suggested by our re-examination of the terrace deposits may be briefly stated as follows. The latest cultural manifestation, represented in Petrocchi's layers A, B and C, is considered to show a significant degree of resemblance to the 'Oranian' coastal culture of the Maghreb. The virtual absence of the dominant type of backed blade from the earlier portion of Layer D, leads me to suspect that it contains traces of an appreciably earlier and less specialized version of the same tradition. Whether this difference is confirmed or not, the basic character of the stone-working traits of both layers remains substantially the same and belong as a whole to a late and evolved style of blade industry showing not the slightest traces of contact with, or derivation from, the Middle Palaeolithic.

RELATION OF THE BRECCIA TO THE TERRACE SEQUENCE

At the time of our work, all that remained of the brecciated formation shown on Petrocchi's plan was a small patch apparently representing the *upper* portion of what had been a more extensive deposit largely removed by our predecessor with the aid of dynamite (see Fig. 19). This formation could be followed south-westwards along the rock face as far as the dry stone wall. On removal of the latter the breccia could apparently be followed without break till it graded into the deposits of Layer D of the terrace. No break or disconformity of any kind could be detected by eye in this lateral sequence, but samples were taken at foot intervals in the hope that chemical analysis might reveal some evidence in favour of or opposition to the existence of such a disconformity. Professor Zeuner kindly undertook examination of this problem but it will be seen from Appendix F that no positive results were obtained.

If, however, it should be accepted, despite these uncertainties, that the breccia containing the 'Mousterian' industry figured by Petrocchi is in fact continuous with Layer D on the terrace, the most plausible theory on purely typological grounds would appear to be that an abrupt change in industrial habits took place in a very short space of time, more probably as the result of ethnic movement, rather than of gradual cultural evolution *in situ* in the manner proposed by Petrocchi in his preliminary report.

CHAPTER XIII

THE EXCAVATION AND ARCHAEOLOGICAL
RESULTS AT THE CAVE OF
HAGFET ED DABBA

Throughout both seasons a sharp lookout was maintained for cave sites likely to contain prehistoric deposits. The great majority of caves actually visited contained no deposits whatever, perhaps for the reasons mentioned at the beginning of the last chapter. An exception was provided by the site about to be described. When first seen it appeared as an inconspicuous opening high up in the steep south-facing slope of a tributary valley of the upper Wadi Cuf.

The region is near the centre of the Gebel district proper, on the main road between the towns of Barce and Cyrene. It is one of the few to preserve to a large extent the natural vegetation of the territory, and may be described as not untypical of Mediterranean mountain environments in general. Topographically it forms part of a limestone plateau, dissected by a network of deep and steep-sided valleys; the altitude of the plateau is here about 500 m. above sea-level. The plateau itself and much of the higher slopes are covered with an evergreen scrub, while lower down in the valleys larger trees occur, including large conifers, wild olive, and scrub oak.

Running water is found only for a short time in the winter season, and for the rest of the year the valleys are perfectly dry. The present—mainly pastoral—population draws water entirely from large storage cisterns of Greco-Roman manufacture, and there are no springs within a distance of ten miles.

I. DESCRIPTION OF SITE AND EXCAVATION

The site consists of an elongated cavity 23 m. long by roughly 7·5 m. wide with its long axis east and west, and two openings facing south divided by a massive rock and stalagmite pillar., At the outset of our investigation the floor was approximately level with a slight slope downwards to the west and south. No obvious signs of talus could be recognized outside the smaller western entrance and the cavity behind was dark, low-roofed, and generally less suitable for habitation than the fairly ample, high-roofed chamber to the east. The general shape of the eastern entrance also suggested that there was greater likelihood of accumulated deposits in this area.

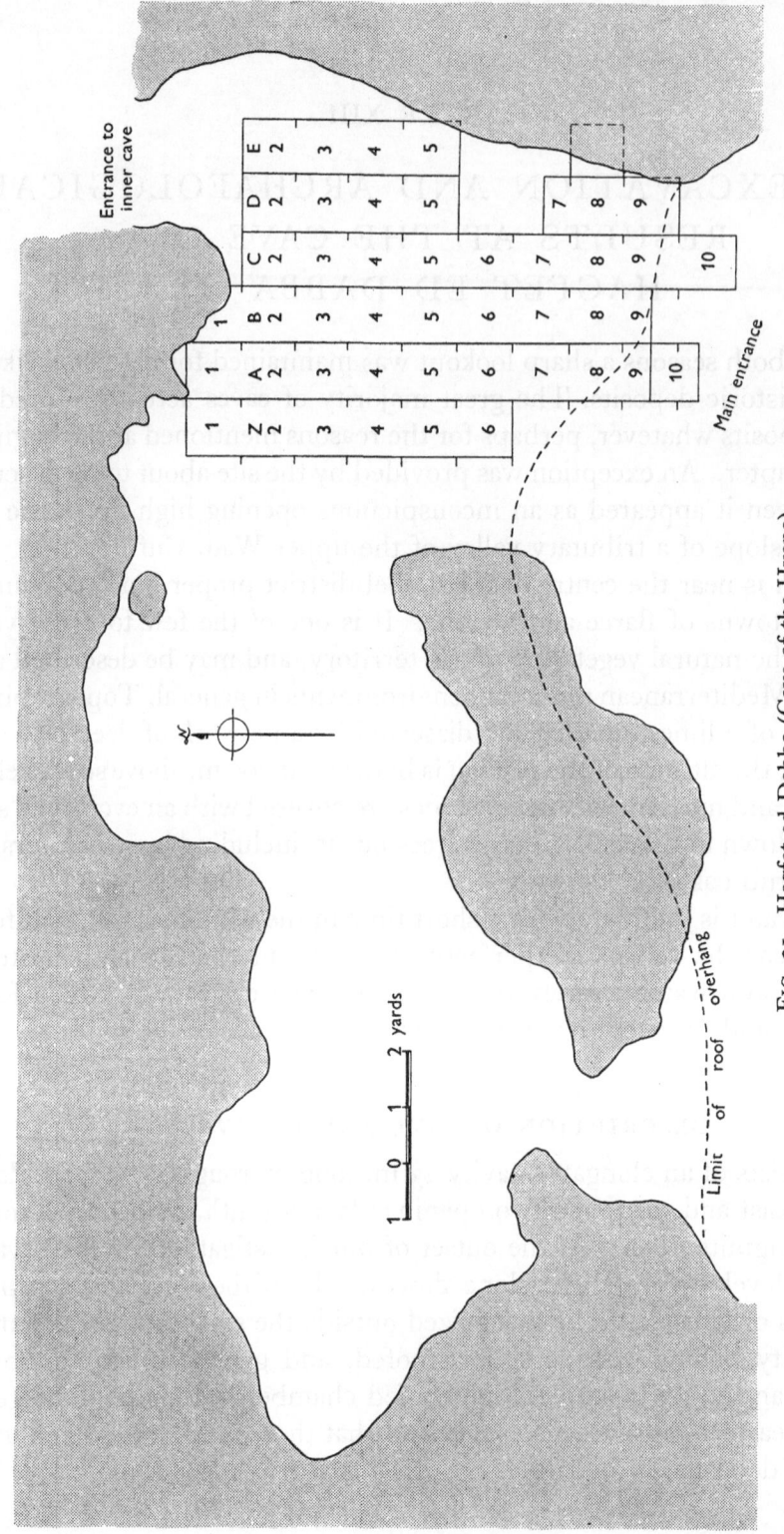

FIG. 21. Hagfet ed Dabba (Cave of the Hyena)

Ground plan showing position of excavation; north-south axes denoted by letters,
east-west axes by numbers.

Massive rock-falls encumbered the floor to some extent against the eastern wall of the cave, and at the entrance to an inner cave consisting of a series of narrow passages much obstructed by curtains of stalagmite. The floor at the entrance to the inner cave lay some 10 ft. above that of the two outer chambers, and the deposits which filled it to this level clearly owed a good deal of their bulk to infiltrations through a small chimney opening on to the plateau above.

After clearing the entrance of loose rock and vegetation the first operation undertaken was a sounding trench extending from north to south from the back wall of the cave through the centre of the entrance in the position of trenches A and B in Fig. 21. As this immediately produced flint artifacts and fossil bones in some quantities, it was decided to enlarge it to a total width of 2 m. and a total length of 7 m. During the first season the deposits in this trench were examined for a distance of $4\frac{1}{2}$ m., in thin spits down to a maximum depth of 1·80 m. below the highest point. A stalagmite floor was encountered at a depth of about 1 m. at the northern end, and was not penetrated during the first season.

These first two trenches were labelled A and B. In view of the somewhat complex stratigraphy revealed, particularly on the east face of B, it was decided during the second season that the supplementary trenches to be dug adjacent to the first two should be removed in natural layers in alternate square metres, the stratigraphy in the three faces exposed being fully recorded before the intervening 'baulk' squares were touched. In this way a nearly complete network of sections at metre intervals was obtained over an area of about 27 sq. m., and the difficult task of following the irregular surfaces of the different natural layers much facilitated. Digging was carried out mainly with trowels; and the spoil from each layer of each square was riddled separately through 4 mm. mesh. Owing to the relatively shallow depth of the layers it was considered advisable on unpacking to assign a code number to all tools and specimens of special interest conveying the horizontal position within the yard grid, as well as the vertical segregations into layers.[1]

2. STRATIGRAPHY

The surface of the hard stalagmite floor on which the various fossiliferous formations were resting lies in a horizontal plane at the north-western end of the area examined, but falls away appreciably towards the entrance to the south, and towards the inclined eastern wall of the cave. The general pattern of the stratigraphy overlying this slightly domed shape may be described as that of slates on

---

[1] A conversion table giving the meaning of each code number has been deposited with the original collections in the Museum of Archaeology and Anthropology at Cambridge.

a roof; that is to say, each layer tends to overlap the next over part, but not all, of its extent. This scheme is twice interrupted—by the penultimate layer (counting from bottom to top) which covers nearly the whole of all the preceding layers and by a final zone of recent disturbance affecting to a greater or less extent about one-third of the region excavated.

Layer IX—Basal stalagmite

The entire floor surface of the area excavated was found to consist of an extremely hard breccia of angular limestone fragments presumably derived from the roof and walls of the cave—cemented together with a red matrix. The resistance of this material to penetration with cold-chisels was hardly if at all inferior to the bedrock. In the end a pit 1 m. deep was made with the aid of four charges of dynamite. The deposit was found to remain of the same consistency throughout this depth, but bedrock was not reached and its total thickness could not be estimated. A few small bone fragments were collected from it, but no traces of artifacts or charcoal were observed.

Layer VIII—Decomposed stalagmite

A lenticular mass of white powdery deposit with a maximum observed thickness of 80 cm. was found to overlie Layer IX at the southern end of C trench (see Fig. 22). A few bones and artifacts obtained from it are considered to be intrusive from the overlying Layer VII; as far as they go they show no perceptible difference from the contents of the latter.

Layer VII

A moderately hard deposit composed of reddish earth with occasional small angular pieces of limestone, artifacts, rare charcoal, and bones. The area occupied by this deposit extended from the northern part of square D6 where it rested in a thin lens directly on the stalagmite, to the southern end of C trench, where it reaches a thickness of 45 cm.

Layer VI

Similar to VII, but darker in colour and containing appreciably more traces of charcoal. To the westward this layer extended from the northern part of transverse trench 5 to the southern margin of the excavation, while on the east it was spread out in a thin sheet over the stalagmite floor at least as far north as D3. In this region also it was almost black in colour and contained a much higher proportion of charcoal. Most of the finds were made in the south-west portion where immense numbers of small artifacts occurred.

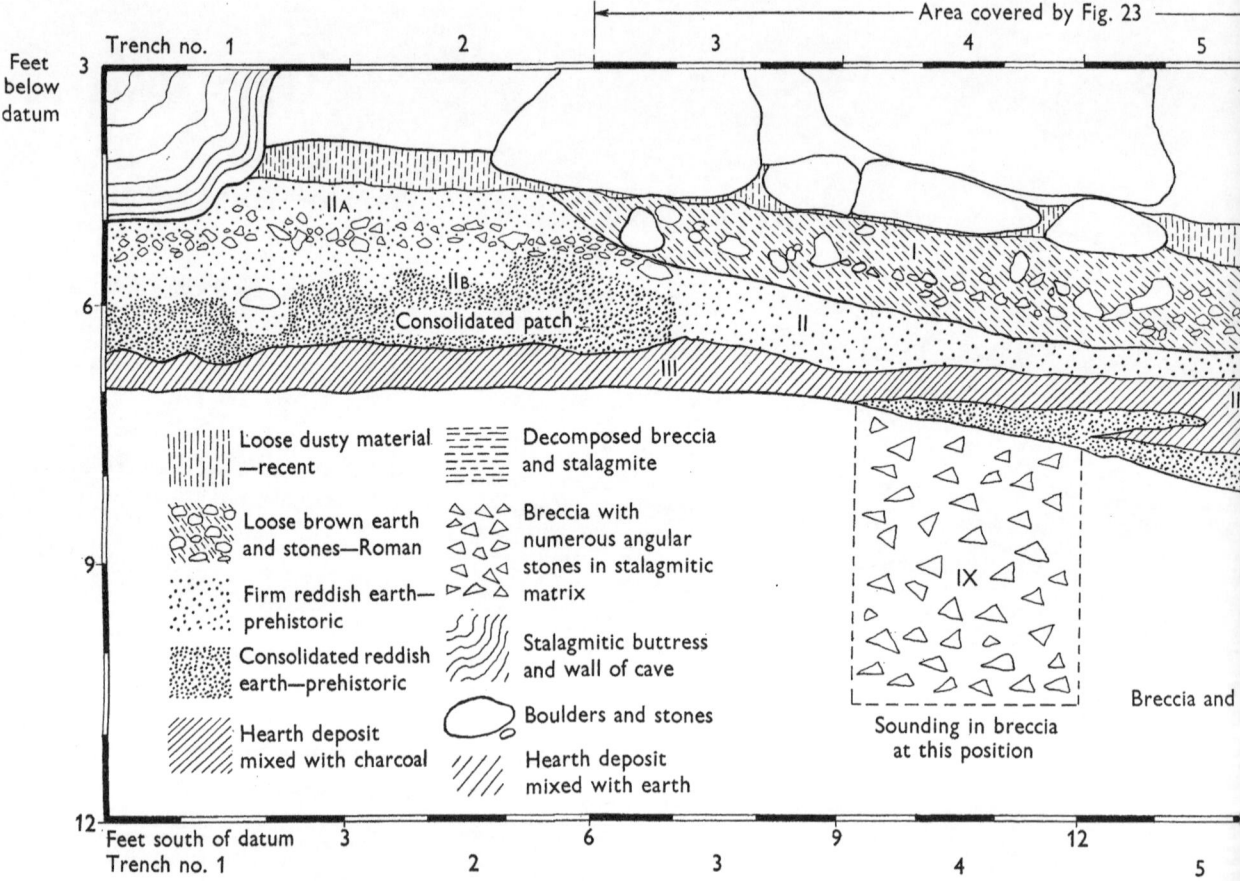

Area covered by Fig. 23

Feet below datum

Trench no. 1   2   3   4   5

3

IIA

IIB
Consolidated patch

6

I

II

III

II

Loose dusty material —recent

Decomposed breccia and stalagmite

Loose brown earth and stones—Roman

Breccia with numerous angular stones in stalagmitic matrix

Firm reddish earth— prehistoric

Stalagmitic buttress and wall of cave

IX

9

Consolidated reddish earth—prehistoric

Boulders and stones

Sounding in breccia at this position

Breccia and

Hearth deposit mixed with charcoal

Hearth deposit mixed with earth

12

Feet south of datum   3   6   9   12
Trench no. 1   2   3   4   5

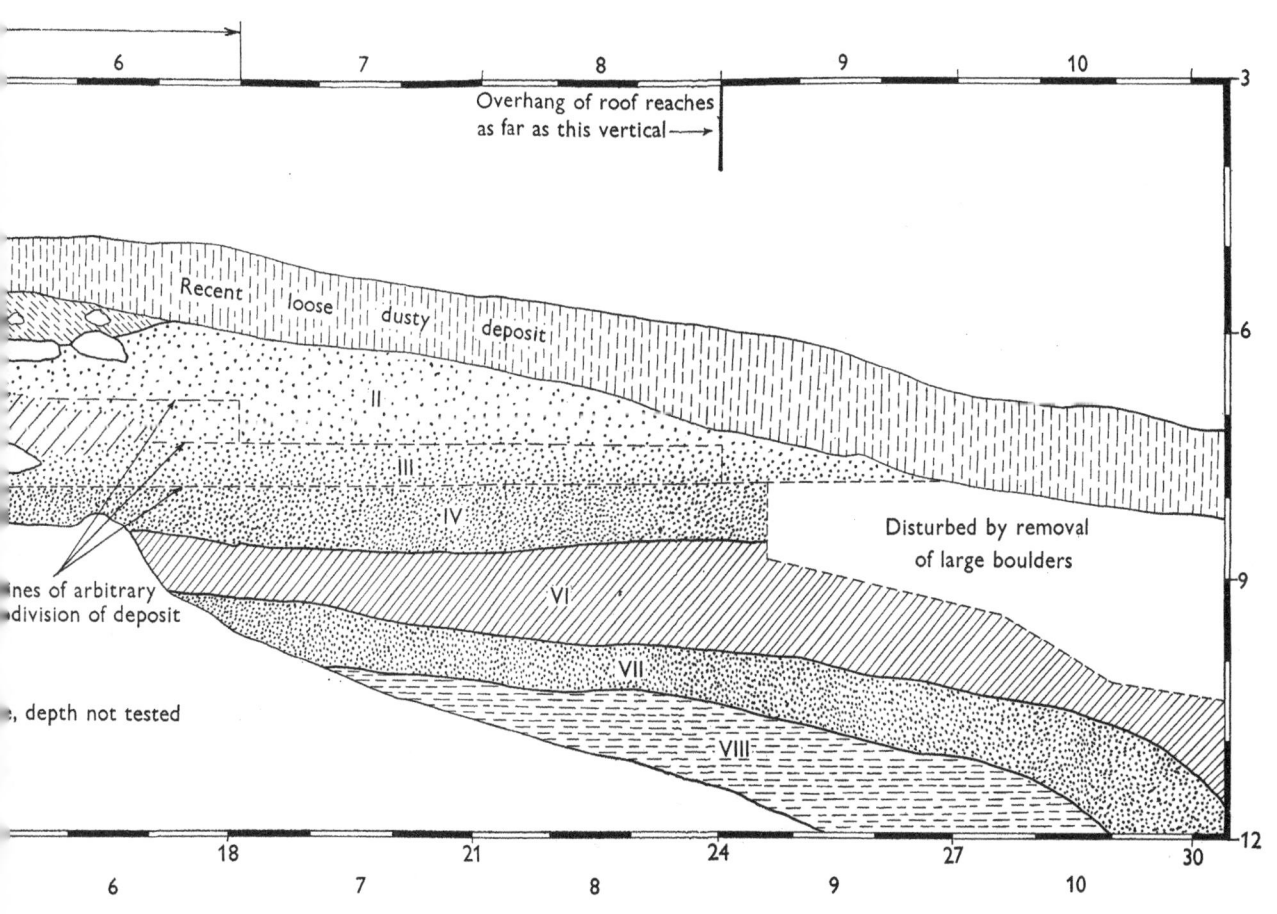

Overhang of roof reaches
as far as this vertical ⟶

Recent loose dusty deposit

II

III

IV

Disturbed by removal
of large boulders

VI

ines of arbitrary
division of deposit

VII

, depth not tested

VIII

Fig. 22. Hagfet ed Dabba: north-south section between trenches B and C

Layer V

This layer, which appeared to consist mainly of ash mixed with brown earth and strongly cemented with calcium carbonate, formed the basal layer in this southern portion of trench Z and squares 4, 5 and 6 in trench A. It yielded only an insignificant collection of stone artifacts but contained a remarkable mass of fragmented bones in the southern portion observed, where it reached a maximum thickness of 30 cm.

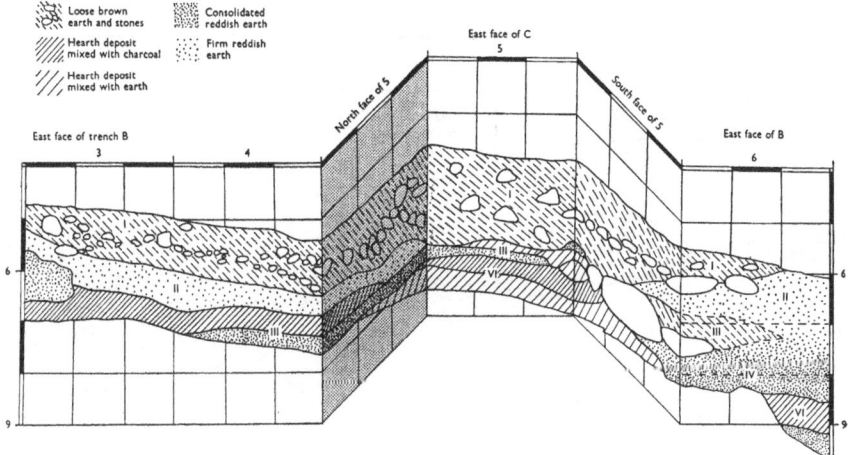

FIG. 23. Hagfet ed Dabba: isometric diagram of portions of sections between trenches B and C, 5 and 6, C and D, 4 and 5

Layer IV

The northern margin of this layer tapered out approximately on the limit between transverse trenches 4 and 5. The upper limit between it and Layer II could not readily be distinguished south of trench 5, where they ceased to be separated by Layer III, and an arbitrary horizontal division was accordingly established in this area. It was, however, considered that the first foot at least of deposit overlying Layer VI in this area (south of trench 5), could be confidently correlated with the deposit underlying Layer III in trenches 5 and 6, on grounds of colour and general appearance. To the east this layer tapered out very gradually till it formed little more than a line of discoloration separating Layers III and VI in the southern face of square D 3. The relationship of this layer to Layer V to the westward was difficult to establish with certainty. No clear-cut line of demarcation could be established and it seems probable that Layer V really represents the lateral facies of Layer IV to the west.[1] This is

[1] This is consistent with the chemical and microscopical data obtained by Professor Zeuner and reported in Appendix F.

13-2

borne out by the nature of Layer IV which consists of a very high proportion of reddish earth and is generally remarkably free from traces of hearth debris, although yielding a particularly rich assemblage of faunal and industrial remains.

Layer III

This is the principal hearth layer. In the entrance to the inner cave a black mass of comminuted charcoal mixed with cave earth forms a continuous layer nearly 60 cm. thick from wall to wall. Westwards, this deposit interdigitates

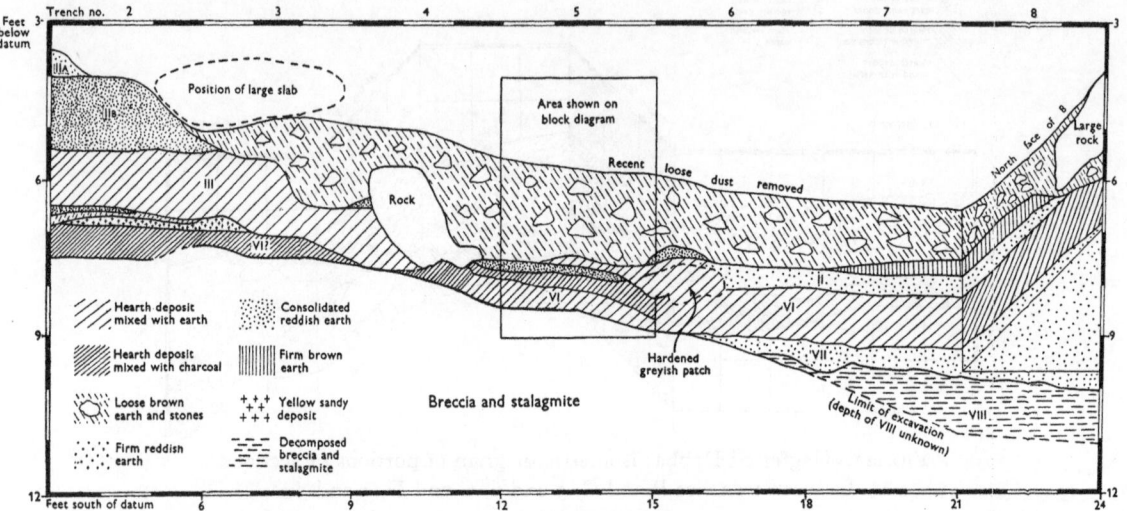

FIG. 24. Hagfet ed Dabba: north-south section between trenches C and D, with isometric addition of part of east-west section between trenches 7 and 8

in squares A 1–2, and to the south in trenches 4 and 5 (see Figs. 22, 23 and 25), with consolidated reddish earth. In the latter square and further to the east and south it has been destroyed by the disturbed zone of Layer I.

Layer II

In consistency this layer closely resembles Layer IV, being composed mainly of reddish earth. In some places, particularly to the north of the excavated area, it assumes the appearance of true terra rossa and is no doubt very largely composed of infiltrations through the chimney. It is cemented into an extremely hard formation approaching the consistency of true stalagmite in places. These cemented patches show no correspondence to changes in bedding or colour and are perhaps to be attributed to purely local infiltrations of water dripping from particular irregularities in the roof. Close to the entrance the concretionary patches are absent, and the deposit assumes a more nearly brown colour. In the northern portion of the area excavated, it was found convenient to divide Layer II into two portions, an upper and a lower, called A and B respectively. The upper portion was petrographically identical with the lower except that it seldom showed a comparable degree of cementing; it was, however, generally of a firm consistency contrasting with Layer I and other recent deposits. Archaeologically

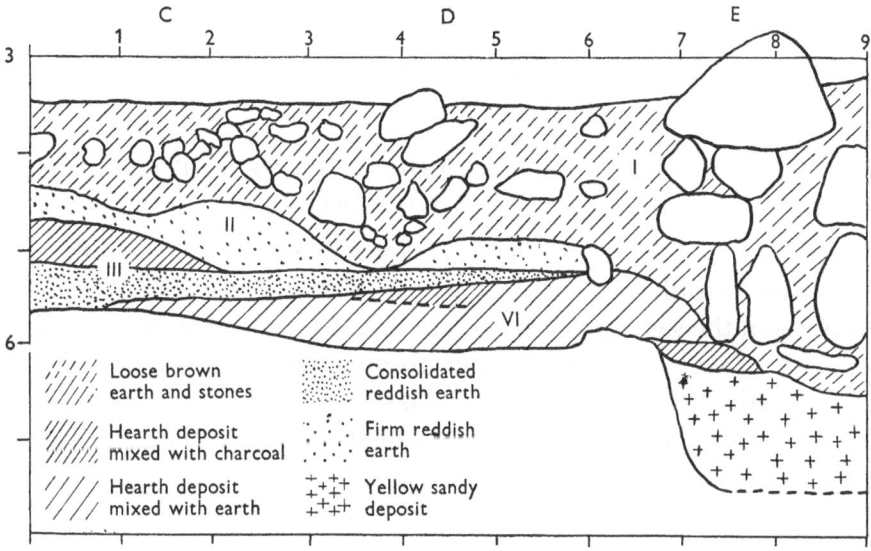

Fig. 25. Hagfet ed Dabba: east-west (transverse) section between trenches 5 and 6

Layer II A differed from Layer II B in that it included the surface of the prehistoric deposits immediately underlying those of modern or historical date, and was consequently more liable to contain intrusive elements later than the period of its formation; as were also the south and east edges of other layers.

A possible agent for such intrusion was provided by root action, and no doubt that of occasional small burrowing animals as well (though the latter would have been effectively held up in Layer II B by the concretions just mentioned). The roots were commonest towards the entrance, where they account for the finds of Roman pottery in prehistoric layers.

Layer I

This layer displayed throughout a loose texture, was a characteristic dark brown in colour, and contained large quantities of angular limestone fragments and Roman pottery of about 100 B.C. described in Appendix G. It was found to form the infilling of a large irregular pit with very gently sloping sides in the east-central portion of the area examined. At its deepest point in squares E4 and E5 (Fig. 21) this pit cut right through all the deposits except the basal stalagmitic floor.

RECENT DEPOSITS

The greater part of the floor of the cave at the time of discovery of the site was covered by 15–30 cm. of loose grey dust. Elsewhere, particularly in the north-eastern area, large falls of rock had taken place at some time subsequent to Layer I, and included immense slabs up to several tons in weight (Fig. 22).

3. THE ARCHAEOLOGICAL CONTENTS OF THE LAYERS: STONE INDUSTRY

Nearly all the artifacts of archaeological importance were of flaked flint. A considerable range of different varieties was made use of, and approximately the same varieties in the same proportions are found from top to bottom. Very little use was made of coarser-grained materials and there is no sign of the peculiar chalcedony used in such quantities at Hagfet et Tera. The sources of the raw materials were not traced, though for geological reasons they can hardly have been nearer than about 10 km.[1]

Layer VII

The total yield of this lowest layer was not large owing to the relatively small area examined—about 4 sq. m. The rate of occurrence can be estimated at about 450 to the cubic metre.

Cores (13). All are blade (as opposed to flake) cores; there are four remarkably neat elongated double-ended specimens and one of a flatter double-ended pattern.

Flakes (624). These show a very high proportion of finely-made blades, often of considerable size. Most of the rougher flakes can be satisfactorily explained as products of the earlier stages of core preparation or rejuvenation. The ratio of flakes to cores is thus about 48:1, or about twice that observed at Hagfet et Tera. While this may be in part an artificial result arising from the removal of a high proportion of partially worked cores by the no doubt nomadic inhabitants of the cave, it is also possible to see in it a reflection of the higher standard of flaking compared with that of the other cave site.

Backed blades (28). The majority of these are made on narrow parallel-sided blades with a straight line of retouch running the whole length of one side. The

---

[1] The main source must have been the chert-bearing layers of the Lower Eocene limestone whose nearest outcrops are near the coast.

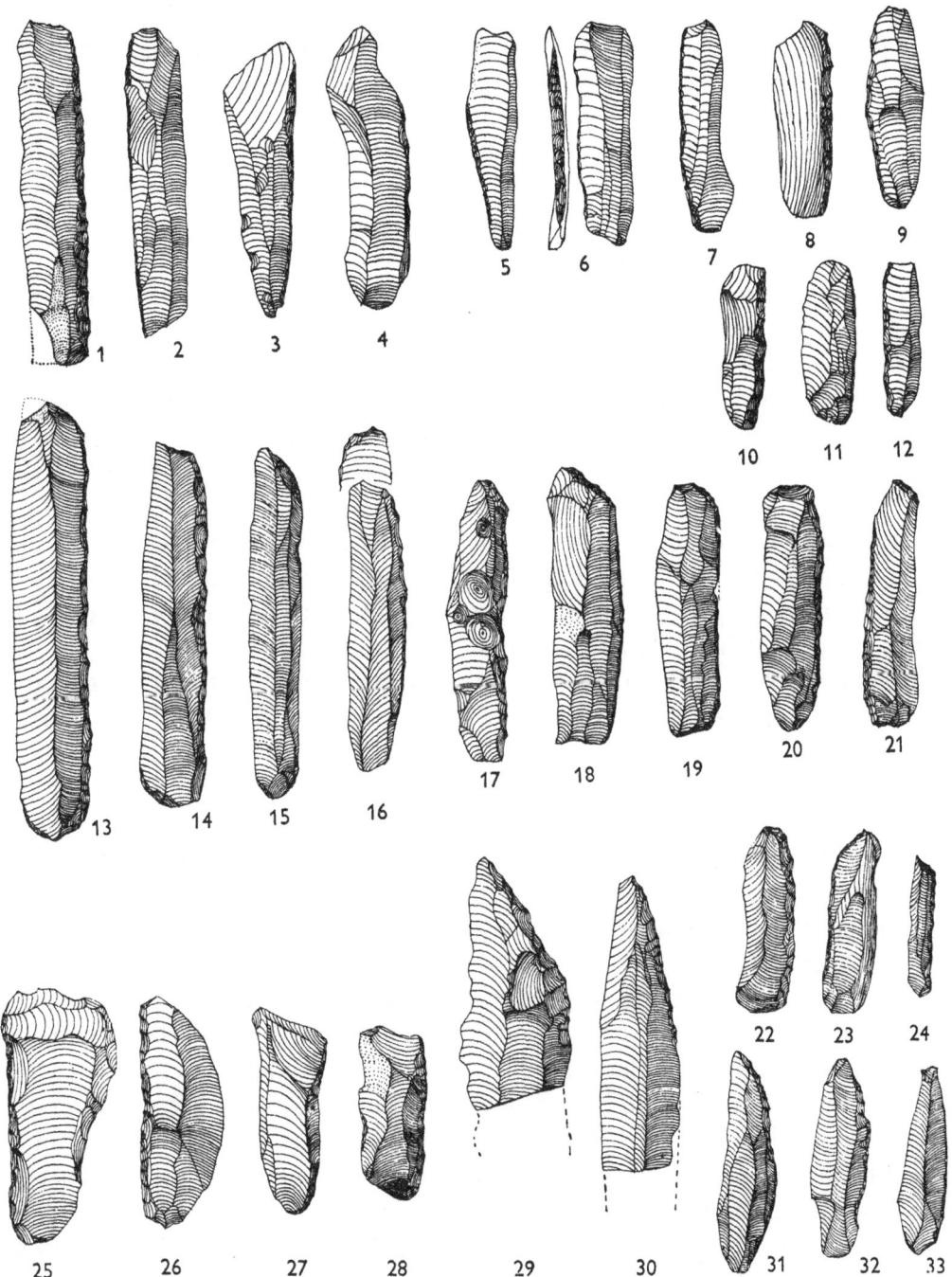

FIG. 26. Hagfet ed Dabba: representative series from Layers IV–VII

Nos. 1–12, backed blades with straight line of trimming; Nos. 13–24, backed blades with terminal—in addition to lateral—trimming; Nos. 25–28, backed flakes, Nos. 29–33, backed blades and bladelets with curved backs. All × 2/3.

striking-platform is frequently present at one end, while the primary transverse edge of the blade forms the opposite end. More rarely the retouch is continued round the tip or the platform or both, but the general shape of the tool remains rectangular as opposed to the curved-backed form often met with on Capsian or Chatelperron blades.

Burins (5). Both transversely truncated angle-burins and bec-de-flute varieties are represented. There are ten identifiable spalls.

Transverse-burins (15). A detailed description of the peculiar and highly characteristic tools included under this denomination will be given in connexion with the much larger collection made in the succeeding layer. Two are of the rare right-handed variety while the remainder conform to the usual left-handed pattern. The unmistakable squills or spalls of this type number 31 in this layer.

Microburins (3). There are only one definite and two atypical pieces of this class. While these are all quite possibly accidental, there are some signs that the technique was occasionally used for finishing off the tips of backed blades.

End-scrapers (5). All are made on blades though poorly worked.

Miscellaneous scrapers (1). A single specimen of a curious naviform shape like a steep 'limace', which recurs in several later layers.

Squamous flakes (1). There is only one possible example of this class of tool which is well represented in later layers and presumably functioned as a struck chisel of some kind.

Layer VI

This layer produced just over 4000 artifacts, the largest single collection to come from any one layer. An area of about 8 sq. m. was examined and the rate of occurrence may be estimated at about 1000 to the cubic metre or about double that of Layer VII.

Cores (37). All are quite evidently the residues of blade production. The majority are more or less elongated and double-ended with only a slight tendency towards lateral flattening. There are no globular or discoidal flake cores whatever.

Flakes (3660). Well-made blades of small to moderate size are common and the flakes can all be reasonably attributed to the early stages of blade manufacture. The high rate of flakes per core is even more pronounced than in Layer VII; including the tools, it appears to be about 100 : 1.

Backed blades (199). This relatively large collection enables the type to be studied in some detail. Except in the matter of size a rather high degree of homogeneity is revealed in the class as a whole. Backing is invariably from one face only and the cross-section is consistently low and wide. Thick, triangular-sectioned sharply pointed pieces backed from both faces, like Fig. 20, no. 34 and

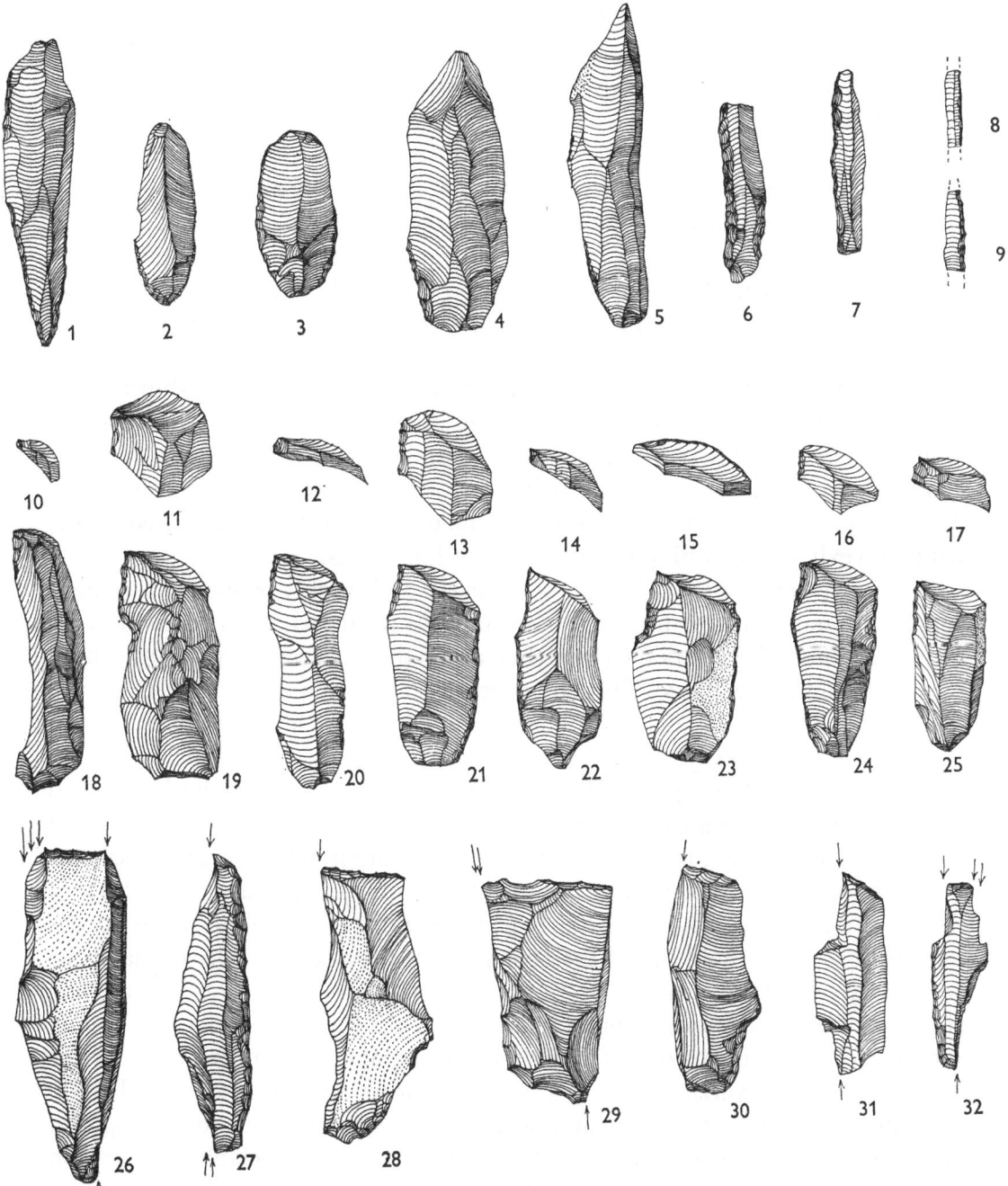

FIG. 27. Hagfet ed Dabba: representative series from Layers IV–VII (*cont.*)

Nos. 1–9, miscellaneous trimmed blades, No. 1 apparently for insertion in terminal handle, Nos. 8 and 9 microlithic; Nos. 10–17 re-sharpening spalls from transverse-burins; Nos. 18–25, transverse-burins; Nos. 26–32, angle-burins. All × ⅘.

nos. 37–9 from Hagfet et Tera are virtually absent (though rare examples occur in other layers). A rare though possibly distinctive form shows retouch for a short distance along both edges of a blade as if to adapt it for insertion in a terminal handle (Fig. 27, no. 1). The remainder conform closely to the range of shapes mentioned in Layer VII. By far the commonest size is one of about 40 × 10 mm.; there is also a fairly well-marked larger class, about 70–90 × 15 mm. Microlithic pieces in the strict sense, measuring about 15 × 3 mm., occur but are very rare (Fig. 27, nos. 8, 9). There are also a few rare large specimens showing curved backing rather roughly executed in the Capsian manner (Fig. 26, nos. 29–31).

Burins (36). Well-made transversely or obliquely truncated angle-burins predominate. They show considerable variation in size from stout specimens made on flakes to remarkably small and delicate examples made on flat blades. A rather characteristic local variant shown in Fig. 27, no. 27, tends to be exceptionally narrow and appears to be made on a fragment of large blunted-backed blade, with convexly curved oblique truncations at both ends.

Transverse-burins (117). These specialized and highly standardized tools supply the most interesting single item of cultural information to come from this cave. In all the layers in which they occur they are remarkably uniform in technique and range of size. They are generally made on a thick blade, the burin facet being struck from a straight line of retouch on the left or right margin of the blade—nearly always the former.[1] The resulting fracture is inclined at an average angle of about 45° to the plane of the bulbar face of the blade, and strongly curved, so that the bulbar surface of the squill is markedly concave and skew. Squills resulting from second and later resharpenings accordingly assume a peculiar concavo-convex form impossible to confuse with any other type of flaking (Fig. 27, nos. 10–17). A fairly constant ratio of just under three squills per burin seems to hold for all the layers where they occur.[2] Occasionally the squill may carry away as much as 1 cm. or more of the blade, but normally the squills are not more than 2–3 mm. thick. A total of 285 squills can be confidently recognized. The dimensions of these tools as estimated from the yield in this layer alone show a variation of 0·9–4·0 cm. in width, and 1·6–6·2 cm. in length. The modal width may be estimated at about 1·7 cm. and the modal length at about 3·6 cm. The dispersal about the median length is appreciably greater than about the median width, a feature which is most easily explained as resulting from the varying effects of resharpening on a fairly homogeneous group of blades.

[1] In Layer VI only 3½ per cent. burins and 7 per cent. squills have been struck from the right. I know of no exact parallel for this curious consistency in manufacture of this kind.

[2] It was also noted that not more than 30 per cent. could be regarded as initial sharpening squills.

As with most classes of flaked tools the precise purpose for which these trans-verse-burins were intended is extremely difficult to guess; so also is the method whereby they were held and used. As far as general form goes many could have quite well served as end-scrapers rather than true burins. In support of this theory attention may be drawn to the much lower proportion of end-scrapers in the layers in which the transverse-burins occur. On the other hand, the propor-tion of true burins accompanying the latter is rather higher in the basal layers than elsewhere. An examination of resharpening squills revealed that about 30 per cent show evident traces of utilization along the intersection corresponding to the cutting edge of the end-scraper.

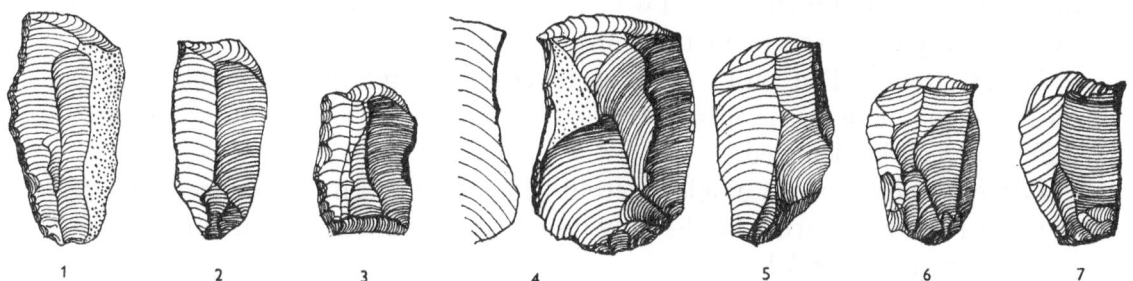

FIG. 28. Transverse-burins from surface site at Nag Hammadi, near
Luxor, Egypt (after Vignard)
For comparison with Fig. 27, Nos. 18–25. All × ⅔.

On the other hand, the complete absence of any effective parallels throughout the vast range of variation of blade industries in Western Europe would tend to make it appear that the class represents something more than a local device for manufacturing end-scrapers, and it seems at least conceivable that some special chiselling or cutting activity is mainly responsible for its invention.

End-scrapers (10). Only three could be regarded as thoroughly characteristic specimens, all extremely small. The rest are merely blades with a little atypical retouch at one end.

Miscellaneous scrapers (4). Three large flake scrapers, and one broken 'limace' type.

Squamous flakes (4). Three fairly characteristic examples on fragments of wide blades. There are also two notched blades showing rather similar treatment.

Layer V

This layer, which, as mentioned above, may be simply the lateral version of Layer IV, produced a total of less than 200 artifacts. As far as this very small series goes, no significant difference from the preceding layer was revealed.

Layer IV

The distinction between this layer and Layer VI was not always possible to establish with certainty, as it depended on a relatively slight difference of colour. In order therefore to afford the maximum opportunity for detecting any cultural difference that might be present, the archaeological contents of a 6 in. zone lying on either side of the plane of contact were separated from both and are not included in the table on p. 217.

The certain contents of Layer IV can now be described as follows:

Cores (14). The majority are typical, well-made double-ended blade cores of moderate to small dimensions. Two show slight signs of wear and there is one exceptionally small specimen 1 cm. in diameter.

Flakes (1034). The ratio of flakes to cores, including flake tools, can be reckoned at about 82:1 or slightly less than Layer VI. No other technical difference can be detected.

Backed blades (70). A single microlithic piece, and three narrow forms re-touched from both faces, are the only types not encountered in about the same proportions in Layer VI.

Burins (14). Polyhedric varieties predominate but there are also two typical angle-burins.

Transverse-burins (13). Two are right-handed and the rest left-handed; there are thirty-two spalls.

End-scrapers (4). All are rather roughly retouched.

Miscellaneous scrapers (2). Both are flake scrapers, and one shows fairly clear signs of intentional denticulation.

Microburins (?1). Possibly the accidental product of blade backing.

Squamous flakes (4). Only two can be regarded as really typical.

No clear distinction can be observed between this and Layer VI other than a slight increase in the number of backed blades and a rather more marked decrease in the number of transverse burins.

Layer III

The hint of cultural change noted in Layer IV becomes fully apparent in Layer III—it is expressed mainly by the virtual disappearance of the transverse burins and a noticeable increase in scrapers of all kinds, especially well-made end-scrapers and large flake-scrapers. The rest of the tool kit, as will be seen from the following details, remains substantially the same.

Cores (16). All are somewhat rough and irregular, apart from one large well-made double-ended blade core. Three seem to have been made from large flakes,

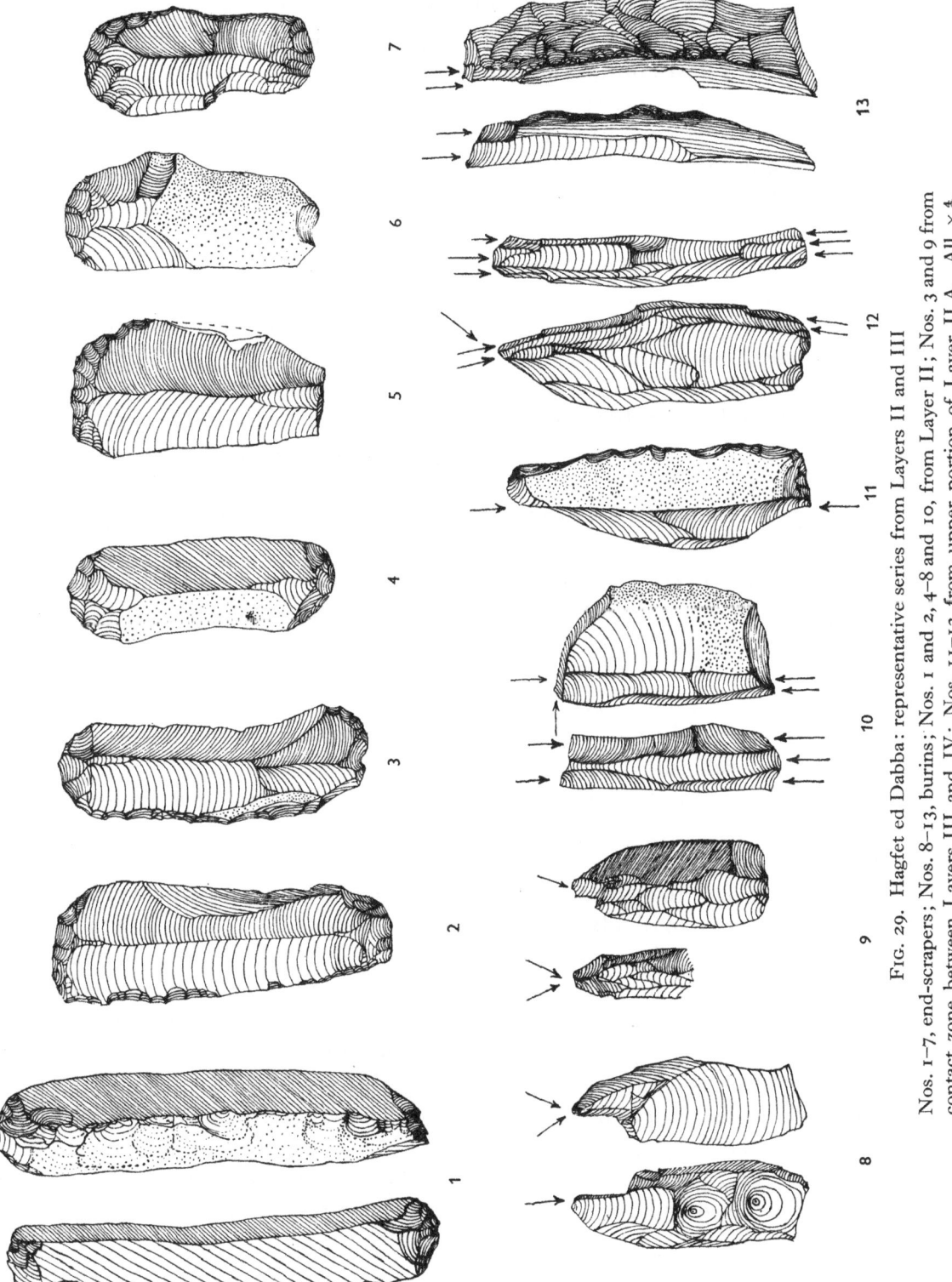

Fig. 29. Hagfet ed Dabba: representative series from Layers II and III

Nos. 1–7, end-scrapers; Nos. 8–13, burins; Nos. 1 and 2, 4–8 and 10, from Layer II; Nos. 3 and 9 from contact zone between Layers III and IV; Nos. 11–13 from upper portion of Layer II A. All ×⅘.

one has been trimmed into a rough core-scraper, and one is clearly intended for the production of flakes on the alternate principle.

A few of the flake-cores (Fig. 31, nos, 1, 3, 5 and 6) assume a curious wedge-shaped section suggesting that they may just conceivably represent a forerunner of the flaked-adze so characteristic of the Neolithic cultures of Eastern Libya described in Chapters xiv and xv.

Flakes (909). Good blades are still well represented and there is no obvious distinction with earlier layers. The ratio per core has dropped (including tools) to about 65:1.

Backed blades (54). The series includes three specimens of the thick narrow form retouched from both faces, associated with the Gravettian in Europe and the Oranian rather than the Capsian in North Africa. The remainder are somewhat rougher and larger than the corresponding tools in the earlier layers, and there is a more marked class of flat broad flakes with a single straight line of retouch (Fig. 30, no. 35).

Burins (7). There are four transversely truncated angle-burins and one typical double burin.

Transverse-burins (2). Both were found in an area where Layer III directly overlies Layer IV and may possibly be derived. The same applies to three spalls.

End-scrapers (16). Most of these are much more characteristic than the earlier pieces, and can be described as quite typical tools of their kind (Fig. 29).

Flake-scrapers (8). These are mostly large and varied in type. There is one example with fairly definite signs of intentional serration, and one (possibly derived) of a Levalloiso-Mousterian type of side-scraper.

Squamous flakes (7). Five are large and typical, and two are more doubtful.

Layer IIB

It will be recalled that in the interior of the cave (north of trench 6) it was considered advisable to divide this topmost prehistoric layer into two halves, an upper portion IIA firm but uncemented, the upper surface of which lay in direct contact with deposits of recent and historical date, and a lower portion IIB extensively cemented and less liable to contamination with specimens of later date. Towards the entrance this distinction could not be made on grounds of appearance or consistency and is merely of an arbitrary division of the deposit into an upper and a lower half. Figs. 29 and 30 show a representative series from this and the underlying Layer III.

Cores (35). The majority of these are small but very regular blade cores, there are also a few on globular and roughly discoid patterns for the production of small flakes.

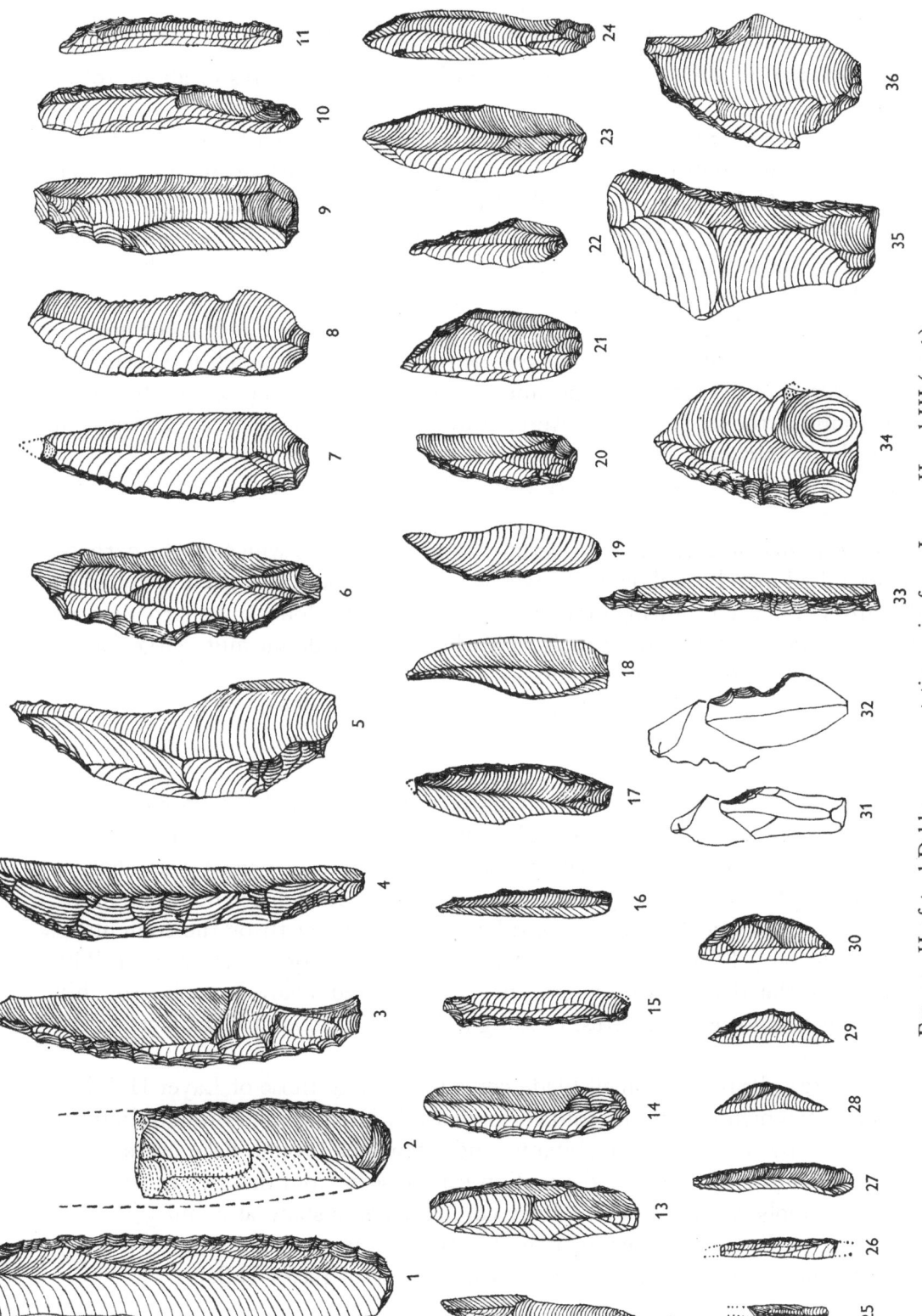

Fig. 30. Hagfet ed Dabba: representative series from Layers II and III (*cont.*)

Nos. 1–27, blunted-backed blades and bladelets; Nos. 28–30, crescents; Nos. 31 and 32, microburins; No. 33, Gravette type backed blade with awl tip; Nos. 34–36, blunted-backed flakes. Nos. 7, 13, 26 from Layer III, remainder from Layer II. All × ⅘.

Flakes (3536). Including the tools, a ratio of flakes to cores of about 108:1 can be estimated.

Backed blades (152). The slight distinctions noted between Layer III and the lower layers are confirmed in this larger series, but no further innovations can be detected. Only two specimens show the narrow form with a thick triangular cross-section, but blunting or partial blunting of a curved back suggests that not a few were intended to be held in the hand rather than mounted in a handle of some kind.

Burins (20). Single-blow burins made on broken blades and flakes are the most numerous, but the angle and polyhedric types are also well represented.

Transverse-burins (2). Both specimens come from the entrance area and none were found in the interior; there were, however, seven spalls. All are considered to be almost certainly derived.

End-scrapers (28). Nearly all are very well made and representative examples of the type.

Miscellaneous scrapers (44). A certain number of flakes which might perhaps be described merely as showing heavy signs of utilization have been included in this group, but even allowing for these there can be little doubt of a sharp increase in large scraping and cutting tools. Nearly all are made on quite large flakes, and serrated edges are not uncommon.

Squamous flakes (7). Some of these are thick and rather like rough 'burins plans.'

Layer II A

In excavating this layer the overlying recent deposit of loose grey dust 15–30 cm. deep was first removed together with the upper 7–8 cm. of Layer II A. The difference between the recent and the prehistoric deposit was sharply defined both by colour and texture, but owing to the occasional penetration of the former in shallow pockets it was considered advisable to jettison the contents of the superficial 7–8 cm. after examination for any pieces of special interest. The volume of the deposit whose contents are analysed below may be roughly estimated at eight cubic metres, yielding a density of about 300 artifacts to the cubic metre.

Cores (42). Principally small blade cores resembling those of Layer II B, but including fragments of two discoid cores, possibly intrusive and brought in from the surface by the ancient inhabitants, and a few small rough flake cores.

Flakes (2130). No obvious distinction can be seen from those of Layer II B. Including tools, the ratio of flakes per core appears to stand at about 57:1.

Backed blades (148). Specimens with a thick triangular section are virtually absent. The remaining types are identical to those of Layer II B.

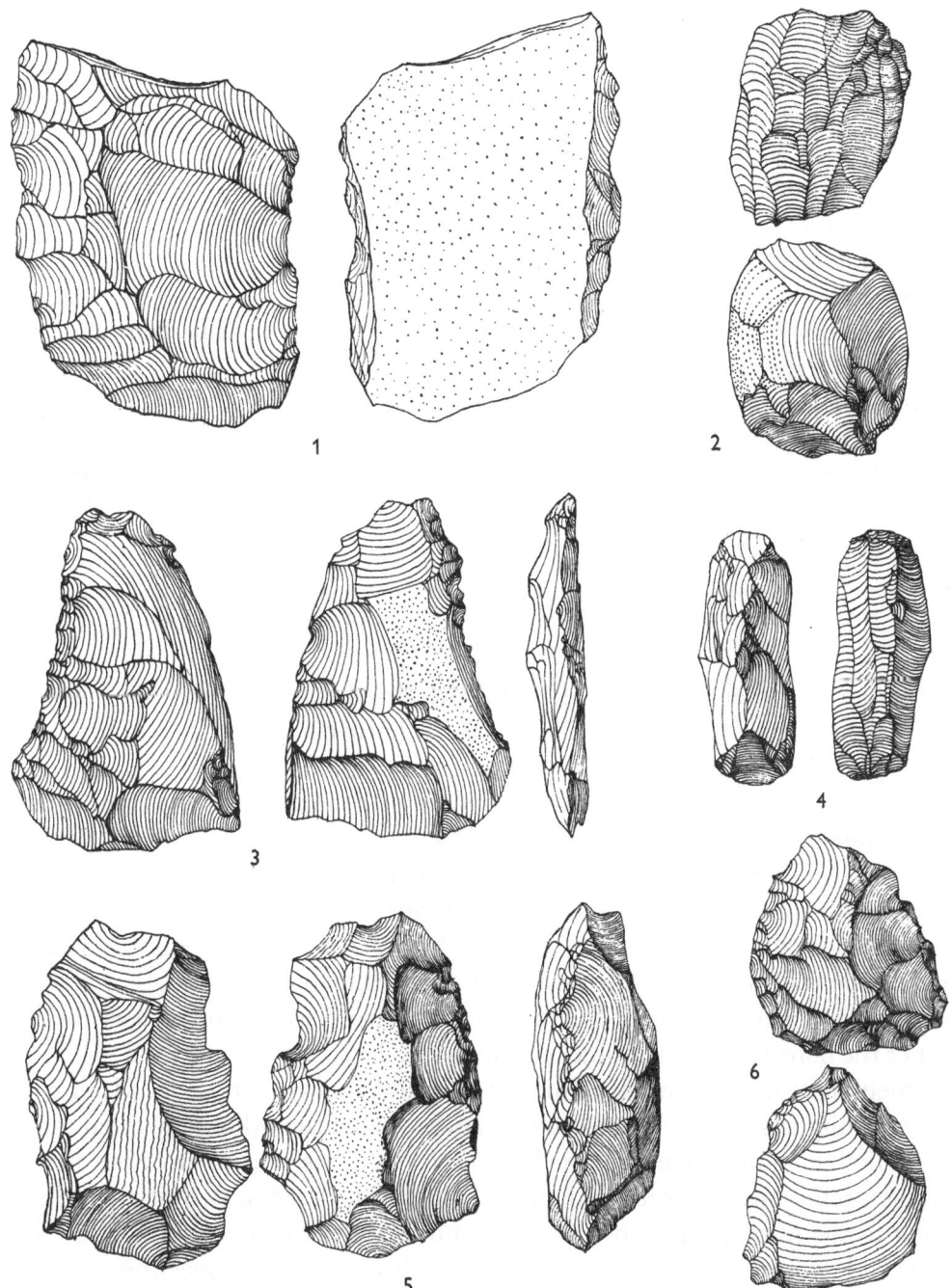

FIG. 31. Hagfet ed Dabba: representative series of nuclear artifacts
from Layers II and III
Nos. 1, 3, 5 and 6, flake-cores, possible prototypes of Libyan Neolithic adze;
Nos. 2 and 4, blade-cores. All × $\frac{2}{3}$.

Burins (20). The majority are angle-burins, but there are also a few single-blow and polyhedric pieces.

Transverse-burins (5). These, together with eighteen spalls, are probably derived from the lower series, since they occurred only in those areas near the entrance where this layer was in contact with or close proximity to Layers IV–VII.

End-scrapers (29). As in the two immediately underlying layers most of the representatives of this class are perfectly typical. In a few cases the terminal retouch may be straight or irregular in outline.

Miscellaneous scrapers (32). The details given for Layer II B apply here also, and no new types can definitely be identified.

Squamous flakes (5). Two are thoroughly typical and the remainder doubtful.

Awls (?2). No tools could definitely be attributed to this type with more than moderate probability, and even these were very rare throughout the whole succession. The two specimens noted here are of this description and made on the tips of blades.

4. THE ARCHAEOLOGICAL CONTENTS OF THE LAYERS: BONE INDUSTRY

Despite an extremely careful watch and the cleaning and examination of many thousands of small fragments of bone, shell, horn-core, etc., intentional workmanship could only be detected on five specimens. The first of these, a small sliver of ivory roughly pointed at one end, is of little interest owing to the fact that it came from the disturbed portion of Layer I and must be regarded as unlayered. Of greater importance are two articular extremities of bird bone severed with a clean grooved cut in the manner common to most ancient stone-using peoples, about 4 cm. from the end.[1] No trace was recovered of the corresponding central portion of the bone, but from the absence of further work on the surviving pieces it would seem that these are wasters and it was the long shaft of the bone that was the object of the work. This idea is confirmed by a fourth specimen, a central portion of a smaller bird limb-bone from Layer II, of which one extremity is severed in the above manner, while the other is worked into an awl or bodkin point by grinding (Pl. 10, no. 2). It is curious to note in view of the extreme rarity of bird-remains of any kind, that the two first specimens belong to the same portion of the right and left ulnas respectively of the same species of large accipitrine bird—probably a vulture. The fifth specimen is a small flattened fragment of bone scored with a criss-cross pattern (Pl. 10, no. 1), that finds its nearest analogy in a series of bone trinkets from the Mesolithic layers at Khartoum.[2]

[1] From Layers IV and VI respectively.   [2] Arkell (1949), pl. 52.

One or two marine shells were collected in each of the main layers, and although none show visible traces of intentional work, the fact that they are insufficient in number and size to have served as an article of diet would seem to suggest that they were collected as ornaments or trinkets of some kind. Among the genera represented are *Dentalium*[1] and *Cardium*.

ECONOMIC AND ENVIRONMENTAL CONTEXT

The evidence for the economic basis of existence of the society responsible for the remains just described and the natural environment in which it functioned, is as follows. Bones and calcareous organic remains are on the whole well-preserved at this site and afford fairly complete information regarding the principal varieties of animal food.

The percentage of specimens belonging to different food animals may be approximately estimated as:

	Layers							
	I	II	III	IV	V	VI	VII	Average
Rhinoceros	—	Traces	—	—	—	—	—	Traces
Ox or buffalo	14	11	8	5	10	5	22	9
Antelope (?)*	—	—	Traces	—	—	—	—	Traces
Zebra	12	13	26	4	5	4	—	11
Gazelle	11	7	2	3	5	4	11	6
Barbary sheep	64	68	63	88	80	88	67	74
Total specimens	(146)	(155)	(65)	(111)	(21)	(82)	(9)	(590)

* Dr A. T. Hopwood has kindly given it as his opinion that one molar can be identified as hartebeeste.

From the above there can be little doubt that the main objective of the hunters (for there is not the slightest trace of domestic animals of any kind) who visited Hagfet ed Dabba from time to time were herds of wild Barbary sheep, with an occasional buffalo or zebra, and still more rarely gazelle. Antelope and rhinoceros were killed only in insignificant numbers. Among small game, tortoise figures fairly regularly, though never in large numbers. Snail shells occur in considerable quantities, mainly *Albea candidissima*, and may perhaps have formed an article of diet. Bird remains, apart from some exceedingly rare bones of a bird about the size of a pigeon, and the vulture bones mentioned above, are virtually absent, so that fowling can have played little part in the economy.

As for the climatic and vegetational elements in the environment, it may be concluded from the presence of such creatures as buffalo or ox, rhinoceros, and

[1] This genus, it may be remembered, was used in enormous numbers by the Natufians of Palestine for decorative purposes on head-dresses.

zebra, that considerable supplies of water were not far distant at least at some times in the year. At the present time the available water supply would hardly permit the existence of the first two in Cyrenaica except in extremely few favoured localities. On the other hand, the appreciable numbers of gazelle, as compared with their virtual absence from Hajj Creiem,[1] would tend to show that although the rainfall may well have been higher than today, it can hardly have been comparable to the maximum during the Levalloiso-Mousterian occupation. A few tentative identifications of plants from small fragments of charcoal are consistent with this view, since they comprise wild olive, lentisk and date palm.[2]

5. CONCLUSIONS: STRATIGRAPHICAL

In its main outlines, the story revealed by the stratigraphy seems clear enough. The first stage is represented by a long phase of uniform deposition of a mixture of red earth and fragments of the surrounding rock cemented by calcium carbonate presumably held in solution by a fairly abundant supply of water. No certain correlation can be attempted between this and geological series elsewhere in the absence of artifacts or useful fossils common to both.

The second stage seems to be separated from the first by a considerable disconformity although uneven deposition of calcium carbonate continued to some extent throughout the various subsequent prehistoric layers. It had however, apparently ceased by historic times and no formation of the kind seems to be going on at present. Layer VIII seems to represent a period of weathering intervening between the two stages, during which little or no red earth was reaching what would then have been the southern margin of the roofed-in area, whereas the subsequent layers represent a return to conditions more nearly resembling those of the first stage, though less marked.

From Layer VII to Layer II we appear to have a fairly continuous record of deposition and human occupation without important break in either. The alternation of red earth and hearth deposits would seem, however, to require some specific explanation. This is probably to be sought in the effects of the piecemeal collapse of the outer edge of the roof which was clearly taking place on a considerable scale throughout this time, and the gradual enlargement of the chimney in the interior. But whether this was the ultimate cause or not, the basic observation seems to be that the favourite site for the hearth was moved at intervals from one part of the cave to another. During the occupation of Layer VII the position of the hearth, which must surely have been present, cannot be deduced,

---

[1] According to Dr A. T. Hopwood a simple specimen can be referred to gazelle.
[2] See Appendix H.

but Layer VI clearly represents a period when fires were constantly lighted at or near the present entrance. If we are right in assuming that Layer V is simply a lateral version of Layer IV, then the main locus of hearths at that time would have been in the region of squares Z–A4, 5 and 6, and perhaps slightly to the west.

While Layers III and II were accumulating, it was clearly customary to make the fires close to the north wall of the cave towards the east, and the final situation for hearths seems to have been in the passage leading to the inner cave, almost immediately below the opening of the present chimney (Figs. 22–4).

It may perhaps be thought that this general interpretation of the stratigraphy conflicts with the archaeological record inasmuch as the latter shows a sudden sharp change in equipment, which it is suggested represents two successive stages of the same industrial tradition. It should, however, be remembered that cultural changes *may* take place quite rapidly under the stimulus of newly-diffused ideas or changing natural conditions, whereas a short hiatus of a few centuries or even a millennium would not necessarily leave any very tangible trace in a formation of this sort.

In a word, although both deposit and cultural sequence are continuous in the sense that the former clearly represents a single phase of sedimentation and there are no sterile layers, it is by no means possible to conclude that occupation took place throughout the year or even every season. It may well have been intermittent for considerable periods, and in fact the displacement of the various hearth deposits would rather favour such a view.

6. CONCLUSIONS: CULTURAL

The evidence for the recognition of two main stages in the archaeological sequence has been given in detail in the description of the layers; it is further summarized in statistical form in Table IV. Briefly, the distinction between the two may be described as the replacement of a dominant and specialized tool found only in a few widely-separated localities east of Cyrenaica by other tool-forms, notably the end-scraper and flake-scraper common to a wider range of blade traditions. It is of some interest to compare this with the faunal evidence. Although the series are too small to be conclusive it may be noted that if Layers II–III are compared *en bloc* to Layers IV–VII an appreciably greater diversity of game becomes apparent, with a consequent slight but noticeable decrease in the relative importance of Barbary sheep. While this may be no more than an effect of random sampling, or perhaps the result of some minor climatic oscillation, a cultural explanation may also be put forward, namely that we have in the lowest layer at Hagfet ed Dabba traces of an early stage in the colonization of the region by communities bringing with them a cultural tradition evolved in a somewhat different environment, and attracted mainly by the ubiquitous source of food offered by the Barbary sheep. Subsequently, closer adaptation to the new environment could have led to exploitation of more varied types of game and concomitantly to certain modifications in equipment.

No doubt future stratified finds will throw clearer light on this problem and show which of the two explanations fits the facts more satisfactorily. The next question requiring some discussion is that of the outside affinities of the industry.

Attention was drawn in the last chapter to the evidence in favour of the assimilation of at least the Upper Layer at Hagfet et Tera to the Mesolithic cultural continuum of the Atlas littoral—the Oranian. The question naturally arises at this point whether the contrasting remains from Hagfet ed Dabba show any perceptible relationship to the only other pre-Neolithic blade tradition known in detail from North Africa—the Capsian.

The curious but very characteristic tool assemblages grouped under this name contain a mixture of traits which both in Europe and Palestine typify the Upper Palaeolithic and Mesolithic respectively. They seem to represent the stone equipment of an apparently uniform culture group occupying a small inland area measuring about 450×250 km., in the Shott el Jerid region of southern Tunisia. At least two developmental stages can be recognized, an earlier stage wholly restricted to the southern region, and a later extending some 100–150 km. farther into the hills to the north around Tebressa.

Thanks mainly to the work of R. Vaufrey, the characteristics of both stages are now known in some detail: the earlier or 'Typical Capsian' of this author is distinguished mainly by the relative abundance of large elements such as normal-sized burins and large backed blades; microlithic forms are rare and on the whole ill-defined. In the later stage—Vaufrey's 'Upper Capsian'—there is an immense increase in true microlithic forms which now fall into sharply-defined classes, while true burins and other large forms have become much scarcer. Among the various classes of microliths a prominent place is now occupied by the trapeze, and microburins—presumably in part by-products of its manufacture—are correspondingly abundant.

Table III below, calculated from figures published by Vaufrey[1] and Gobert to correspond as nearly as possible with the classification used at Hagfet ed Dabba, will suffice to convey the main distinguishing features of both stages of this tradition and the Oranian.

[1] In Vaufrey (1934), and Gobert and Vaufrey (1932), Gobert has recently contributed greatly towards more detailed knowledge of these cultures.

Comparison with the two lithicultural stages at Hagfet ed Dabba analysed in Table IV, p. 217, suggests as the closest parallel the Typical Capsian, and the following as the most important points of resemblance and dissimilitude respectively.

(i) Presence of abundant angle-burins and fairly numerous large backed blades in both.

(ii) Absence of the transverse-burin in the Capsian (or indeed in any other assemblage of the Maghreb).[1]

(iii) Complete absence of the trapeze at Hagfet ed Dabba.

The differences are probably sufficient to indicate a significant cultural divergence, while the features held in common are mainly those of any industry in process of emergence from a normal Upper Palaeolithic substratum.

TABLE III. *Percentage of different tool classes in the principle blade industries of the Maghreb*

	CAPSIAN (Ain Metherchem)	UPPER CAPSIAN (Rhilane)	ORANIAN (Ouchtata)
Large and medium backed blades	34·5 %	4 %	— %
Microlithic backed blades	36	34·5	94
Trapezes and triangles	3	15	1
Lunates	2	3	—*
Rectangles	—	—*	·25
Scrapers, etc.	7	9	1·5
Awls	—*	—*	·25
Microburins	9	28·5	1
Burins	8	4	·25

* These items, although absent from these particular sites, are represented to a small extent in other sites of the same facies elsewhere.

If we turn to the east, however, an interesting analogy to Hagfet ed Dabba can be found in a somewhat unexpected quarter, namely in the remarkable surface industry collected by A. E. Vignard at Nag Hammadi near Kom Ombo in Upper Egypt, a description of which was published in 1920. At this site the following tool types were found apparently in mutual association in a sample of about 2000 specimens:

(i) Transverse-burins, forming about 20 per cent of the total and identical to those of Hagfet ed Dabba except for a more or less even distribution between those struck from the left and those from the right (Fig. 28, nos. 1–7).

(ii) Normal burins, forming about 20 per cent of the total and again identical to those of Hagfet ed Dabba, with a dominance of angle burins.

(iii) End-scrapers, very numerous and typical, similar to those of Hagfet ed Dabba.

[1] Information kindly supplied by Dr E. Gobert.

(iv) Transversely sharpened bifacial axe-heads, providing about 15 per cent of the total (unknown in Cyrenaica).

(v) Bifacial sickles, very rare; two specimens only (unknown at Hagfet ed Dabba).

The association of the sickles with the rest of the assemblage may conceivably be accidental but that of the previous group—the transversely sharpened or 'side-blow' axes—must be regarded as highly probable, and is extremely interesting from several points of view. The type is a highly specialized one most unlikely to have been subject to convergent development. Its occurrence has been carefully studied by S. A. Huzzayin, who shows that as far as Egypt is concerned it is found only in a relatively restricted area of Upper Egypt, where it is regularly associated with a number of other elements mainly attributable to the Middle pre-Dynastic (Amratian) continuum, and perhaps indicates a regional variant of that stage.[1]

No pottery, it is true, was noted at Nag Hammadi, though similar sickles, and above all angle-burins, occurred with abundant Middle pre-Dynastic pottery at Armant. The absence of pottery at Nag Hammadi may be explained by supposing the site to be a chipping-floor rather than a normal habitation site, and in fact we know certainly at Armant that the bulk of the chipping was carried out at some locality other than the settlement itself.

Accordingly there is strong evidence for the regular use of transverse and other types of burins in at least one region of Upper Egypt during the Middle pre-Dynastic. At first sight the survival of these two significant traits to so late a period in Egypt might seem to afford some grounds for assigning a comparable date to Hagfet ed Dabba also, were it not that both traits have recently been identified in an entirely different cultural context of great antiquity in Syria, namely in a well-characterized and satisfactorily dated Lower Aurignacian industry at Abou Halka.[2] The main elements of this industry may be quoted for comparison in Table IV.

The analogy between the lower zone at Hagfet ed Dabba and Abou Halka is best seen in the relative frequency of burins of both kinds, end-scrapers, and squamous flakes. The most significant divergence is probably in the backed blades. Apart from relative frequency the Cyrenaican product unquestionably represents a far more fully integrated tool idea. Some difference in basic flaking practice also deserves a mention; the Syrian blade technique shows signs of being considerably coarser, and flakes struck in a more or less Middle Palaeolithic manner with faceted butts are reported to be not uncommon. This, combined with the presence of more or less Mousteriform flake points, imparts to the Syrian industry an undeniably archaic general appearance. It is hard to believe

[1] Huzzayin in Mond and Myers (1937), p. 223.   [2] Haller (1946).

that the highly-evolved assemblage from Hagfet ed Dabba is not separated from the Syrian industry by a long period of development. On the other hand, the important common features cannot easily be ignored either, and it is probable that the whole question of the relationship of Hagfet ed Dabba to these two finds of widely differing date must be left in suspense until further data, in particular C 14 readings, are available from both areas.

TABLE IV. *Percentage of tool classes at Hagfet ed Dabba compared with the Lower Aurignacian (or 'Emiran') of Syria*

	Abu Halka	Hagfet ed Dabba Layers IV–VII	Hagfet ed Dabba Layers II–III
Backed blades, etc.	12 %	56 %	56 %
Burins, normal	19	10	13
Transverse-burins	15	27	2·5
End-scrapers	3	3·5	11
Scrapers, miscellaneous	32	2·5	14
Squamous flakes*	1	1	4
Flake points	18	—	—
Tabelbalat points	1	—	—
Total no. of observations	444	582	315

* These are not specifically mentioned in Haller's text but are shown in his illustrations, Haller (1946), pl. II, nos. 22 and 23, for instance.

One fact alone seems to emerge with tolerable clarity, namely, that the transverse-burin is an eastern trait whose origin must be sought on the east Mediterranean coast rather than in Africa itself. This conclusion, if it comes to be generally accepted, is clearly not without a certain bearing on the wider problem of relations between the early blade industries of the Maghreb and the Levant. That biological intercourse had been established between the two areas at or before this time has long been inferred from the vertebrate palaeontology, and is now further confirmed as far as Cyrenaica is concerned by the discovery of the fossil vole described in Appendix A.

Unfortunately none of these considerations throw any light on the other problem of greatest local interest, that is to say, the relative antiquity of the Hagfet ed Dabba and Hagfet et Tera blade industries. The only hint that we can obtain in this connexion is provided by the nature of the stratigraphy at the latter site. Since no difference was disclosed by Professor Zeuner's analysis of the interior breccia containing a purely Mousterian assemblage at Hagfet et Tera, and the extension of the same formation yielding a pure blade industry a few feet away, it seems tolerably likely that at that site the Oranian style of work succeeded the Mousterian after no great lapse of time. Only the discovery of a stratified site or the close absolute dating of the two facies by other methods can be expected to settle the point effectively.[1]

[1] These lines were already in the press when further evidence became available which appears to offer a solution to this problem. The analysis of material from a sounding in the newly discovered site of Haua Fteah near Appollonia affords indication of an industrial horizon very similar to that of Hagfet ed Dabba, stratified below a second horizon with an industry typologically close to that of Hagfet et Tera. Carbon readings suggest absolute ages of 14,000–10,000 and 10,000–6,000 B.C. respectively for these two horizons.

While further data are desirable for confirmation, these results must certainly be regarded as highly probable. The carbon readings—but not the cultural diagnoses—are published in Suess, (1954). A general description of the site appears in McBurney, Trevor and Wells (1953).

CHAPTER XIV

PREHISTORIC FINDS FROM NEIGHBOURING TERRITORIES AND THE SOCIAL BACKGROUND TO THE NEOLITHIC IN NORTHERN LIBYA

Chapter XIII was concerned with the description of a hitherto unknown industrial tradition which, in its later stage, came to resemble in many respects the earliest manifestations of the Capsian of south-east Tunisia. This, together with a second, distinct blade tradition from Hagfet et Tera, provides all that we know of the more immediate predecessors of the Neolithic in Cyrenaica where indeed the latter stage is as yet very imperfectly documented. No important further discoveries have been made since the publication of a series of surface finds from the coastal plain near Tocra and Benghazi in 1947.[1] In fact, any attempt to sketch the part played by northern Libya in the growth and dissemination of Neolithic arts and crafts must of necessity be based largely on the general pattern of the distribution of comparable finds from the neighbouring territories of Marmarica and Tripolitania.

But before starting on this discussion, complicated as it is by recent discoveries and new hypotheses in Egypt and the Maghreb, it will be convenient to give some account of finds throwing light on the earlier cultural stages in Tripolitania and Marmarica also since these, after all, provide the social background required to see the later events in their proper perspective. The present chapter is designed to fulfil this purpose.

I. THE EVIDENCE FOR THE PRE-NEOLITHIC CULTURES OF THE MARMARICAN COAST AND THE SIWA DEPRESSION

Apart from the oases of Kharga and the Fayum in the immediate vicinity of the Nile Valley, indications regarding the early hunting communities of the region east of Cyrenaica are of extreme rarity. Some specimens of Lower and Middle Palaeolithic typology are understood to have been collected from surface sites along the northern fringe of the Sand Sea south and south-east of the Gebel Akhdar.[2] These are apparently sufficiently characteristic to indicate fairly definitely an occupation of this region at that time, though nothing further can be said until the publication of the specimens in question. A few specimens of Middle Palaeolithic character were obtained by the writer on the surface of a

---

[1] See p. 269n.   [2] Verbal information kindly imparted by Professor O. Myers.

gravel spread near the Gulf of Ghazala in 1942,[1] where their fresh condition can probably be taken to indicate that they are subsequent to the formation of the deposit. A cursory examination of one or two localities further to the east gave no indication of finds in that area.

Inland to the south the only data available come from the large surface collection formed by Dr C. Willett-Cunnington in 1918, described in full in the next chapter. At this point attention need only be drawn to one possible example of an unfinished hand-axe of Acheulean style in chert, a group of large discoid cores in an extreme state of mechanical and chemical disintegration, and what appears to be a heavily sand-blasted fragment of an Aterian point. A number of coarse flakes and cores in the same collection, judging by their fresh state and by the occurrence of similar specimens in undoubted association with mining activities at Kharga and elsewhere during Neolithic and later times, need not belong to an earlier period at Siwa either.

2. PRE-NEOLITHIC FINDS IN SIRTICA

In the 650 km. of deeply indented coast intervening between Cyrenaica and Tripolitania proper, the only prehistoric finds to be described in detail, so far as I am aware, are those included in my survey of war-time observations published in 1947. Two well-marked habitation sites were identified which both yielded samples of a highly characteristic microlithic industry of an evolved kind.

MARBLE ARCH AERODROME

This site lies some 250 km. south-west of Hagfet et Tera and the western extremity of the Gebel Akhdar. The country between the two consists of a flat and arid plain with a line of mobile dunes along the modern shoreline. Along the southern shore of the Gulf of Sirte these give way for the most part to a well-developed ridge of dune limestone, perhaps of the same date as the corresponding formation in Cyrenaica. On the crest of this, immediately behind the present shore and close to the aerodrome, a closely concentrated working-floor was discovered in 1942. The remains comprised very large numbers of microlithic chippings, minute implements, a few small fragments of bone, marine shells, and fragments of ostrich eggs. Apart from one or two sherds of Roman or later pottery there was little reason to suspect the intrusion of objects foreign to the original settlement. The area occupied measured about 15 × 6 m.

The main features of the industry were as follows:[2]

Cores (30). The specimens collected are all very small, ranging from 2–3 cm.

---

[1] In the Wadi et Tmimi, see McBurney (1947).
[2] Further details of this and the next two finds are given in McBurney (1947).

in the residual state. Two types seem to be represented: a roughly globular or polyhedric variety resulting from the production of oval flakes 1–7·5 cm. long by alternate flaking from a curved ridge, and a rather rarer conical type for the manufacture of rather rough blades. Judging from surviving fragments the maximum length of these can seldom have been much over 5 cm. Taken as a whole the general standard of flaking was distinctly of a low order, though no hints of faceted platforms or other archaic features were present. At a rough estimate the ratio of flakes to cores is about 50 : 1.

Blunted backed bladelets (20 per cent). There are no traces of large blunted backed knives, and all the surviving examples are of microlithic proportions. Two classes can be distinguished—a simple bladelet 2–3 cm. long by 0·5–1 cm. wide with a straight line of steep backing down one side (frequently directed from both faces in the ordinary Gravette manner), and a rather narrower implement with bilateral trimming to produce a sharp point at one end, averaging somewhat smaller than the foregoing.

Geometric microliths: (a) Transverse arrow-heads (4 per cent). Clearly-defined geometric forms are lacking, but a rather variable class of artifact, D-shaped or triangular in outline, formed by steep backing leaving a chisel edge on one side can probably be interpreted as serving this purpose. The range in maximum dimension is about 0·5–1·5 cm.

(b) Lunates (3 per cent). A few probable examples of this class are rougher in finish than is usual in industries where they form an important element, and there is no sign of the microburin or notch technique in making the points. The range is from about 1·5 to 2·5 cm.

Round-based points (5 per cent). This implement, which provides the most distinctive feature of the industry, consists of an elliptical piece of blade averaging about 2 cm. long by 1 cm. wide, with one carefully-rounded extremity produced by fine trimming, and the opposite end coming to an extremely sharp triangular-sectioned point, formed as a rule by the intersection of a negative 'microburin' facet with the primary edge of the blade. The trimming at the butt is mainly unifacial but usually includes a few squills to remove irregularities from the bulbar surface. The whole device may thus be said to represent a remarkably novel and specialized production of its kind, and although the microburin technique is of course frequently used to produce sharp points on a wide range of microliths, the writer knows of no exact parallels (apart from the derivative form at Siwa to be described in the next chapter).

Microburins (28 per cent). These show no features to distinguish them particularly from the corresponding class in most microlithic industries. The range is from about 0·5–2 cm. in length, and it may be noted that blade tips predominate over butts, and both are far more numerous than central portions showing two facets, though the latter do occur. Negative fractures occur only on the round-based points, and the close correspondence between the mean angle of incidence and length of the fracture in the two forms strongly suggests that the microburins are merely a by-product of the manufacture of the round-based points. The discrepancy in frequency can presumably be explained by the character of the site as a working floor from which a high proportion of finished tools were removed for use elsewhere.

Scrapers. A number of relatively thick flakes show coarse secondary work along about a quarter of their periphery, which may indicate their use as some sort of rough scraping tool, though they are by no means typical of the class and there are no definite end-scrapers.

SIRTE

All the above forms are represented by characteristic examples obtained *in situ* in a mobile dune deposit exposed in a small excavation at the side of the road some 8 km. west of Sirte, over 200 km. west of the preceding site. Apart from the inclusion of a few rather larger backed blades, and a fragment of a double-backed point no significant difference could be seen between the two. Organic remains at both sites included marine shells and fragments of ostrich egg shell, one at Marble Arch showing traces of deeply-engraved parallel lines.

Briefly, there can be little doubt that these finds demonstrate the existence along the whole southern shore of the Gulf of Sirte of food-gathering communities, dependent to some extent on the marine environment, and practising a homogeneous and not unspecialized material culture, at some time subsequent to the last low sea-level. As for the origin of their technological tradition, it would seem that this may well be in part local and in part connected with the pre-Neolithic tradition of the Maghreb, if account be taken of the presence of the somewhat attenuated version of the transverse arrow-head (of which this is the most easterly record so far in Libya) and the engraving of ostrich eggs. For this tradition the name of the Sirtican Culture is now proposed.

3. THE PRE-NEOLITHIC CULTURE STAGES OF TRIPOLITANIA

In Tripolitania the sequence is rather more complete since finds of both the Lower and Middle Palaeolithic as well as a late stage of the Mesolithic are now known. From a topographical point of view the whole of northern Tripolitania may be said to consist essentially of a wide tableland separated from the sea by a varying width of coastal plain, and marked by two main systems of drainage, one directed eastwards into the Gulf of Sirte, and the other northwards into the waters immediately east of the Gulf of Gabes. The first two sites to be described belong to the eastern watershed.

GRAVEL TERRACE NEAR BIR DUFAN

The finds were made on the surface of a coarse alluvial, perhaps fluviatile, deposit near the head of a small tributary of the Wadi Merdum, itself a tributary of the Wadi Sofeggin. The latter rises far inland to the west and empties into the salt flats of Taorga on the north-east margin of the Gulf of Sirte. The site lies about 100 km. from the coast, in a region of considerable aridity at the present time. The collection brought back from this locality[1] comprises a representative selection of the many thousands of large artifacts which strew the site over an area of several acres. With a few insignificant exceptions the forms are all those of the Middle Palaeolithic and earlier stages. Preservation varies from a patinated but unabraided state to one of such extreme chemical and mechanical disintegration, that it is likely that a proportion of the artifacts originally present have since been rendered totally unrecognizable.

This wide variation in state, coupled with the signs of water transport in the deposit, led to the classification of the series into four arbitrary categories of preservation, the assumption being that a part at least of the decomposition was due to the ancient climatic events responsible for the formation on which the series lay.

The first group contained all specimens showing heavy wear and deep chemical corrosion, stained a dark red-brown. Typologically these comprised a series of large cores made on rounded cobbles flaked alternately on either side of a ridge in a manner essentially comparable to the European Clactonian, massive flakes with wide plain platforms at a high angle, and a few crude hand-axes.

The second group showed a degree of chemical change that might be described as 'deeply patinated' rather than corroded, but was still considerably abraided and stained. It included a number of flakes with unmistakably faceted platforms at between 90° and 100° inclination to the bulbar surface, considerably more elaborate dorsal patterns, and one or two cores showing clear traces of tortoise-core technique.

A third group, entirely free from mechanical abrasion, largely unstained, and only moderately patinated, contained a considerable series of much smaller and more carefully worked cores comparable in technique to Mousterian disc-cores, one or two in particular being less than 5 cm. in diameter with the characteristic cortex patch on the base. Some of these were made on reworked specimens showing flaking in one of the earlier states so that there could be no doubt that

[1] McBurney (1947), pp. 61-9.

223

possible differences in the chemical constitution of the specimens could be ruled out as a possible cause of their differing states of preservation.

Finally, the freshest group, comprising pieces which had undergone only a slight degree of patination, included similar cores plus neatly-made side-scrapers and points.

These results combined with the observation that specimens in the earlier states were many times commoner than those in the later and fresher states, suggested that the former were at least as old as a period of greater chemical and mechanical activity in the area, that is to say a period of greater rainfall. Granted that on typological grounds the freshest series can hardly be much less than 10,000 years old (since it is difficult even on the shortest chronology to imagine the replacement of the Middle Palaeolithic by the earliest blade industries much later than 8000 B.C.), the small amount of change produced in them by exposure for this length of time would seem to imply a disproportionate difference in age for the earlier specimens, without the intervention of climatic conditions differing appreciably from those of the last 10,000 years.

It is interesting to compare this final result with that deduced by Miss Caton-Thompson from her observations in the oasis of Kharga in southern Egypt, some 1800 km. to the east-south-east. Here an archaeological sequence showing a gradual transition from late Acheulian to full Middle Palaeolithic started in a period of markedly greater rainfall, but the final stage of the Middle Palaeolithic cycle was not reached until after a prolonged period of increasing desiccation. It is true that the different groups at Bir Dufan are all noticeably cruder in workmanship than the corresponding finds at Kharga, but although it would be difficult to demonstrate the point satisfactorily without elaborate experiment, it seems not unreasonable to explain this mainly as the result of the difference in raw material. The chert used at Bir Dufan is coarser-grained and certainly flakes far less easily than the fine flints of Egypt and eastern Libya.

WADI BEI EL KEBIR

Further evidence of the survival of a later Middle Palaeolithic type of industry in Tripolitania into a period of relative climatic stability comparable to the present, is provided by a small but very characteristic series in the Levalloiso-Mousterian style from the surface of a gravel terrace in the upper reaches of the Wadi Bei el Kebir, 160 km. south-east of Bir Dufan, since these also were perfectly fresh and free from any traces of abrasion.

It will be noted that no mention has been made in the above record of the Aterian. The presence of this stage over a wide area in northern Tripolitania is in fact attested by a number of loose finds of typical tanged points, though no more can be said of its relative stratigraphical position than in Cyrenaica. The remarkable closed find from the Wadi Gan on the northern watershed, about to be described, does, however, yield a detailed picture of the lithic elements in at least one local version of this tradition.

WADI GAN—THE ATERIAN STATION

A collection of material from this locality, together with explanatory manuscript notes from which the following details are abstracted, were presented to the National Museum of Southern Rhodesia by their discoverer, Mr K. S. R. Robinson, in 1946. They were subsequently presented to the Museum of Archaeology and Ethnology at Cambridge, and Mr Robinson has kindly allowed the following description to appear in the present work.

The locality is one of a number of wadis on the northern slopes of the Gebel Tarhuna, some 15 km. south of the village of Garian. Two implementiferous horizons were identified in close proximity on opposite sides of the wadi, exposed in alluvial deposits by recent flood-water erosion.

The deposit occurs in the left (north) bank of the wadi. The specimens were excavated from a well-defined cultural horizon 1·25 m. above the visible base of an alluvial deposit, and overlain by up to 10 m. of the same formation. No organic remains are reported.

It seems clear both from the nature of the collection, its state of preservation, and mode of occurrence, that we have here the remains of a single chipping floor or camping site.

The collection as received at Cambridge contains the following elements:

Cores (11). Two are characteristic miniature tortoise-cores of proportions usual in the Levalloiso-Mousterian, e.g. 5·5 × 5 cm. (Fig. 32, no. 8) and 4·8 × 4 cm., with main scars 4·2 × 3·6 cm. and 4·3 × 4 cm. respectively. A third specimen on the same pattern is unusually small, 3·2 × 2·9 cm., with main scar only 2 × 1·9 cm. (Fig. 32, no. 9). Since among the smallest tools are some only 3·1 × 2 cm. and 3·2 × 1·5 cm. there seems no reason to doubt that this is simply a core rather than a primary product of some kind. Two fairly characteristic examples of simple disc cores are available, measuring only 3·7 × 3·7 × 1·3 cm. and 3·0 × 2·8 × 1·8 cm., together with three larger more irregular pieces, one subsequently trimmed into a rough awl.

Clearly distinct from the above are two well-defined prismatic cores from which thick but narrow 'flake-blades' (to use the convenient phrase coined by Miss Caton-Thompson) have been struck apparently by direct percussion with a hard hammerstone (Fig. 32, no. 7).

There is also a small globular alternate core (Fig. 32, no. 14). Only two cores

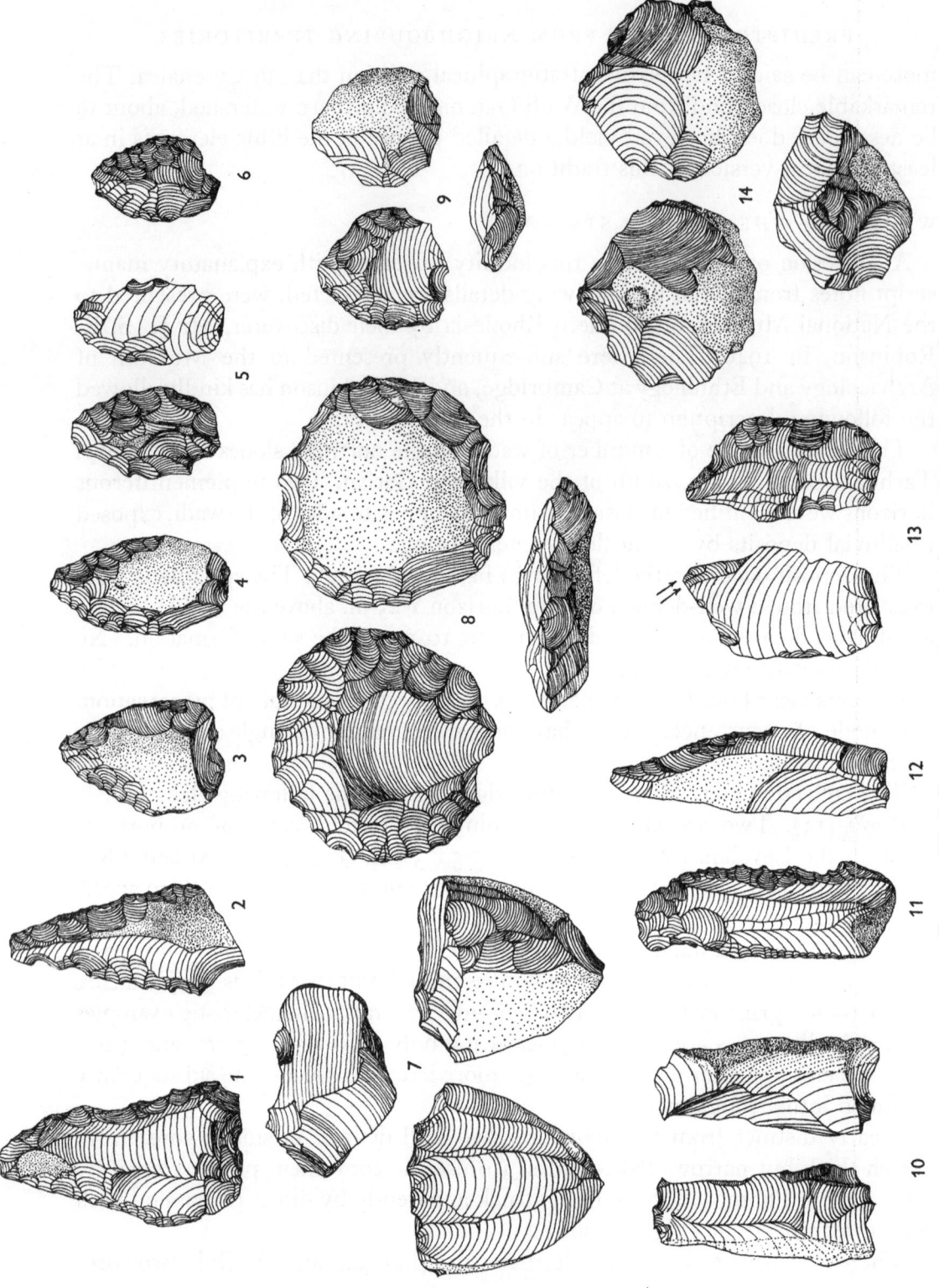

FIG. 32. Wadi Gan: representative series from main terrace deposit

Nos. 1–6 points, No. 5 with bifacial working; Nos. 7 and 10, flake-blade cores; Nos. 8 and 9, miniature tortoise-cores; Nos. 11 and 12 trimmed flake-blades; No. 13, burin; No. 14 flake-core. All ×⅔.

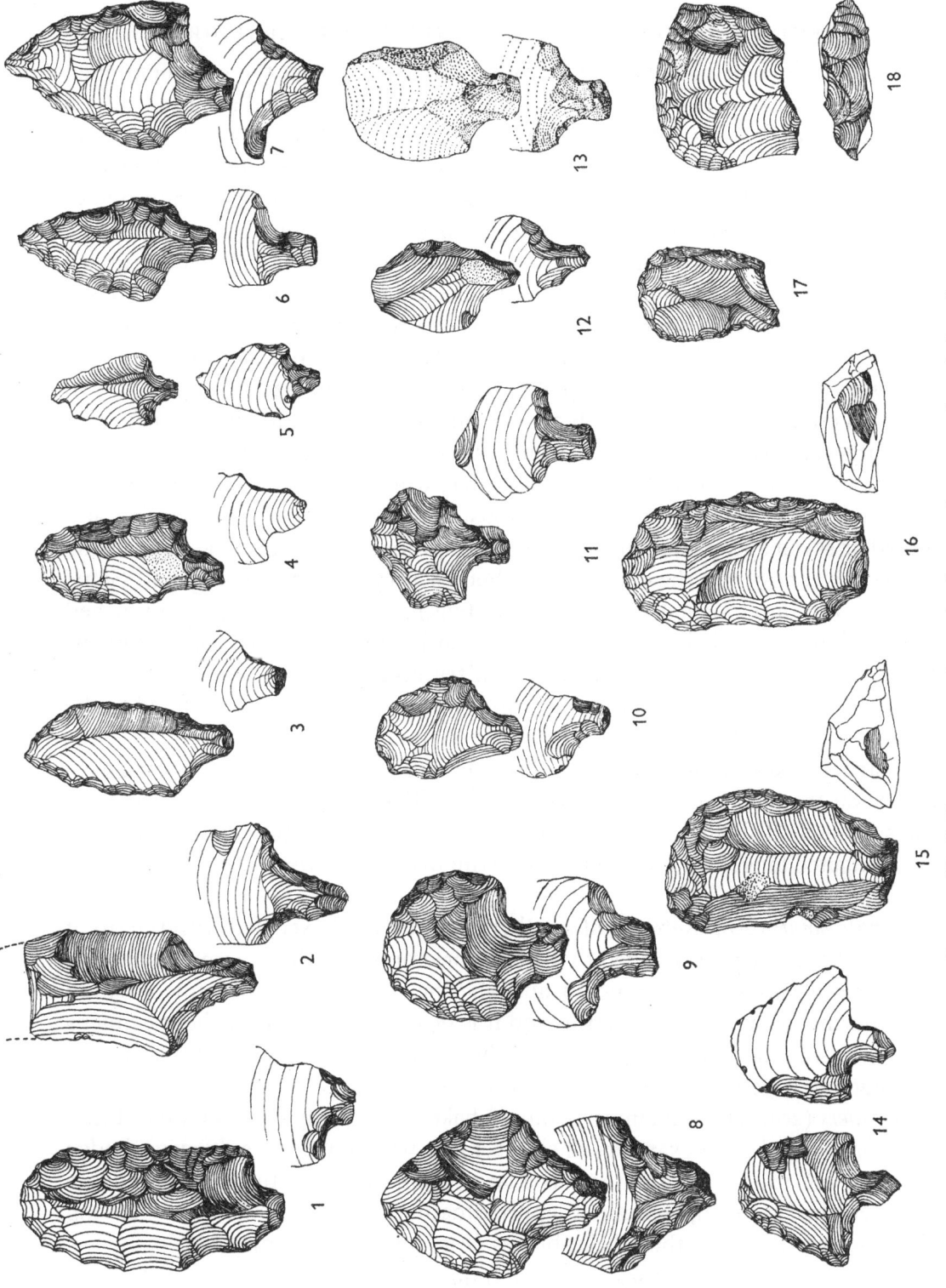

FIG. 33. Wadi Gan: representative series from main terrace deposit (*cont.*).

Nos. 1–14, tanged implements; Nos. 15 and 16, steep-scrapers; No. 17, end-scraper; No. 18, side-scraper. All ×⅔.

227

show clear signs of having been intended for use as tools, the one mentioned above and a flake-core trimmed to resemble a rough carinated scraper.

Flakes—unretouched (26). The general indication of at least two basically distinct methods of flake production suggested by the above details of the cores, can be confirmed by a study of the flakes themselves. Six have wide finely-faceted platforms and typical convergent dorsal preparation showing that they were derived from normal disc cores, while a single example with eleven dorsal scars clearly results from a true tortoise-core technique. Nine elongated flake-blades were evidently struck from prismatic cores, but the remainder are too irregular to afford a clear impression of the flaking system aimed at.

Tanged implements (26). Complete specimens vary in size from $3 \cdot 0 \times 1 \cdot 8$ to $5 \cdot 5 \times 3 \cdot 8$ cm., with average length $3 \cdot 5$ cm. and average breadth $2 \cdot 5$ cm. (Fig. 33). The largest (broken) piece must have measured over 4 cm. in width and 6·9 cm. in length. Six have lateral and terminal secondary working evidently designed to give the distal portion the general form of a Mousterian point (Fig. 33, nos. 6 and 7), and a further seven specimens, though damaged, probably belong to the same class.

In twelve, however, the outline of the body is roughly oval or polygonal. Some of these are unretouched (except for the tang) and may be simply unfinished pieces of the pointed class, but in two cases at least this rounded shape is manifestly intentional and produced by careful trimming (Fig. 33, no. 9).

There are no true bifacial examples and the retouch is usually confined to the margin, revealing that the majority of tanged tools were made from tortoise- or disc-struck flakes. In general the outlines tend to be irregular and the workmanship is on the whole of rather inferior quality. There is no positive evidence for the so-called 'Tabelbalat' class of tool and only one possible representative of the elongated tanged end-scraper class distinguished by Miss Caton-Thompson[1] (Fig. 33, no. 1).

Simple points (20). Retouch of a rather irregular kind executed with the tip of a pointed percutor was applied to produce or emphasize the triangular shape of a number of flakes (Fig. 32, nos. 1 and 2). The class is less well-defined than in most normal Mousterian industries and it is possible that some of the specimens are unfinished tanged points, though others seem unsuitable for this purpose. Average measurements are $5 \cdot 0 \times 3 \cdot 0 \times 0 \cdot 5$ cm.

Scrapers (20). The remaining trimmed flakes are mainly of two kinds. First, there are several good examples of side-scraper with a wide, evenly curved edge (Fig. 33, no. 18). Secondly, there is a large group of small, thick, stubby flakes with terminal retouch, sometimes approximating to true end-scrapers—among the latter are a few on thick flake-blades (Fig. 33, nos. 15–17).

[1] Caton-Thompson (1946*b*), p. 36, n. 3.

Saws (6). A few flake-blades have lateral retouch that appears unsuitable for scraping and may perhaps have been intended for filing or sawing (Fig. 32, no. 11), but there are no cases of the regularly denticulated edge found in some variants of the Capsian, for instance.

Burins (1). The only specimen included in this collection can be classified as a 'burin ordinaire', and probably accepted as intentional (Fig. 32, no. 13).

Bifacial tools (1). A single roughly-worked foliate of very small size—4·2 × 2·7 cm.—apparently unfinished, is the only representative of this class, which is fairly common elsewhere in the Aterian (Fig. 32, no. 5).

While the assemblage just described is probably too small to provide reliable statistical information, from its general character and composition it is clear that it corresponds closely enough to the last main stage of development proposed by Miss Caton-Thompson in her general study of the culture.[1] (The highly specialized facies so far only known from the Mughharet el 'Alyia in Tangier[2] would presumably provide a still later and more evolved stage in that region, though the extension of its geographical distribution is as yet entirely conjectural.)

WADI GAN—THE INTERGETULO-NEOLITHIC SITE

On the opposite side of the wadi Mr Robinson discovered the only other closed find so far known from this territory. As will appear from the details given below, it seems probable that this collection provides us with our first glimpse of the immediate predecessors of the local Neolithic; as such it is of not inconsiderable interest in the present connexion.

The remains are contained in a clearly defined cultural layer exposed in the vertical side of a small artificial cutting to accommodate a path. The layer consists of a black earthy deposit about 30 cm. deep, containing considerable quantities of charcoal, ostrich egg shell and flint implements, resting on a reddish sandy alluvial deposit apparently representing the same formation as that containing the Aterian remains on the other side of the wadi, and covered by about 30 cm. of light-coloured earthy deposit containing boulders, probably a hill wash or slope deposit.

The character of the industry is conveyed by the following critical inventory, though the writer is not certain, from the details supplied with the collection, how far it can be relied upon to provide a true statistical picture of the relative frequencies.

Cores (8). All except one are clearly the by-products of prismatic blade production with fairly definite signs of the use of a punch. Two in particular show a high standard of technique, and apparently started from small cylindrical

---

[1] Caton-Thompson (1946*b*).   [2] Howe and Movius (1947).

nodules using a single platform and a few preparatory squills after the method described by Barnes.[1] One is rather coarser and seems to have been partly intended for small flakelets.

Raw flakes (34). These are all more or less blade-like. Some showing cortex are rather thick and wide but there are several others very narrow and regular (though the whole industry seems to be on a small scale). Nearly all show slight traces of utilization.

Scrapers (16). This category includes four short end-scrapers on stubby flakes (Fig. 34, nos. 23–26), one small double-ended end-scraper, one small limace (Fig. 34, no. 27) and a number of blades and flakes with irregular or slightly concave trimmed edges unevenly spaced along the periphery (Fig. 34, nos. 11, 13 and 18–20).

Saw (1). A small blade shows an evenly denticulated convex edge similar to those well known from various Maghrebian sites and generally referred to as 'scies' in the French publications.

Backed blades (21). All are very minute microliths except two fragments of possibly rather larger specimens. One of the latter, backed from both faces, is very narrow and has a triangular cross-section (Fig. 34, no. 15). The remainder are all of the evolved Capsian type, 1·5–2·5 cm. long by 0·4–0·8 cm. wide, frequently coming to a very sharp tip (Fig. 34, nos. 5–9). A single specimen showing pressure flaking, is iron-stained and deeply patinated (as opposed to the rest of the collection, which is perfectly fresh) and is doubtless intrusive.

Awls (2). Both on the tips of blades, rather coarsely worked, and only just recognizable.

Ostrich egg shell beads (2). Among several fragments of egg shell are two somewhat burned, one a fragment of a finished bead and one a small piece of irregular outline with a small bored hole—presumably an unfinished example of the same type.

Typologically similar surface-finds of identical materials, in the same physical condition and from the immediate vicinity, confirm the above description. Isolated pieces of this kind, made of the same type of raw material and in a similar state of preservation, were found loose at two other localities in the district. No other finds that can be attributed with any confidence to a pre-Neolithic phase of the blade tradition have so far been discovered in Tripolitania, so that the question of the outside affinities of this assemblage is of some importance. Comparison with Professor Vaufrey's well-known summaries of the different facies of blade industries in the Maghreb[2] suggests that only one assemblage offers a satisfactory parallel, namely, the so-called 'Intergétulo-Néolithique'.

[1] Barnes (1947), p. 109.   [2] Vaufrey (1933), p. 469.

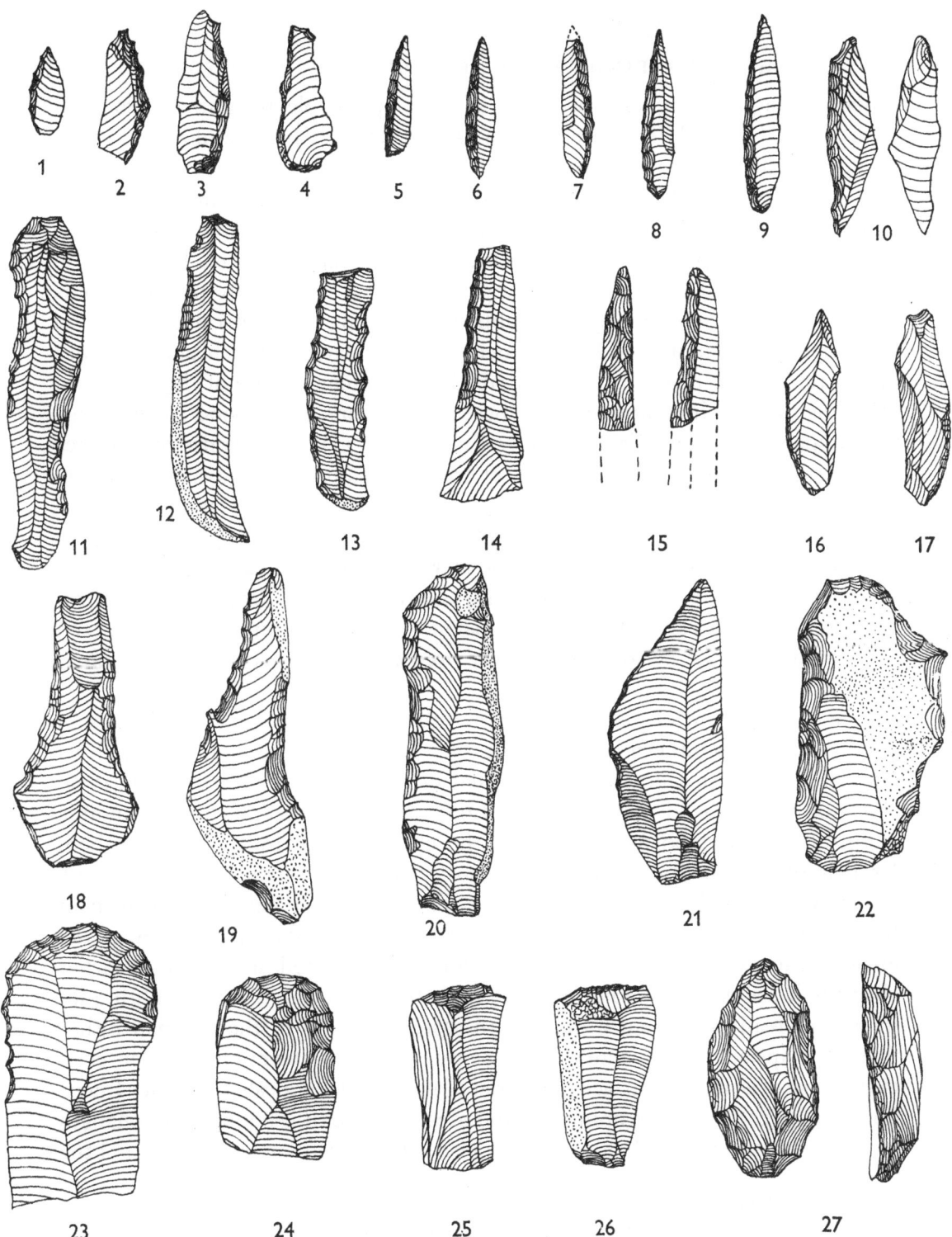

FIG. 34. Wadi Gan: representative series from late deposit overlying
main terrace formation, Intergetulo-Neolithic

Nos. 1–9, microlithic backed blades and flakes; Nos. 11–14 and 18–20, miscellaneous unilaterally
and bilaterally trimmed blades; No. 15, fragment of Gravette type blade; No. 21, Capsian type
backed blade; No. 22, flake-scraper; Nos. 23–26, end-scrapers, No. 27, limace. All actual size.

Gobert's original diagnosis of this tradition, applicable to a small group of sites in the Shott el Jerid area, reads as follows:[1] 'L'Intergétulo-néolithique est un Gétulien (Capsien) évolué où la lame à tranchant abattu est devenue rare (sauf sous sa forme microlithique) et le burin absolument exceptionel: c'est quelque chose comme un Azilien barbaresque.' Vaufrey adds in the course of a description of Lala—taken to be a typical site of this kind: '...en effet, si les grattoirs, generalement larges et courts, sont encore plus ou moins nombreux, les grandes lames à dos rabbatu et les burins lateraux ont disparu ou se comptent par unités.' The presence of denticulated retouch and irregular lateral retouch on blades is also described as a regular feature of this industry, and illustrations show how closely these descriptions correspond to the equivalent specimens from Wadi Gan (see Fig. 34, no. 13). 'Les vraies microlithes géometriques, triangles équilateraux et trapèzes, sont rares, les scies peuvent-être nombreuses.' Such differences as do occur between the Tripolitanian and Tunisian site appear to be of a less significant kind—namely the greater rarity of microburins (only one atypical example at Wadi Gan (Fig. 34, no. 10)) and the virtual absence of true 'pointes scalènes', and well-defined 'lames-à-troncature-oblique'—and scarcely affect the conclusion of a close cultural affinity between the two areas in the period immediately preceding the emergence of the Neolithic.

It may not be out of place at this juncture to point out that the title chosen for this facies in southern Tunisia is in some respects misleading. Although its late date is stratigraphically demonstrated in at least one instance,[2] there is no evidence whatever to suggest association with the appearance of economic culture-traits for food-production, in the characteristic form just described. Its relationship to the Capsian is at most that of a distant collateral to the Upper Capsian, since by far the most distinctive and persistent feature of that tradition in all its stages, the trapeze, is virtually absent. In view of the greatly extended distribution of the Oranian suggested in this and earlier chapters it seems reasonable to inquire whether the 'Intergétulo-néolithique' might not really represent a final south-eastern version of the Oranian rather than the Capsian. At any rate we have as yet no hint whatever of the latter, even in the form of scattered surface finds, east of its area of classical occurrence (mentioned on p. 214).[3] In the opinion of the present writer it is not until the changes in technique and material culture associated with the economic revolution of the

---

[1] Quoted in Vaufrey, *op. cit.*   [2] Gobert (1912).

[3] Unless, as Dr Gobert has suggested in a letter to the writer, the upper zone at Hagfet ed Dabba is to be regarded as ancestral to the Capsian. The difficulty is the complete absence of trapezes both in this deposit or anywhere else east of the Maghreb as far as the Nile Delta.

Neolithic that we have any solid evidence of the spread of peculiarly Capsian ideas, notwithstanding much that has been said and written to the contrary over the last fifty years.

CONCLUSION

At this point it may be convenient to summarize the conclusions regarding the industrial traditions of pre-Neolithic character in northern Libya as a whole. During the Lower Palaeolithic the fragmentary observations obtained supply further confirmation for the idea suggested by the evidence from Kharga, and also by the general distribution of finds of this type in the western Sahara, namely, that the Acheulian must be associated with one or more periods of considerably increased rainfall. It is clear that during this period or periods an appreciable degree of cultural uniformity was established throughout North Africa, among communities inhabiting all the more fertile regions of the present day, and many that are now wholly arid.

At Kharga, as mentioned above, some evidence has been recorded which tends to show that the subsequent Middle Palaeolithic tradition was the product of gradual evolution *in situ* from the Acheulean; as far as it goes, the new evidence from Wadi Merdum in Tripolitania seems to suggest a similar phenomenon. In Cyrenaica a later more evolved phase of the Middle Palaeolithic, the Levalloiso-Mousterian, can be accurately dated to a period of pluvial increase apparently to be equated with the first cold oscillation of the Last Glacial maximum in Europe, and displays close affinities to a Palestinian equivalent of perhaps comparable date. How far west this facies may have extended, or whether it is in fact typical of North Africa as a whole we do not yet know for certain, though as stated above, the Wadi Bei el Kebir finds can almost certainly be assigned to it.[1]

The final phase of the Middle Palaeolithic cycle, the Aterian, provides an indication of activities at once so specialized in character and so unvarying throughout their geographical range of occurrence, that it is possible to speak with confidence of a true cultural continuum extending across the whole of North Africa from Morocco to the southern Egyptian oases. In this connexion, the new evidence from Tripolitania and Cyrenaica serves merely to emphasize the ubiquitous nature of this spread throughout the area in question, rendered in any case a practical certainty by the character of the remains at Kharga.

---

[1] The suggestion that a number of stations in the Maghreb can also be assigned to this facies, needs, I think, to be accepted with caution, owing to a certain amount of confusion regarding the true nature of the traits which differentiate the Levalloisian from the Mousterian in Europe, on which the new appellation is based. The criteria, as pointed out in connexion with Hajj Creiem, though perfectly objective, require to reach a definite statistical standard before they can be regarded as conclusive. Different criteria for this purpose proposed by F. Bordes (1950) will be discussed elsewhere.

At the same time it is interesting to note how closely the new site from Wadi Gan follows the well-known pattern of previous finds in the Maghreb down to almost the last detail. The limits of this distribution are almost equally significant, for the number of discoveries is now such that the chances of future research revealing its presence either south of the Sahara or eastwards in Palestine, would now seem to be small.

In North Africa, as in Europe, a third broad industrial epoch can be recognized in which the whole pattern of stone-working and the design of tools were profoundly modified by the new technological device of true blade-making. Although much experimental work still remains to be done on this subject (and it is more than doubtful whether any living civilized man can reproduce at all accurately the process, which only a few millennia ago was practised almost throughout the inhabited world), investigation of records of recently extinct primitive peoples has produced some highly suggestive observations. The upshot of these[1] is to suggest not merely that the removal of true blades from a core cannot be performed without a carefully developed manual skill, but that further prerequisites are an intermediate tool or punch with definite characteristics, and finally that appreciable skill is required in the preparation of the core before the process is even started.[2]

That a device of this nature should have been independently discovered on a number of different occasions is no doubt conceivable, as with every important invention, but that it should have arisen simultaneously throughout a wide area as a natural development of ordinary flaking methods seems, as a result of these new data, to be distinctly improbable. Whether the spread of the new type of equipment was accomplished by a gradual process of diffusion, or resulted from large-scale ethnic migration, is of course another matter, though in many areas a complete break with earlier traditions seems to favour the latter alternative. In any case the main concern of the present discussion is with the date at which the change took place rather than with its exact nature. In Cyrenaica we have seen that a *post quem* date is supplied by the Sahel alluvium, apparently corresponding to the second maximum of the Würm, while the latest radiocarbon results make it nearly certain that the Dabba variant was fully established there by the fourteenth millennium B.C. It is likely that an interval in the order of 10,000–20,000 years separates these two events, but within these limits

---

[1] This, at any rate, is how the writer would interpret the data collected by the late A. S. Barnes.

[2] Among American Indian tribes it was customary for some men to specialize in this part of the work and some in the actual removal of the blades. In any case it was all man's work, whereas the pressure-flaking of arrow-heads, often practised by the women, was regarded as a much simpler process, carried on with the aid merely of a piece of leather and a hand-held pressure-flaking tool.

we are as yet unable to give the date of the latest Middle Palaeolithic[1] or the earliest Upper Palaeolithic in Cyrenaica.

In the Maghreb the Upper Capsian according to the latest radio-carbon readings available,[2] was being practised during the period 6400–5000 B.C.; how much earlier it began or later it lasted we have as yet no means of telling. Nor is anything yet known of the date of the replacement of the Aterian by the Typical Capsian, since the latter has not yet been recorded from a geologically datable deposit of Pleistocene character. On the whole the balance of probability, having regard to the typological character of both phases and their close evolutionary connexion, may perhaps be said to favour a slightly later date than that of the Dabba variant in Cyrenaica, though nothing short of a proper stratigraphical sequence with radio-carbon dates can settle the point.

In general the climatic conditions accompanying the blade industry of early hunting peoples in North Africa, seem to have been characterized by progressive and marked desiccation. Since this would inevitably have led to gradual isolation of communities and restriction of movement and intercourse, it is not surprising that the archaeological record begins at this time to show unmistakable signs of regional specialization. Thus, although a rather undifferentiated version of the Oranian coastal culture seems at first to have extended as far east as the Benghazi plain, it would appear to have been ultimately replaced in Tripolitania by the South Tunisian Intergetulo-Neolithic culture, and by the newly-discovered microlithic Sirtican Culture along the south shore of the Gulf of Sirte.

In Cyrenaica there is clear indication that the transverse-burin is of purely eastern origin, and never penetrated further to the west, while on the other hand there is a striking absence of certain extremely characteristic western traits such as geometric microliths. Further east in Egypt the microlithic equipment appears, as will be described below, to have been simpler still, while some Palestinian mesolithic traits which penetrated the area are not known west of the Nile.[3]

To generalize broadly, the sequence in North Africa of the various industrial traditions practised by the early hunter-gatherer communities, seems to present a picture of gradually increasing regional specialization. Starting with the inter-continental uniformity of the Lower Palaeolithic hand-axe equipment, there

---

[1] Suess (1954). [2] Kulp, Tyron, Eckelman and Snell (1952).

[3] This does not, of course, preclude their diffusion up the Nile to the south. It is not always realized how sharp a contrast is offered by the enormous predominance of the lunate in the Natufian—well over 50 per cent. of the total tool output in both layers at Mughâret el Wad—and its regular, but very restricted role in the Maghreb. In the latter region, there is no single recorded instance of its contributing more than 7 per cent., and less than 1 per cent. is the norm. To find an African analogy to the Palestinian state of affairs we must look to the Upper Nile at Khartoum, or to East Africa.

follows the full Middle Palaeolithic, when some continuity seems still to have been preserved with western Asia, while later still in the closing phases of the same epoch the continuum, though clearly represented, extends no further than the limits of Libya and the Maghreb. Finally, the blade industries, taken as a whole, present still more significant signs of locally restricted technological developments.

While it may be argued with some justice that the above scheme is in part the effect merely of the quantity and nature of the surviving evidence, and in part also, the direct result of climatic change; it is at the same time reasonable to see in it some reflexion of a third factor, namely that of growing economic efficiency with a tendency towards more complete adaptation to the different ecological opportunities offered by various territories. As an example one might cite the ostrich and mollusc remains associated with the Sirtican microlithic sites, as compared with the Barbary sheep, zebra, buffalo, etc., and even rhinoceros, which seem to have formed the mainstay of human life in the Gebel Akhdar.

CHAPTER XV

NEOLITHIC MATERIAL FROM NORTHERN LIBYA AND THE PROBLEM OF THE SPREAD OF NEOLITHIC ARTS AND CRAFTS ALONG THE NORTH AFRICAN LITTORAL

This final chapter is devoted to a description of two important collections of Neolithic type material from northern Libya, and a discussion of their bearing on the diffusion of Neolithic arts and crafts throughout the more habitable portions of the northern margin of the Sahara. At the present stage of the investigation, with the material available, conclusions regarding the spread and interrelations of these earliest food-producing societies in the area are necessarily of a restricted kind. The semi-desert conditions obtaining along the greater part of the littoral are for the most part unsuited to the preservation of a detailed picture of social and economic activities, and little evidence of the kind has yet been published. Only two sources of information are to be found in any quantity— rock art, and the form and technique of stone implements.

The first of these, the rock art, has been exhaustively examined by R. Vaufrey as recently as 1939,[1] and later discoveries have as yet done little to modify this aspect of the evidence. The same is by no means true of the study of flaked stone objects. In Egypt important fresh observations have been made and radically new interpretations have been put forward to explain them, while in the Maghreb one new discovery at least deserves discussion in the present connexion.

The main problem at issue may be defined thus: how far can the observable traits in the two areas of most intensive settlement—the Nile Valley and the Maghreb—be regarded as autochthonous, and how far they are best explained as the result of diffusion from one territory to the other, or from some third source? In order to assess the contribution of the new Libyan material to this problem it will be convenient to begin with a summary of the present state of knowledge and hypothesis in the two main (and geographically remote) areas.

[1] Vaufrey (1939).

I. THE TRANSITION FROM MESOLITHIC TO NEOLITHIC
IN THE MAGHREB

Mention has already been made of the earlier stages in the succession of the blade-industries of the Maghreb, in connexion with Hagfet ed Dabba and Hagfet et Tera. The two facies in question are comparatively fully studied and belong to two geographically distinct groups—an inland variant, the Capsian, and its derivative, the Upper Capsian, confined to a maximum area of some 120,000 square km. on the south-eastern slopes of the Atlas massif; and the Oranian, a strictly coastal culture, identified from Morocco to the Gulf of Gabes, contemporary at least with the Upper Capsian.

Finally, concentrated mainly in the eastern part of the Capsian province *sensu stricto*, is the stratigraphically late facies of the Intergetulo-Neolithic referred to in connexion with Wadi Gan.

Overlying all these types of tool assemblage, wherever stratigraphical juxtaposition occurs, are traces of a strikingly different industrial tradition of far wider geographical extension, the so-called Neolithic-of-Capsian-Tradition. Generally speaking, this facies shows a remarkable degree of uniformity throughout the mountainous areas of the Maghreb, though some regional peculiarities become apparent among the more southerly finds in the desert proper. The two most significant innovations of this new tradition are probably the tanged arrow-head, and a new and much improved technique of manufacture—pressure-flaking with some sort of hand-punch. This latter, which forms so universal a feature of the technology of most early food-producing societies, and has spread among many modern or recent food-gathering communities throughout the world, was quite clearly unknown to any of the earlier blade traditions just referred to.

In the Neolithic-of-Capsian-Tradition, however, it was applied not merely to the production of tanged arrow-heads, but to a variety of other forms as well. Some of these, such as the peculiar chisel-shaped arrow-head, clearly derive from the Upper Capsian; indeed, virtually all the tool types current in the earlier blade-traditions survive to some extent, and provide eloquent testimony for a gradual development of the new tradition, rather than for any sudden break with older practices.

Accompanying these changes in flint work in some of the more complete samples of archaeological material, are polished stone wood-working tools and fairly frequent traces of crude pottery. Of the economic developments to which these technological changes may have been correlated, very little has as yet been satisfactorily established in the Maghreb. On the other hand the survival of so many characteristic features of the material culture of the earlier hunting communities certainly suggests that whatever social and economic modifications took place, whether as a result of internal development or of diffusion from outside, they were of a gradual character unlikely to have been caused by sudden widespread immigration or invasion. The skeletal evidence as far as it can be relied upon apparently points to the same conclusion, though it is reported to offer some slight indication of a slow change in the physical nature of the population.[1]

It is unfortunate in this connexion that the archaeological record should be so particularly incomplete and confused in the final stages immediately preceding the historic period, since it makes it difficult to know how far the social and economic conditions at that time, as recorded by written documents, may have reflected those of prehistoric times. Thus we have at present no picture at all of the process whereby the stone equipment was eventually ousted by cheaply made objects of iron, presumably during the first millennium B.C., nor have we even a complete pottery sequence to afford some clue of break or continuity in the cultural tradition. For what it is worth, however, it may be noted that the earliest written descriptions of the last few centuries before the Christian era describe the economic status of the aboriginal population along the Maghreb littoral as that of pastoral nomads, raising an occasional catch crop of wheat or barley, and still depending to some extent on hunting.

At any rate this general picture fits well enough with that suggested by the rather meagre palaeontological observations—made for the most part many years ago—in cave deposits in the coastal regions of Tunisia and Algeria containing Neolithic type stone tools and pottery. Further to the south, however, south of the mountains, the higher proportion of arrow-heads may well be taken to indicate a greater reliance on hunting. One of the few faunal assemblages to be found in this latter region in association with Neolithic material—that of El Arouïa, near Figig—showed only a few species which may conceivably have been domesticated, while the great mass of remains quite certainly belong to wild animals of the chase.

In view of Miss Bate's demonstration of the great importance of *Ammotragus* to the larder of the early Libyans, it may perhaps be questioned whether full weight has been given to the possibility that the famous ritual representations of ovids in southern Oran may be intended for this wild species rather than the domestic *Ovis longipes* to which they are generally assigned. Again the identifications at Mugharet el 'Aliya and elsewhere in the Maghreb of (wild) *Bos taurus* and *Bos primigenius* make one wonder whether some at least of the figurations of

---

[1] Wulsin, F. R. (1941), chapter x.

cattle so confidently assigned by most authorities to domestic varieties may not also have been wild. Admittedly, in a certain proportion of sites the grouping of the animals, and especially the attitudes of the human figures, do certainly support the former alternative, but this by no means applies to all instances.

At this point it may be pertinent to re-examine the traditional view that tanged arrows and pressure-flaking are the indubitable indices of a pastoral and partly agricultural society in this area. Two considerations in particular give grounds to doubt this long-accepted notion.

The first of these is the distribution of the early blade industries outlined above, that is to say that the Capsian in both its manifestations occurs only in a limited area in the eastern half of the Atlas massif, while the Oranian is confined almost without exception to the immediate vicinity of the coast. In the whole vast hinterland of the Western Atlas innumerable finds of blade-industries attest only a hybrid tradition in which one or both of the traits in question are combined with devices of later Capsian origin. In a word there is throughout this area no evidence of any intervening cultural stage between the latest Aterian, and the earliest version of the Neolithic-of-Capsian-Tradition.

A second observation adds considerably to the force of this first; namely the recent discovery in the Mugharet el 'Aliya in Tangiers, and subsequently in Morocco, of a final stage of the Aterian in which the two traits of pressure flaking and tanged missile heads are combined to produce implements indistinguishable from those hitherto taken to be type-objects of the Neolithic.

Of course, before the conclusion can be drawn that the Maghreb Neolithic did indeed derive these two important elements locally from culture contact with a belated Aterian, much additional data will have to be accumulated under various heads. In the first place, reliable chronological evidence based on carbon analysis or stratigraphical overlap is required to show that the two generically different types of industry were in fact practised contemporaneously in close proximity. Secondly, some indication is needed that the highly peculiar newly-discovered variant of the Aterian, known so far only from a narrow coastal zone in Tangier and Morocco, really penetrated far inland to the southeast; in many coastal areas the Oranian preceded the Neolithic.

Nevertheless, it can scarcely be denied that a local origin for many of the material aspects of the Maghreb Neolithic is now a distinct possibility; and this being so, it may be inquired whether the same is not true also of some of the economic aspects. While the investigation of this last problem is as yet in its infancy it is of interest to note that a possible—though perhaps not altogether satisfactory—wild prototype for the barley, that formed (and still forms to this day) the stable crop in the region, is known both from the Maghreb and

Cyrenaica.[1] Again, it is perhaps not altogether inconceivable that the domestic cattle also may have been developed from the wild species of *Bos* mentioned above, though according to Hilzheimer the sheep and the goats must certainly be derived from Asiatic species.[2]

The nature of the contribution to this general problem which can be expected from the archaeological record of stone-working and other traits in Libya, may be said to reside largely in the distribution and stratigraphy of the more specialized devices. A long-continued process of diffusion from east to west or vice versa might be expected to result in a fairly well-marked geographical gradient in the number and complexity of traits, with few traces of positive local peculiarities. Conversely, well-defined provinces of material culture might be interpreted as the result of prolonged elaboration of *a few* basic ideas which, because of their functional importance, spread rapidly but were not followed by any considerable degree of regular intercourse.

From the foregoing it may be seen that remains amenable to this type of treatment are by no means lacking in the Maghreb; and a similar analysis will now be attempted of the comparable evidence in the Egyptian sequence.

2. THE STONE-WORKING TRADITIONS OF THE EARLIER FOOD-PRODUCING SOCIETIES IN EGYPT

Owing to factors of surface geology and climate, the Nile Valley throughout Egypt, though admirably suited to illustrate the sequence of the broad cultural subdivisions of the Palaeolithic, offers few facilities for recording the finer details required for the proper reconstruction of the later periods. In default of micro-stratigraphy based on pollen or other organic remains, the best method of obtaining a reasonably complete cultural classification is usually provided by cave deposits. Unfortunately, the conditions are such that there can be little hope of obtaining even this type of evidence anywhere between the Gebel Akhdar and Mount Carmel.[3]

The principal remains of pre-Pharaonic agricultural communities so far made in Egypt fall into two main geographical groups, the one near the root of the Delta in northern Egypt, and the other between Assyut and Edfu, separated from the first by over 150 miles of territory little explored from this point of

---

[1] See Childe (1952), p. 25.

[2] Quoted in Baumgartel (1947), p. 24.

[3] What is wanted, of course, is a site like the Belt Cave in Persia where the transition from latest Mesolithic to the earliest Neolithic can be followed in a continuous or nearly continuous sedimentary record. See Coon (1951).

view. In both groups the cultural material of most sites is contained in very thin deposits where only the vaguest stratigraphy can be inferred and the separation of the products of two or more periods is often a difficult matter, depending to a large extent on the sporadic occurrence of reasonably intact tombs, small patches of deposit sealed in by later structures, and the like.

THE SUCCESSION IN LOWER EGYPT AND THE FAYUM

Probably the best isolated succession[1] is that obtained by Miss Caton-Thompson and Miss Gardner in the Fayum depression where they were able to associate settlements of different ages with the shorelines representing successive stages in the shrinkage of the Pleistocene Lake Moeris down to its present level. Here an initial industrial tradition of generically Middle Palaeolithic type lasted from the highest (Pleistocene) level at 34 m. above sea-level to one at 23 m. above sea-level. The earliest Neolithic was probably associated with a level at 18 m. and certainly with a level at 10 m. Thereafter no further certain correlation can be made till early Dynastic times, which were shown to start with a lake level of 2 m., though a pre-Dynastic settlement can be intercalated on typological grounds between the two.

Of the cultural events and duration of the possibly considerable period intervening between the latest Middle Palaeolithic remains and the earliest Neolithic we know as yet virtually nothing. Abundant remains of primitive food-gathering societies practising an evolved blade industry of microlithic proportions occur in the same area, and there is evidence of extensive culture contact between these and the *later* of the two phases of the Neolithic. Slight traces of similar contact on a much smaller scale during the *earlier* Neolithic stage are perhaps also perceptible. From this it seems not impossible that some at least of the microlithic chipping sites represent the dwelling-places of an aboriginal population in possession of the territory before the arrival of the Neolithic intruders—the latter hailing presumably from the Delta, though perhaps ultimately from further afield. At any rate the phenomenon of an immigrant neolithic population to a new territory gradually assimilating some of the culture traits of more primitive aboriginal inhabitants, is well attested elsewhere, for instance in Western Europe.[2]

If this reading is correct it would follow that at least four stages can be recognized in the closing phases of flint-working along the western edge of the Delta and adjacent regions. Without going into the highly controversial subject of the

---

[1] Though even here there was occasional uncertainty regarding the possible inclusion of stray surface specimens.

[2] Childe (1947), p. 106, for instance.

origin and mutual relationship of the industrial practices in question, the distinctive features of each stage may be listed as follows:[1]

Presumed original microlithic tradition in the Fayum

It is unfortunate that although sites representing this type of industry have long been known from several areas on the western edge of the Middle and Lower Nile, so little attention has hitherto been paid to their systematic description. One of the few published accounts is that of a site in the Fayum provided by Miss Caton-Thompson.[2] The site in question was regarded as roughly contemporary with the later stage of the local Neolithic on typological grounds, though if the view be taken of the cultural history of the area just outlined this argument loses much of its force. At any rate, there seems no doubt that the outstanding characteristic of this and other sites of the same description is the immense predominance of small, flat, blunted-back blades, coupled with a complete absence of the evolved geometric types of microlith such as the trapeze and scalene triangle which typify the latest Capsian, and Neolithic-of-Capsian-Tradition in the Maghreb. The absence of microburins is undoubtedly significant in the same connexión, and it may be noted that there is little or no trace of normal burins either.

On the other hand, links with the main east-to-west continuum across the whole North African littoral are provided by the double-backed blade (judging by later parallels this may well be part of a composite drilling or boring tool quite unconnected with the obvious cutting function of the single-backed blade), and that ubiquitous North African implement, the pressure-flaked trihedral rod. The latter seems to have gained popularity among most of the later blade-working communities from Egypt to Spanish Morocco.[3] At the same time its absence among communities of comparable status south of the Sahara, in Europe, or western Asia, would emphasize its significance as a specialized culture trait.[4]

---

[1] It is necessary at this point to pay tribute to the brilliant field-work of Miss Caton-Thompson and Miss Gardner, whose thoroughness and system have resulted in what must surely be one of the most significant single contributions to field archaeology ever made—*The Desert Fayum*—without which no analysis of this kind could be attempted.

[2] Caton-Thompson and Gardner (1934), p. 67, pl. XLIX; in general the Beduin Microlithic of Kharga may be assumed to belong to a more evolved facies, or at any rate one with more evident signs of Neolithic influence. It contains true geometric microliths in common with more southerly microlithic industries, see for instance Caton-Thompson (1952) p. 159–162 and Arkell, A. J. (1949), pl. 15.

[3] Almagro Basch (1946), p. 174, for instance.

[4] Miss Caton-Thompson's suggestion that all or most of the specimens are made on a rare thin faceted form of wind-eroded pebble—the so-called 'dreikanter'—though ingenious, does not, to my mind, carry conviction.

16-2

The basic blade-making technique seems to have been of a normal fully-evolved kind, using well-developed single-platform conical cores.

The earliest Neolithic or Neolithic 'A' Tradition

Apart from rare blunted-back blades comparable to the above[1] the main characteristic of the stone-working tradition associated with the complex economic traits of the first food-producers is that of bifacial pressure-worked tools exploiting the tabular form of the local raw material. Various types of fine knife-, dagger- and lance-heads were produced in this manner, and when a suitable naturally occurring shape of nodule was not available, the process was supplemented by preliminary polishing and grinding. This latter technique is of particular interest as forming one of the many important resemblances linking this culture to the other site of discovery of Neolithic typology at Merimde Beni Salame some 100 km. north-east of the Fayum on the fringe of the Delta.[2]

Pressure-flaked bifacial arrow-heads were abundant but represented by two forms only, both without exact parallel outside the Nile Valley—a deeply indented hollow-based shape, with curved sides, and a stouter, more triangular form with slightly convex anterior edges. Tanged varieties are altogether exceptional, if indeed they are really associated.

Two other bifacial implement types are also characteristic of this stage—elongated chisels, and the denticulated elements of composite sickles. Polished axes and adzes of both local and imported rocks are another well-represented feature. Pottery is undecorated but sufficiently distinctive in shape to be easily separated from the products of later periods, and is obviously allied to that of Merimde.

The later Neolithic or Neolithic 'B' tradition

Observations at this stage are considerably less complete than those affording evidence of the earlier type of settlement, though data are sufficient to enable a number of important distinctions to be established. Bifacial pressure-flaking was still extensively used, but a striking innovation is the incorporation of a large microlithic element of the type described above. Both earlier variants of arrow-head have become rare; the ubiquitous foliate and narrow-tanged shapes widely distributed throughout the littoral of North Africa,[3] make their first appearance.

No polished axes of imported stone are found and two specialized wood-

---

[1] Caton-Thompson and Gardner (1934), p. 30, section 36.

[2] In Upper Egypt, according to Dr Baumgartels' new analysis, this trait first certainly appears in the earlier of the two settlements, at Nakada.

[3] Caton-Thompson (1934), p. 73, n. 2.

working tools appear instead—the 'gouge' and the 'plane' of Miss Caton-Thompson's classification. The latter is a highly characteristic tool which gains considerably in significance, as will be seen below, from its restricted and continuous distribution.

The absence of polished stone axes and pottery, while requiring confirmation, seems to indicate an appreciable degree of cultural impoverishment, although the presence of sickles and a closely similar siting of the settlements with regard to the new lake shore may be taken to argue at least some survival of the former way of life. The main factor for this change is thought to have been climatic, but in any case it is evident that many links with the Delta tradition had been broken and extensive borrowing and cultural hybridization had taken place with the local hunter-gatherer industries which, as already noted, seem to be related to the main North African cultural continuum.

The later Predynastic stone-working tradition of Lower Egypt

The later Predynastic industrial succession is as yet less fully documented in Lower than in Upper Egypt, owing to the smaller number of sites to receive adequate publication. Above all, a pottery sequence comparable in detail to that of Upper Egypt has yet to be established. Apart from loose surface finds, only one definite settlement has so far been reported from the Fayum, at a site called Qasr Qarun, which could be attributed to this period on the strength of resemblances to larger settlements bordering the Nile Valley at Maadi, Gerza and Harageh, and to certain sites in Upper Egypt.

Among the distinctive features of the material *in situ* at Qasr Qarun was the manufacture of large blades which reached a high standard. Narrow blunted-back forms are prominent, frequently with a rounded trimmed base, a straight back and a convex cutting edge. Petrie's 'three-faced twisted blades' are also well represented. Both forms recur at Maadi where they are accompanied *inter alia* by true burins. Sickles, as at Maadi, are made of sections of narrow blade with one or both ends squared with trimming and a denticulated edge showing heavy straw polish. Large, flat flakes were trimmed into oval or elliptical scrapers somewhat resembling the 'cutting-flakes' of early dynastic assemblages.

Bifacial implements form a much smaller proportion of the whole output than in the two Neolithic industries just described, but include several highly distinctive types such as the 'fish-tailed lance', the asymmetrical 'sickle-shaped knife', a handled knife, and a tanged arrow-head.

All the above tool classes are represented at one or other of the late Predynastic settlements in Lower Egypt, though, with the possible exception of the arrow-head, entirely lacking in primary association with the Neolithic settlements of the Fayum. Another striking difference is the rarity or (at Maadi) complete absence in the later Predynastic sites of ground axes and adzes of stones other than flint, in contrast to their characteristic occurrence in the Neolithic sites of the Fayum and at Merimde. The ceramic evidence as far as it goes endorses the cultural classification suggested by the flints, and although the Qasr Qarun settlement cannot be directly dated in terms of lake shorelines there cannot be any reasonable doubt that it represents a tradition closely allied to that of dynastic times[1] and widely removed both in date and social context from that of the Neolithic A and B sites.

THE STONE-WORKING TRADITIONS OF THE PREHISTORIC AGRICULTURAL COMMUNITIES OF UPPER EGYPT

Although the number of sites investigated in the southern group is considerably larger than in the northern, the complete cultural succession is given at no one single site, and can only be reconstructed from a series of overlapping stratigraphical observations. Despite fairly extensive geographical dispersal, the total number of finds is insufficient as yet for the confident delimitation of the distribution of many important tool types; the whole picture is further complicated by the undoubted survival of some of these in the more southerly sites long after they had become obsolete in the northern sites of the group.[2]

The first classification of prehistoric Egyptian village settlements, elaborated by Sir Flinders Petrie and others, subdivided these into three periods: Amratian, Gerzean, and Semainian, which were believed at the time to apply virtually to the whole of Egypt north of Edfu. Subsequently two earlier stages, the Tasian and Badarian, were described by Brunton and Caton-Thompson from Upper Egypt, while accounts of the Neolithic-type settlements from Lower Egypt (outlined above) were published by Caton-Thompson and by Junker.[3] Finally, a new classification has been proposed by E. Baumgartel,[4] based on a thorough re-examination of the material collected by Petrie from the two sites at Nakada in Upper Egypt, together with a critical revision of the crucial evidence from both Upper and Lower Egypt.

The new scheme proposed by Dr Baumgartel would involve the subdivision of the Upper Egyptian sequence into four stages: the Tasian, Badarian and

---

[1] Caton-Thompson and Gardner (1934).

[2] Discussed in Reisner (1911) and elsewhere. A comprehensive analysis of the dating of a wide variety of flint types is given by Huzayyin in Mond and Myers (1937), pp. 191 *et seq.* Many of these are further discussed in detail in Caton-Thompson (1952).

[3] Junker (1929, 1930, 1932, 1940).   [4] Baumgartel (1947).

Nakada I—all representing closely related evolutionary phases of a single tradition—and Nakada II, representing a sharp break in the cultural evolution, heralding the main features of the historic civilization and attributable to large-scale ethnic immigration. In addition, Dr Baumgartel suggests an entirely new correlation between Upper and Lower Egypt, differing substantially from that envisaged by Caton-Thompson and by Junker.

Detailed criticism of the new hypotheses is, of course, outside the scope of the present discussion, even if the present writer were qualified to undertake it, and it is as yet too soon to say how far Dr Baumgartel's views will meet with general acceptance among Egyptologists. Nevertheless, some assessment of the various opinions offered is necessary if any attempt is to be made to interpret the desert material described below.

If then the considerable body of observations arising from the re-examination of the Nakada collections is interpreted in the manner Dr Baumgartel suggests, the outline of the Upper Egyptian sequence as far as stone-working is concerned would be as follows.

During the Badarian, among the most distinctive shapes so far identified (apart from some crude core tools perhaps used in wood-working or tanning) are bifacial sickles, essentially similar to those of the Fayum Neolithic, and equally similar hollow-based arrow-heads, as well as a leaf-shaped form. Although the woodland environment attested along the fringe of the swampy valley floor at this time would certainly seem to demand efficient wood-working tools to meet the needs of any largely agricultural people such as we know the Badarians to have been, according to Baumgartel no ground axes can be associated with their tradition with absolute certainty, though their use seems likely on general grounds. In general the manufacture of all the flint equipment seems to have been done on the spot, and we have no evidence of specialized factory sites such as characterize the later Nakada I and II periods. Observations bearing on the nature of the flint-work during the preceding Tasian period are said to be somewhat inconclusive but as far as they go show no positive differences with the Badarian.

The main developments during the Nakada I[1] period include characteristic triangular arrow-heads identical in every way to those of Fayum A, together with leaf-shaped varieties, celtiform hoe-like implements, 'planes', and other types characteristic of Fayum B, all associated with the famous fish-tailed lance-heads, which in the Fayum are unknown till the very much later pre-Dynastic settlement at Qasr Qarun. A further characteristic of Nakada I is the frequent use of the rather peculiar technique of grinding before flaking, also noted in the Neolithic of the Fayum, and possibly already known during the Badarian.

[1] Baumgartel (1947).

The most striking innovation during Nakada II seems to be the far greater preponderance of unifacial tool forms coupled with a very high standard of blade manufacture. In addition, forms characteristic of most blade traditions such as true burins of all the usual types, transverse-burins of the Hagfet ed Dabba type, and end-of-blade scrapers, now appear in great numbers. Sickles are now more usually made of segments of blades with terminal retouch at right-angles or pointed ends, and denticulated edge, instead of the older bifacial form. At the same time some new bifacial classes make their first appearance, such as the highly evolved ripple-flaked knives, and a new form of flaked axe with a cutting edge formed by a characteristic lateral stroke.[1] In some sites also a number of older forms continued in popularity, such as the 'plane' and the celtiform flaked implement, abundantly represented, for instance, at the Nakada II site at Armant.

The acceptance of the above scheme would of course in no way prejudice the issue of the correlation between Upper and Lower Egypt, which remains an entirely distinct problem. In the original publication on the Fayum material (some years before Dr Baumgartel's study), Miss Caton-Thompson stressed the analogies between Fayum A and Merimde on the one hand, and between both and the Tasian and Badarian stages of Upper Egypt on the other. Dr Baumgartel would now prefer to separate Fayum A from Merimde; correlating the former with Nakada I and the latter with Nakada II.

It must, I think, be conceded that Dr Baumgartel's correlation is on the face of it open to a number of objections which are not fully dealt with in her original exposition. To begin with, it is clear from the above summary, that in Upper Egypt a large number of traits were in use simultaneously which, in the Fayum at any rate, occur successively in quite distinct periods. Such, for instance, are the special type of hollow-based and triangular arrow-heads virtually confined to Fayum A; the planes and celtiform bifaces which are the typical, if not exclusive, property of Fayum B, and finally, the fish-tailed lance-heads unknown before the very late cultural stages represented at Maadi, Gerza, and Qasr Qarun. If Nakada I was really the contemporary of Fayum A and the source from which the latter derived all its most important features, how is it that so many of the Nakada I traits which ultimately reached the Fayum in abundance only did so after so long a delay? It can hardly be questioned that the quantity of material of Fayum A is sufficient to provide a reliable picture of the objects in everyday use, and at the same time the widespread nature of the barter connexions of this culture with distant areas are established beyond question by the

---

[1] Huzayyin in Mond and Myers (1937), pp. 191 *et seq.* According to Caton-Thompson (1952), p. 173 this type also probably occurs as early as the Badarian. Satisfactory prototypes are found in the Neolithic of Jericho, where are also transverse-burins, see Crowfoot (1936).

raw materials used for the ubiquitous stone axes, adzes, and shell ornaments.[1] On the other hand, if a northern origin be allowed for some at least of the traits in question, there would be nothing at all strange in their local survival to a later period in a somewhat different culture complex in the south.

In seeking to separate the Merimde civilization of Lower Egypt from Fayum A and to equate it with the much later Nakada II phase of Upper Egypt, Dr Baumgartel lays particular stress on the stratigraphical distribution of blade-making in the various prehistoric Egyptian cultures. As mentioned above, this trait is one of the characteristic features of Nakada II, only sparsely represented in Nakada I, and doubtfully present in the Badarian.[2] Though rare in the Fayum A its presence can probably be accepted with a good deal of confidence;[3] at Merimde, however, it is apparently somewhat more prominent. As Dr Baumgartel regards this observation as the decisive argument in favour of a Nakada II date for this last it is necessary to consider its significance a little more closely in the light of some of the conclusions recorded above.

In effect, the picture that Dr Baumgartel offers of the origin of the domestic arts of the Neolithic in Egypt is that of a culture complex introduced, presumably by ethnic immigration, from a centre of dispersal in Upper Egypt (ultimately deriving either from some unspecified region still further to the south, or from Arabia via the Wadi Hammamat)[4] and gradually extending northwards along a narrow fringe of woodland and savannah fringing the swampy flood plain of the river.

Both in the Fayum and at Badari some significant evidence was obtained concerning the nature of the natural environment; little attention on the other hand has so far been paid by most investigators to the social aspects of this environment. Dr Baumgartel[5] indeed seems to assume that no such aspect to the environment existed. To the present writer the assumption that the Nile Valley and neighbouring regions, teeming with game and eminently suitable for hunting occupation, remained virtually unpopulated at the time of the first appearance of agricultural societies, seems not merely unsupported by any positive evidence, but on the face of it quite unthinkable. In the Maghreb the antiquity and vigour of the Mesolithic blade tradition is sufficiently demonstrated by the stratigraphical results quoted above; our own observations have now extended the distribution of many of these traits as far east as the Gebel Akhdar, where again it is clear that they are part of a long established hunting culture.[6] Finally attention has been recalled earlier in this chapter to the large number of undated finds of

[1] Caton-Thompson and Gardner (1934), pp. 87–88.   [2] Baumgartel (1947), pp. 87–88.
[3] Caton-Thompson and Gardner (1934), pp. 23 and 30.   [4] Brunton and Caton-Thompson (1929).
[5] Baumgartel (1947), pp. 18–19.

a microlithic industry of apparently Mesolithic character in the Fayum. While the hunting societies which these betoken seem to have established effective contact with the agricultural tradition only at a late stage of the latter, their earlier limit in Egypt is quite unknown. It is interesting to note that the situation at Kharga some 500 km. to the south, is essentially comparable, since there also widespread traces of a closely similar microlithic tradition were discovered, quite distinct from the characteristic products of the earliest agricultural population.

Finally, mention may be made of the oft-quoted surface finds at Helwan on the east side of the 'root' of the Delta. Here, it will be remembered, large numbers of tools showing unmistakable analogies to the Upper Natufian or final stage of the Palestinian Mesolithic, were found loose on the surface. These may or may not be associated with remains of late date found in the same area, but at least it is fair to emphasize that subsequent discoveries concerning the microlithic tradition of Egypt and North Africa generally, strikingly endorse the specialized character of the Helwan finds, and hence the validity of their eastward connexion with the blade tradition of Palestine.

On the whole therefore, though it would be wrong to suggest in default of more precise chronological evidence that the case is as yet complete, there are in fact quite substantial grounds for suggesting that the earliest agricultural communities in Egypt, so far from subsisting in a social vacuum, were in close proximity on the west, north-west and north-east with hunting societies practising a typical evolved blade industry. It follows that little reliance can be placed in slight variations in the number of blades found in various early Neolithic assemblages as an indication of date, since these could quite easily have been acquired in haphazard fashion by barter or imitation from the surrounding hunters.

As to other arguments based on pottery forms and decoration, for the late date of the Fayum and Merimde cultures in relation to the Upper Egyptian sequence, the writer is not qualified to speak.

Some mention must however be made if only in passing, to another aspect of Dr Baumgartel's conception relevant to the same problem, namely that the character of the flint work in general associated with the Badarian, Tasian, Fayum, and Merimde cultures, is such as to preclude a west Asiatic origin, and strongly suggest an Upper Egyptian or Nubian centre of diffusion.

The features regarded as particularly significant in this connexion are three; the rarity of well-worked blades, the predominance of pressure-flaked and bifacially worked tools, and the occurrence of coarse core-tools.

As to the last feature, reference has been made more than once in this report to the difficulty of distinguishing between cores, and cores that have been used as tools. In the case at issue the writer must frankly admit that he can find no proof whatever that the majority of the rougher nuclear specimens referred to are anything more than the cores from which usable flakes and blades have been struck.

With regard to the predominance of pressure-flaking and bifacial work generally, it should not be forgotten that these are extremely widespread characteristics of primitive agricultural societies of many different ages and regions. The disappearance of such ubiquitous hunters' tools as the backed-blade, end-scraper, and burin, at this stage, and their replacement by other and quite different forms of tools, is after all likely enough to find a simple functional explanation. While it may be that this tendency is unusually pronounced in the equipment of the earliest Egyptian farmers and herdsmen, and can perhaps be used as an argument for a local origin, other at least equally weighty evidence points clearly towards Western Asia. That has long been admitted to be the source of two of the domestic animals at least—the sheep and goat—and probably the domestic plants as well. The absolute dates suggested by carbon-readings in both areas, as far as they go accord well with the notion of a west Asiatic origin, since they imply the not unreasonable interval of 1000 years between the first known emergence of the basic ideas of agriculture, animal husbandry, pottery, and ground stone tools in Northern Iraq and their occurrence in primitive form in the Delta.

Finally, the recent discovery of the later stone age sequence in the Sudan, leaves little room for any theory of Egyptian origins in that quarter. In a word far more cogent indications than those afforded by the flint-work are required before an Asiatic origin for the Egyptian Neolithic is abandoned and with it the implied direction of diffusion from Lower to Upper Egypt, rather than vice versa. It is in the light of this conclusion that the collections from Siwa and Tripolitania will now be presented.

SURFACE COLLECTIONS FROM SIWA OASIS

Reference has already been made to the slight traces of the earliest hunting cultures to be discerned in the large surface collections made by Dr C. Willett-Cunnington in this area in 1918. The circumstances under which the material was recovered may now be described in a little more detail, as recorded in unpublished notes lodged with the collections of the University Museum of Archaeology and Ethnology at Cambridge.

Geographically, the oasis of Siwa lies near the north-western angle of a roughly triangular system of widely dispersed depressions eroded in the desert plateau west of the lower and middle reaches of the Nile. The southernmost point of this area may be said to be formed by the Kharga depression, some 1000 km. south of the Mediterranean coast; the north-eastern end by the Wadi Natrun some 60 km. south of the coast, while Siwa itself lies about 300 km. from the coast. The largest single depression is the Qattara, measuring some 300 × 150 km. but others, though smaller in extent, are of greater importance economically owing to their supply of perennial water either in the form of relatively shallow wells, springs, or—in the case of the Fayum—a permanent lake.

The Siwa depression is somewhat narrow and elongated in shape with the main axis east and west and 80 km. long, and a maximum width of about 30 km. at the eastern end. A low ridge separates it on the east from the Qattara depression, whose main axis runs south-west to north-east and reaches to within 80 km. of the coast at the northern end.

Immediately north of Siwa the desert plateau forms a fairly uniform tableland at a considerably higher altitude extending unbroken for over a hundred miles. The important inhabited oasis of Bahriya lies 300 km. to the east-south-east, and the Fayum just under 500 km. due east. The small oasis of Jaghbub lies about 70 km. to the west-north-west of Siwa, while to the southwards the main Libyan sand sea stretches almost continuously for over 1300 km.

Topographically, the Siwa depression presents a relatively flat floor, below sea-level in altitude, and delimited by a well-marked escarpment along the northern margin rising up to the northern desert plateau. Traces of a similar escarpment may possibly occur along the southern margin as well, but if so, it is obscured at the present time by large accumulations of dune sand.

The collections come mainly from two sources—the slopes and terraces of the northern escarpment, and the desert plateau about 30 km. to the north. The collector stresses the fact that, despite repeated search, the rest of the plateau to the north of the zone bordering the oasis proved completely barren of traces of human occupation. When they do occur the plateau sites are concentrated round small depressions that may have been subject to temporary flooding under favourable conditions with a slightly increased rainfall.[1] In the main these sites must represent mere temporary hunting encampments. Within the oasis depression, however, the presence of heavy equipment such as grinding slabs suggests that here many of the sites are connected with some degree of agriculture. They occur on terraces which break the escarpment and would appear suitable for small-scale hoe agriculture under sufficient rainfall—not unlike in fact the catch-crop agriculture practised sporadically along the coast at the present day.

Although the great mass of the material was obtained from these two sources,

[1] The siting of the very similar Beduin Microlithic chipping floors and camps at Kharga is much the same (Caton-Thompson (1952), p. 32).

a few specimens were also obtained further afield, from Jaghbub, and even from Qara in the east on the western edge of the Qattara depression.

On the whole, however, apart from the general indications just given, the absence of precise data regarding location makes a detailed distributional analysis impossible, and the series as it stands is best regarded as a loose scatter from an area with a radius of some 35 km. with little or no known association between the different elements. At the same time, it may be repeated that the difference in state of preservation between the majority of implements and the one or two specimens of Aterian and Middle Palaeolithic typology,[1] is such as to suggest that the former were made over no very extended period, and may quite conceivably belong in great part to a single phase of occupation.

With these preliminary remarks the inventory and description of the various tool classes in the collection (other than those dealt with at the beginning of the previous chapter) may now be given:

Blade cores (5). These are small, ranging from 2·5 × 2·5 × 3·0 cm. to 3·0 × 3·5 × 3·5 cm., and are all of the single-ended conical form for the production of very small blades of the type used for the manufacture of microliths. Other much larger specimens must certainly have existed judging from the size of the blades, and the presence on the sites of fair numbers of cores is mentioned in the manuscript notes.

Unretouched blades (numerous). This series includes some twenty pieces 10–13 cm. and as little as 2 cm. wide. They complete the evidence of the cores by demonstrating the presence of a fairly highly developed blade technique, comparable, say, to that of Hagfet ed Dabba. 'Pièces-à-crêtes' are not uncommon, but the majority of pieces seem to have been struck from fairly highly convex cores with a single platform.

Backed blades (94). These, although forming the most numerous single class in the collection, are all of microlithic proportions. There is virtually no trace of the larger type of blunted-back knife found in the Cyrenaican Gebel, and still more characteristic of the Capsian. At most, one or two small to medium-sized pieces might perhaps be compared to the very prominent corresponding type in the lower series at Hagfet ed Dabba, with narrow shape and straight line of backing. A few large blades do, on the other hand, show slight traces of bilateral nibbling as if for insertion in a terminal haft or handle.

The great mass of the specimens belonging to the microlithic class fall into two main categories about equal in numbers: those which are extremely narrow, with rather thick backing from both faces in a straight line producing a needle-like point at one or both ends (Fig. 35, no. 20), and those of wider proportions with flatter section backed from one face only, in a curved or slightly angulated

[1] See p. 220.

line to produce a rough variant of the 'pointe scalène' of the French authors (Fig. 35, nos. 18 and 19). Apart from these last there is little sign of a geometric tradition; lunates are barely represented, and *trapezes altogether unknown*.

Drills or 'double-backed blades' (25). This highly characteristic tool form consists of a small section of relatively thick blade, ranging for the most part from 1·5–4 cm. long by 0·5–1 cm. wide by 3–5 m. thick (Fig. 35, nos. 25 and 26), produced by steep 'backing' along both edges converging to form a blunt and frequently heavily-worn point. The backing is roughly executed apparently by pressure against an anvil. The same type of point may be made at one or both ends and is occasionally made on blades up to 6·5 or even 9 cm. long (Fig. 35, no. 27). This type of implement with precisely the same range of size and technique is a characteristic feature of all the stages of the inland blade tradition in the Maghreb down to the Neolithic-of-Capsian-Tradition. In the Oranian, judging by published accounts, it is rare or absent. In Egypt it occurs in great abundance with the microlithic assemblages of the Fayum and Neolithic 'B' stages and in the 'Beduin Microlithic' of Kharga, and apparently remained in use until well into the pre-Dynastic or Nakada II stage in the Nile Valley.[1] The type seems to be unknown to the Natufian in Palestine.

Shanked points (11). A clear type seems to be indicated by a series of small bladelets from 2 to 5 cm. long and 1 to 1·5 cm. wide, with a sharp point at one end formed by primary scars sometimes improved with a little trimming, and a well-trimmed triangular shanked base. 'Shanked blades' are reported from the Fayum, but do not appear to form an equally standardized class.[2] Exactly similar pieces, however, occur in the Beduin Microlithic of Kharga[3] (Fig. 35, nos. 6–9).

Round-based arrows (11). The occurrence of characteristic examples of this very distinctive type, previously only known from the microlithic assemblages of the Sirtican littoral, is of much interest. That it does not, however, represent an extension of this tradition in its original form to Siwa, seems to be fairly clearly indicated by the fact that although the tips are finished in the characteristic manner with a microburin blow and the outline and dimensions are closely the same, yet at least half the specimens show traces of true pressure-flaking exactly like that normally found on tanged arrows and other 'Neolithic' forms (Fig. 35, nos. 1–4).

Microburins (2). Only two certain specimens are included in the present collection, apparently by chance, since the type was not widely recognized among prehistorians at the time the collection was made (Fig. 35, no. 5).

---

[1] At Hierakonpolis, for instance—see specimens in the University Museum of Archaeology and Ethnology, Cambridge.

[2] Caton-Thompson and Gardner (1934), pp. 60 and 80, and fig. xviii, no. 3.

[3] The so-called 'Ounanian' shanked type (Caton-Thompson (1952), p. 33 and pl. 98, no. 3).

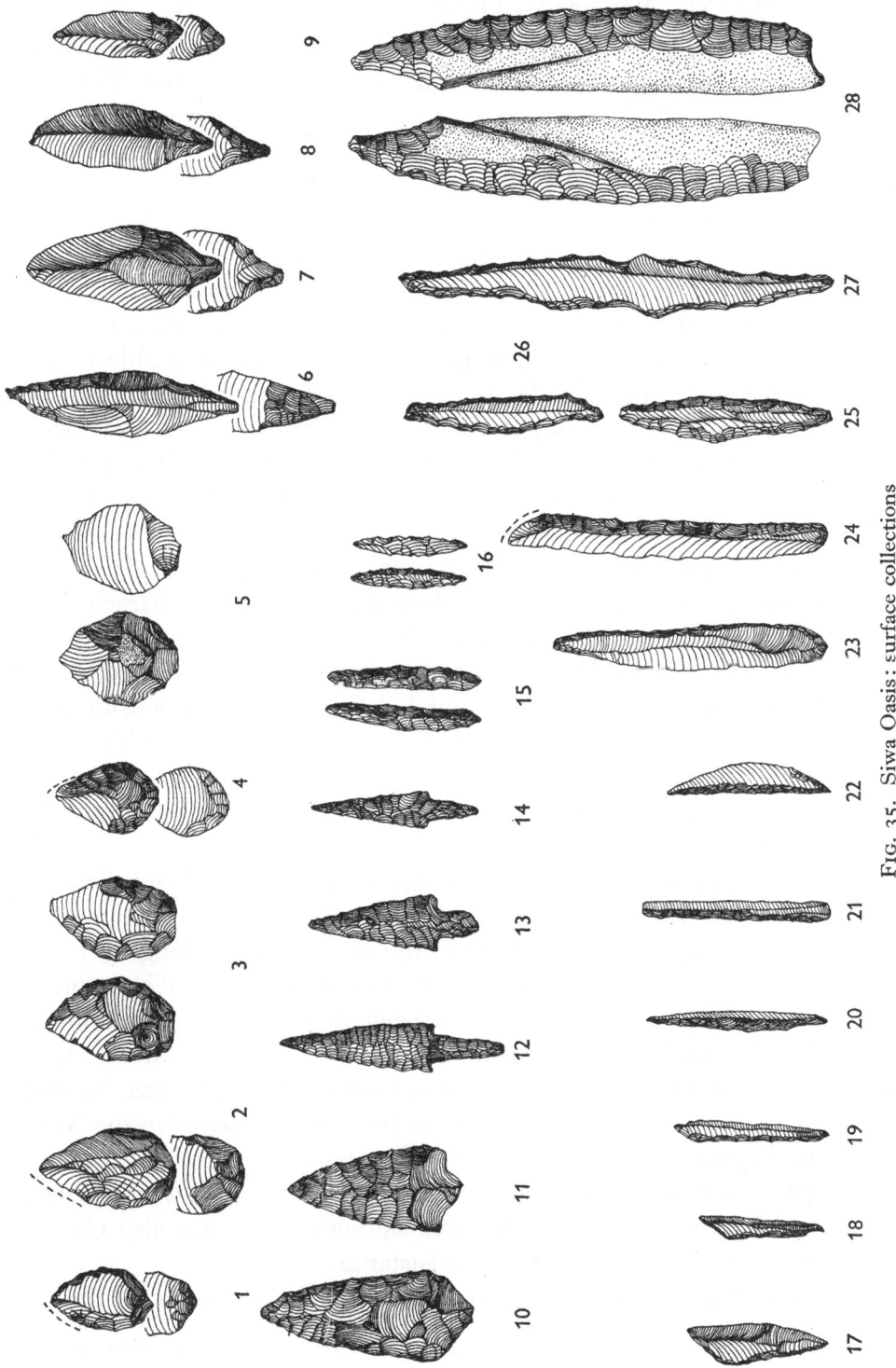

FIG. 35. Siwa Oasis: surface collections

Nos. 1–4, round-based arrow-heads of Sirtican mesolithic type; No. 5, microburin; Nos. 6–9, shanked points of Fayum type; Nos. 10–14, pressure-flaked arrow-heads; Nos. 15–16, small trihedral rods; Nos. 17–19 and 22, microliths; Nos. 20, 21, 23 and 24, Gravette-type backed blades; Nos. 26 and 27, bilaterally trimmed blades; No. 28, bifacially pressure-flaked knife with cortex back. All × ⅔.

True burins (8). Though relatively few in number these specimens are quite typical and amply suffice to establish the use of this tool in the district. They are made on fragments of blade from 3 to 7 cm. long. One is single-blow, one with a trimmed angle, and the rest are 'bec-de-flute' with multiple resharpenings. A significant feature is the occurrence of true pressure-flaking on the base of four examples (Fig. 35, nos. 10–11).

Adzes, etc. (15). Four of these belong to the rather specialized type made with one perfectly flat face formed by a thermal or other flat surface, and described from various Egyptian sites under the heading 'planes'.[1] In the opinion of the present writer, there can be little doubt that the function of these objects was essentially the same as others made in a less distinctive fashion from flat tabular nodules, or from more rounded nodules requiring flaking on both faces, and that all were merely variants of the adze (Pls. 11 and 12). It will be recalled that the first-mentioned class was one of the characteristic features of the 'B' stage of the Neolithic in the Fayum as opposed to the 'A' stage where it is not certainly known, and survived till a late period in Upper Egypt.

Concavo-convex scrapers (2). Two side-blow flakes with the characteristic scraper edge included in the collection, are in every way identical to specimens believed by Miss Caton-Thompson to be attributable to one or other of the two Neolithic stages in the Fayum, and to others found in considerable quantity at Kharga, where they also appear to be associable with a Neolithic type assemblage. As far as the writer's information goes, Siwa is the most westerly locality in which this type is known to occur, and none have yet been reported from the Nile Valley proper (Pls. 13 and 14).

Pressure-flaked arrow-heads, etc. (*a*) *Tanged* (60). The predominant shape is very small and narrow with relatively slight differentiation between tang and head (Fig. 35, nos. 12–14); dimensions are in the order of 3 cm. long by 1 cm. wide, with the longest at about 4 cm. and at least 30 per cent of specimens less than 2 cm. long. This seems to bear out the suggestion made by Dr Willett-Cunnington that the small plateau sites are camps for fowling or hunting other small game, owing to their orientation on the margins of the mud pans. From a comparative point of view all the Siwa types can be closely matched in the Neolithic 'B' of the Fayum and at Kharga, but differ in the most striking fashion from the Fayum 'A' series with its massive triangular and U-based weapons, not a single one of which has yet been identified at Siwa. They are also closely similar to some Tunisian series, Redeyef for instance.[2]

(*b*) *Leaf-shaped* (20). This category grades insensibly from minute specimens

---

[1] Caton-Thompson and Gardner (1934), pl. xxxv, nos. 11–22, for instance, from the Fayum.
[2] See Gobert (1912).

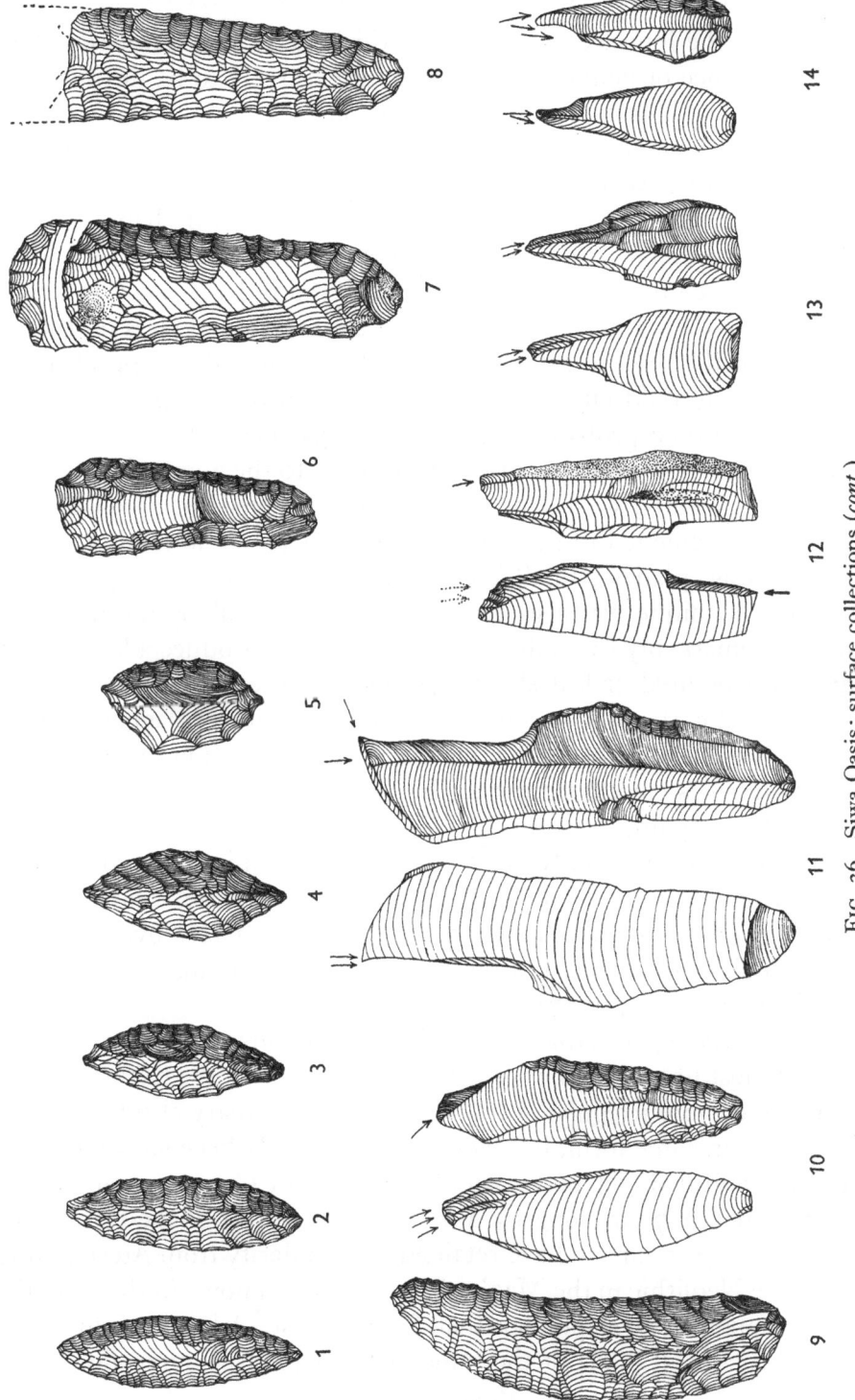

FIG. 36. Siwa Oasis: surface collections (*cont.*)

Nos. 1–4, pressure-flaked bifacial foliates; No. 5, possible double awl; Nos. 6 and 7, pressure-flaked chisels; Nos. 8 and 9, possible sickles, bifacially pressure-flaked; Nos. 10–14, burins. All × ⅔.

symmetrical about the long axis only, with a thickened ridge-like base which, apart from the absence of defined constriction, are scarcely to be distinguished from some of the tanged forms, to larger more regularly leaf-shaped forms, some of which may have been lance- or dart-heads (Fig. 35, nos. 1–4), or side-elements in some large composite weapon.

(c) *Elongated* (5). An extreme variant of the smaller leaf-shape is worth mention in which the length may be as much as eight times the width and the whole object is reduced to a needle-like pressure flaked rod 1·5–2 cm. in length.

(d) *Trihedral rod-shaped* (2–3). Some of the last-mentioned class on the one hand, and some of the more carefully made drills on the other, approach this well-known shape referred to in the summary of Fayum B, though only two can be regarded as more than probable examples (Fig. 35, nos. 15–16).

Miscellaneous pressure-flaked tools (40). In addition to the above characteristic pressure-flaking—executed with a punch—can be detected on a considerable variety of other implement forms, including most of those normally made by other means such as 'planes', small leaf-shaped knives and points (some unifacial), etc. The technique was also applied to both faces along the margin of large plaques of remarkably thin tabular flint in order to produce a long curved cutting edge, or a pointed or leaf-shaped outline. In the latter case, despite a large area of natural cortex, the result may be a large lance or dagger blade no less effective than those with completely flaked surfaces. The general effect created by this series is closely similar to that of pieces made from comparable raw material in the Fayum.

A class of objects that seems only to have been made by this technique are long ribbon-like bifacial knives up to 9 cm. long by 1–1·5 cm. wide. Some of these are cortex-backed (Fig. 35, no. 28) and others completely flaked over both surfaces. Finally, a single example of a unifacial chisel resembling a Fayum type may be mentioned (Fig. 35, no. 7).

Other bifacial forms (70). Apart from the implements showing the above technique in its most unmistakable form are others in which the flaking consists of much larger and more irregular scars, which it is customary also to attribute to pressure-flaking in some form, but which it has recently been shown can also be produced by skilful direct percussion.[1] Here, as in Cyrenaica, an oval or slightly pointed tool, 3–7 cm. long by 2–3 cm. wide, is one of the commonest surface finds. The type seems to have retained its popularity from Aterian times until well into the Neolithic in the Maghreb and is not unknown in the Neolithic of the Fayum, though less conspicuous than further west. A hoard of thirty-one specimens of this sort all made of the same peculiar pink-coloured flint, in all

[1] See Knowles (1947).

258

probability by one craftsman, were found at Shiyata. They are of some interest as affording good evidence of the range in size and outline of a single contemporary group of these implements.

Sickles (1). Owing to the fairly common effects of some degree of sand-blast, it is difficult to affirm that some of the foregoing implements may not originally have displayed straw polish demonstrating their use as sickles. In only one case, however, is the evidence conclusive, namely, on a small specimen of bifacial cutter, strongly reminiscent of the sickles of Fayum and other Egyptian sites, whose denticulated edge shows an unmistakable degree of localized silica abrasion.

Grinding before flaking (1). A single example of this peculiar technique, apparently well represented from the earliest agricultural stage onwards in Egypt, is afforded by a small fragment of a knife or spear-head made by grinding to a very thin plaque less than 3 mm. thick before trimming a sharp bifacial edge —probably by percussion (Pl. 10, no. 4).

Querns and grinders (numerous). A high proportion of the better-defined flaking sites on the escarpment terraces are reported to have contained large flat querns up to 30 cm. in diameter, showing heavy signs of wear, and accompanied by flat circular pebble rubbers or grinders. One of the latter is included in the present collection. Similar objects are reported from all stages of the Capsian in the Maghreb, where they are generally heavily stained with ochre and were apparently mainly used in grinding this substance. At a later period no doubt the same equipment came into service for grinding grain, and since no mention is made of ochre at Siwa it seems likely that this was their purpose here.

Maces (1). A fragment of a bored sphere of calcite, or alabaster of some kind, is exactly matched in form and technique by some surface finds of unknown date from the Fayum.[1]

Imported stone (1). A large cobble of speckled green crystalline rock, with a few flakes removed, is unlikely on geological grounds to be indigenous and is presumably an import. It is the only well-attested example of an exotic material.

Pottery (2). A single sherd of a thick hand-made vessel is of interest as showing bands of an impressed ornamentation made apparently with the articular end of a small bone (Pl. 10, no. 6). Close parallels occur at several sites of Neolithic-of-Capsian-Tradition in the Maghreb.[2] A second sherd of a very thick-walled, round-based vessel though hand-made and probably prehistoric, is unornamented.

In conclusion, the outside affinities of this series of culture traits may be summarized as follows. As compared with the nearest area examined to the east, the

---

[1] Caton-Thompson and Gardner (1934), pl. XXIX, fig. 12.
[2] Redeyef in Tunisia, for instance, see Gobert (1912); see also Vaufrey (1939), figs. 42 and 44.

17-2

Fayum, it will be noticed that virtually all the characteristics of Stage 'B' are represented, but none of those which are the exclusive property of 'A', such as the peculiar arrow-head forms (which it will be recalled were commonly and widely employed elsewhere in Egypt), or the ground stone axes. No specimens displaying traits peculiar to the later Pre-dynastic or early Dynastic flint working traditions have yet been noted in Siwa (with the possible exception of a fragment of dagger-handle showing broad and flat pressure scars suggesting a slightly different technique to that of the remainder of the series.)

One highly specialized class—the side-blow or concavo-convex scraper—has previously been reported only from Kharga and the Fayum (where it is believed to be associated with either 'A' or 'B' stages),[1] and is apparently unknown either in the Nile Valley proper or to the west.

Of specifically western traits in flint-working only two are noticeable; of these the most interesting are the round-based arrows with microburin finish at the tips in the Sirtican style, though the traces of typical Neolithic pressure work on them suggest that they were an integral part of a tradition of the latter type rather than indicating the presence of the Sirtican craftsmen at Siwa. The other tools that may be considered as 'western' are the oval bifaces. These, while not unknown in Egypt, are undoubtedly far more abundant in Cyrenaica, and form a characteristic element in the Neolithic-of-Capsian-Tradition in the Maghreb. Burins may perhaps also come under the same heading; though specimens have been recorded from Kharga, none are apparently yet known from the early stages in the Fayum.

On the whole, the absence of certain typical western traits is a good deal more striking than the presence of those which do occur, especially the absence of true geometric implements or the derivatives which form the dominant feature of virtually all Maghrebian traditions from the Upper Capsian onwards. True lunates are extremely rare in Cyrenaica and Siwa, and apparently unknown in the Fayum or Kharga. Triangles and above all trapezes and transverse arrow-heads are completely lacking from collections from Cyrenaica, Siwa, and the Fayum, and only make a very tentative appearance in Sirtica.

While the existing collections from these areas cannot, of course, be regarded as providing *exhaustive* information on the types of tools that may or may not have been used there, it is difficult to avoid the conclusion that they already provide some statistical basis for assessing the commoner types. When it is realized that trapezes normally contribute 10–15 per cent, and the remaining geometric forms 5–10 per cent in the Maghreb, it can hardly be denied that a significant difference between that area and eastern Libya has already been

---

[1] Caton-Thompson and Gardner (1934), p. 21.

demonstrated. A few trapezes belonging to a peculiar highly-evolved variant have, it is true, been found in Kharga.[1] I must admit, however, that their occurrence in one instance with a handful of small, rather undifferentiated, flakes seems to me insufficient reason for Miss Caton-Thompson's far-reaching theory of their local derivation in this district from the Middle Palaeolithic.[2] The presence of transverse arrows in the full Pre-dynastic of Nubia (and, of course, in the well-known degenerate form of Dynastic times throughout Egypt—a mere untrimmed sliver of flint embedded in a cement mount) together with true lunates, and probably also in the late Mesolithic of Khartoum and along the whole southern and western margin of the desert is, however, well demonstrated.

Such a geographically continuous distribution of this and other western traits suggests that diffusion took place along the south Saharan savannah zone and perhaps by a north-west to south-east route via the Hoggar-Tibesti ridge as well.

Briefly the writer would suggest that the upshot of these observations, both positive and negative, is to stress the existence of a considerable degree of cultural autonomy in eastern Libya, at least from the late Mesolithic or hunter-gatherer stage onwards. Although the Neolithic in this area shares a few traits such as the tanged arrow, trihedral rod, and pressure-flaking itself, with the Neolithic in the rest of the North African littoral, it seems to be fairly clear that the main affinity is with the Neolithic 'B' stage of the Fayum. Thereafter cultural intercourse seems to have had no appreciable effect on this aspect of material culture until the replacement of stone by other raw materials, though, as I have pointed out above, borrowing of native Libyan methods by the higher cultures of the Nile Valley may well have taken place. No doubt the restricted life of the oasis dwellers and desert nomads fostered extreme conservatism then, as it does to this day, and offered little scope for the incorporation of new technological devices once the basic features of the pastoral and primitive agricultural economy had been implanted.

---

[1] Caton-Thompson (1946), p. 117: '...a single mound-spring site yielded in conditions of apparent contemporaneity, small cores and plain thin flakes of epi-Levalloisian style (though not Khargan), associated with three long-stemmed transverse arrow-heads. The find provided no relative stratigraphy, and requires support before the obvious possibility of its "transitional character" can be claimed'. With the last proviso I entirely agree but I must add that the 'plain thin flakes', when Miss Caton-Thompson was kind enough to show them to me, did not, to my mind, afford conclusive evidence of 'epi-Levalloisian' affinities; they appear to me to be the products of free-hand chipping not unlikely to occur in blade industries of many dates.

[2] A similar claim was made by Vignard on the strength of his 'Sebilian' sites at Kom Ombo. To the present writer, the accounts of this much quoted locality are so completely lacking in the type of information required of present-day investigations, that it is virtually impossible to draw any reliable conclusions from them, let alone the far-reaching theories to which they have given rise, as for instance in Caton-Thompson (1946a, p. 118).

As regards the cultural boundary to the west, as far as Siwa is concerned its close affinity with Kharga and the Fayum is probably sufficiently explained by the fairly uniform nature of the oasis and plateau environment which, as pointed out at the beginning of this chapter, forms a not ill-defined geographical unit whose westerly extremity lies at Jaghbub. (The next important area of possible settlement westwards lies at Augila just over 300 km. to the west-south-west, though some intercourse between the two in prehistoric times seems to be indicated by reports of sporadic surface finds along the northern fringe of the Sand Sea.) Communication may perhaps have occurred to some extent along the coast with Cyrenaica, though a second boundary would seem likely to have been formed by the wide arid zone along the southern shore of the Gulf of Sirte. Some reflection of the effect of this last boundary is perhaps to be detected in historical accounts as far back as the second millennium.[1]

'NEOLITHIC' SURFACE SITES OF TRIPOLITANIA

During 1945 and 1946 three officers of the British Military Administration of Tripolitania, Colonel Perry, Major Sandison, and Major Balfour-Paul, made a series of archaeological reconnaissance trips across a roughly triangular region of country, bounded by the coast for some 145 km. to the east of Tripoli on the north, and extending up to 240 km. south to Mizda, on the Upper Wadi Sofeggin. The geographical context of this area may be analysed as follows:

The northern margin of the desert plateau in northern Tripolitania and eastern Tunisia is formed by a crescentic range of hills starting near the Gulf of Gabes, where they are known as the Gebel Ksour, and extending first south-east and then curving east and north-east to reach the coast near the town of Homs, some 145 km. east of Tripoli, where they are known locally as the Gebel Tarhuna. The central and eastern sectors are often called collectively the Gebel Nefusa, and form the main part of the range, reaching an average altitude of between 300 and 450 m. The segmental area of plain to the north of the range, measuring some 480 km. along the chord and 200 km. across at the widest point, is known as the Gefara.

The climatic effect of the hills, especially those near the east such as the Gebel Tarhuna, is to attract a relatively greater rainfall than the surrounding territory, and so form a fertile zone and supply the source of water for oases and wells in the Gefara at their foot. The past and present drainage system in this district is represented by two well-defined groups of wadis—one facing northwards across the Gefara, and the other facing eastwards across the north-western end of the Sirtican Desert. In the latter region the margin of the inland plateau offers a

[1] Bates (1914).

considerably more gradual slope than the bold escarpment bounding the Gefara, and follows a roughly south-easterly line for some 1000 km.

Thus, apart from proximity, it is clear that climatic and topographical considerations also serve to connect the Gebel Tarhuna region more intimately with the main province of the Maghreb than with such habitable regions as lie to the east. Mention has already been made of the extension into this area of at least one late hunting culture, the so-called Intergetulo-Neolithic, located in central Tunisia, and it is interesting to recall that the administrative patterns of early historic cultures show the same relationship, which remained in force up to the end of the Byzantine epoch.

On the other hand, it is evident that some recent changes have taken place in the local ecology. Considerable areas of the Gefara and even some parts of the Gebel are covered at the present time by fields of mobile dunes. The old land surfaces exposed between the ridges of the dunes prove to be by far the most prolific in ancient finds of all dates up to the Roman period, though virtually unoccupied today. Again, whether owing to natural causes such as climate change, or to artificial factors such as exhaustion of the soil or over-grazing, the greater extent of the cultivated area inland to the south in Roman times, as compared with the present, is a striking fact that cannot be overlooked. On the whole it seems reasonable to assume that climatic change has played some part in producing this effect, and consequently to infer that the Gefara as a whole would have been more habitable and an even more effective link with the west in late prehistoric times.

In a word, the area covered by the reconnaissance may be described as the fertile eastern horn of the Gebel Nefusa crescent, and immediate vicinity. The sites discovered, including those previously reported, fall into three main groups —a northern group among the dunes of the Gefara close to the coast, a central group extending eastwards along a line starting at the foot of the escarpment south of Tripoli and ending at a point on the plateau some 25 km. south of the extreme north-eastern tip of the Gebel, and finally, a group of sites centred round an area some 150 km. south of the coast at the southern foot of the Gebel Nefusa, with a few more northerly outliers.

The sites were numbered by the finders from 1 to 26, and the report forwarded to the Museum of Archaeology and Ethnology with a type series of specimens, including a map showing the location of each, and an analysis of the numbers in each class at each site. While the present writer does not agree in every point with the original classification, this very systematic presentation of the field observations enables the results described below to be offered with considerable confidence.

The bulk of the important material comes from the northern group—1450 pieces—with a further contribution from sites of the central group at the foot of the hills or on the plateau near their northern limit. Material definitely assignable to the Neolithic from the southern sites, though sufficient to demonstrate some occupation at that date, is too insignificant to afford any real indication of cultural facies.

Detailed descriptions of the more important types will now be given and followed by a short analysis of their occurrence at the different sites:

Unretouched blades. A sufficient number of these are included in the present collection to throw some light on the technique current. The specimens are medium to small in size and apparently punch-struck with a fair degree of skill from single platform conical cores; the larger pieces reach about 1×5 cm.

Backed blades. (*a*) *Medium.* Only a few of these bordering on the microlithic class occur, and none are of any size. Three pieces in the collection measure about $4 \cdot 0 \times 0 \cdot 5$ cm., have a straight back, pointed base, and are trimmed from the bulbar face only, except for a little possibly intentional squilling near the working edge (Fig. 37, no. 8).

(*b*) *Microlithic.* Two varieties occur, one more or less gibbous, resembling a rough version of the 'pointe scalène', and the other needle-sharp and extremely minute, with bilateral retouch at one end (Fig. 37, nos. 4–6). Similar implements are common in the Sirtican Microlithic but rare or absent from the normal Neolithic-of-Capsian-Tradition in the Maghreb.

Double-backed blades or drills. One is fairly similar to the larger varieties at Siwa and in Tunisia, but the remainder more nearly resemble very rough trihedral rods (Fig. 37, no. 25).

Burins. Only one undoubted example is included, made on a small thick flake, with three burin strokes down one edge (Fig. 37, no. 19). The class as a whole is exceedingly rare throughout the whole territory explored.

Squamous flakes. A single small specimen was found with the characteristic crushing at both ends.

Flake points. Judging by their fresh state and constant association with the other elements, this type, which is otherwise not unlike the Middle Palaeolithic tool of this description, can be certainly attributed to the Neolithic complex. It consists of a rather narrow unifacial point made on a very thin triangular flake frequently with cortex on the back, with two convergent lines of neat trimming, mainly by percussion, more rarely by pressure (Fig. 37, no. 32).

Lunates. Only two examples received, both quite characteristic but very small, trimmed by crushing against an anvil without trace of microburin finish (Fig. 37, nos. 1 and 2).

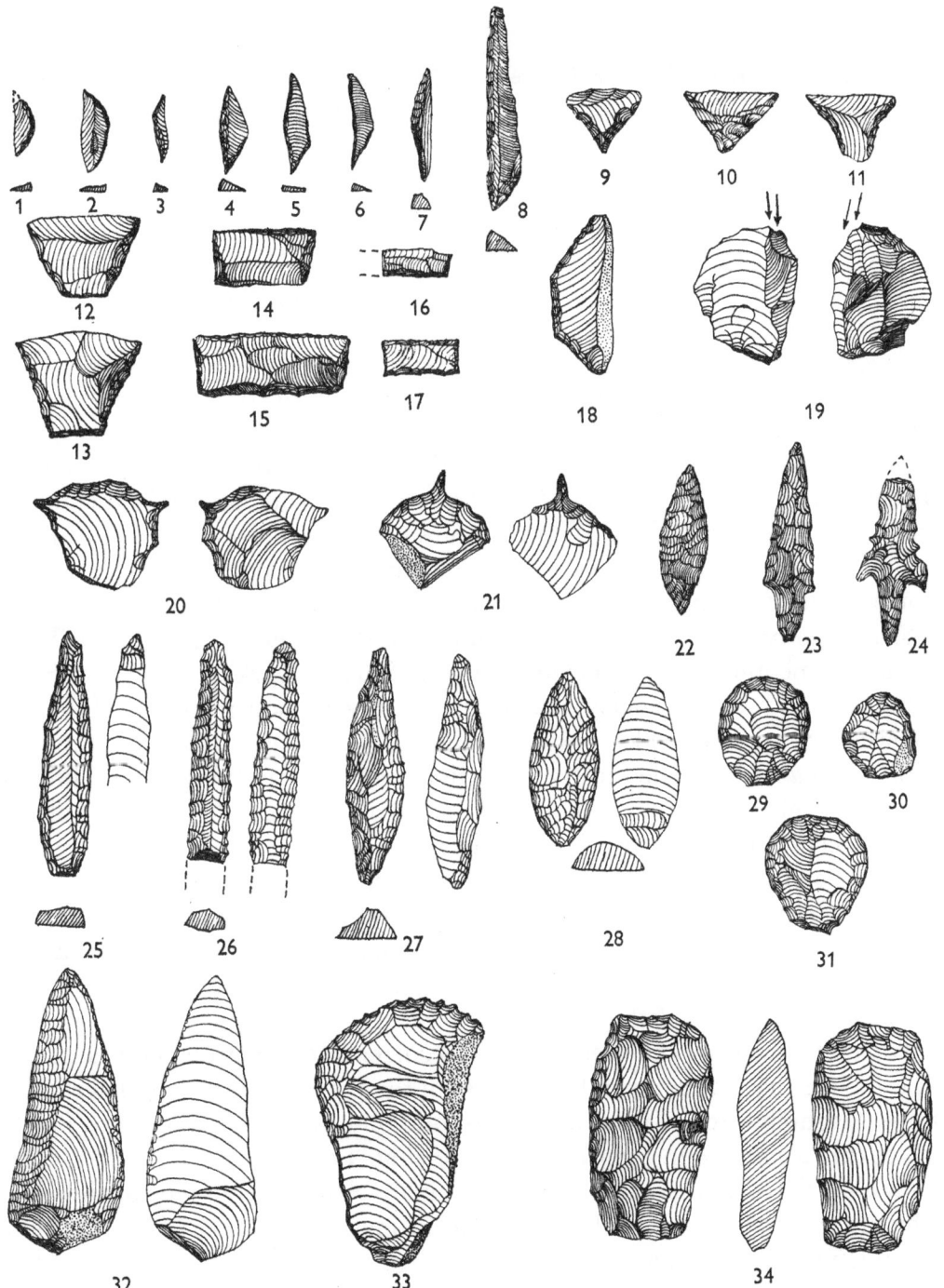

FIG. 37. Northern Tripolitania: representative series of commonly associated types from surface habitation sites

Nos. 1–7, microliths, Nos. 1 and 2 lunates, No. 3 scalene triangle, Nos. 4–6 shanked bladelets; No. 8, Gravette type backed blade; Nos. 9–13, transverse arrow-heads; Nos. 14–17, rectangles; No. 18, trapezoidal blade; No. 19, burin; Nos. 20 and 21, fine awls; Nos. 22–24, various arrow-heads; Nos. 25–27, bilaterally trimmed blades, Nos. 26 and 27 approximating to trihedral rods; No. 28, unifacial foliate; Nos. 29–31, thumbnail scrapers; No. 32, pressure-flaked point; No. 33, denticulated scraper or saw; No. 34, flaked edge. All × ⅔.

Rectangles. These are numerous and absolutely identical in size, proportions, and technique to those of the Neolithic-of-Capsian-Tradition (Fig. 37, nos. 12–17). It may be remarked here that Vaufrey's derivation of this type from Egypt is not altogether convincing, since the supposed prototypes in the latter area are normally much larger, coarser objects with mere terminal truncations and seldom, if ever, have the characteristic finish of 'backing' along three edges invariably featured in the Maghreb.

Scrapers. Somewhat smaller than usual with tools of this sort, the specimens available all seem to belong to a single type, made by steep trimming on a thick flake. The trimming generally extends round the greater part of the periphery to produce an oval or nearly circular outline, varying from 1·5 to 2·5 cm. in diameter. The smaller specimens (Fig. 37, nos. 29–31) are not unlike the minute 'thumb-nail scrapers' of the Wilton culture of South and East Africa. Instruments of this character seem to be a fairly regular accompaniment of the Maghrebian facies of the Neolithic, though less common to the east.

Small oval bifaces. Objects exactly similar to those described from the Tocra plain, and mentioned in connexion with the Shiyata working floor near Siwa, occur here and, as previously stated, are not uncommon in the Neolithic-of-Capsian-Tradition as well.

Bifacial adzes. Included in the collection are three carefully-worked bifacial implements about 4 cm. long, not unlike small adze heads, though much flatter than the characteristic form of eastern Libya described above. One shows possible traces of polish near the cutting edge (Fig. 37, no. 34).

Chisels. A rectangular tool 3·8 × 1·3 cm. worked by pressure over both faces is similar to the specimen placed in this category at Siwa. It belongs to a type not unknown in the Maghreb, though considerably smaller and less definite in form than the chisels of the Fayum.

Trapezes. In form and frequency of occurrence the representatives of this type are in every way comparable to those of the Upper Capsian and later stages in the Maghreb, except in the absence of microburins or other evidence for the use of this technique in their manufacture (Fig. 37, no. 11).

Pressure-flaked transverse arrows. This highly characteristic type of the Neolithic in the Maghreb is well represented in most of the sites of the northern group (Fig. 37, nos. 9 and 10).

Pressure-flaked tanged arrows. These range in size from 1·5 to 2·0 cm. long by 0·7 to 1·6 cm. wide. The shapes normally show a well-detached tang; the extremely elongated type of Siwa seems to be rare, and the minute rod-like form altogether lacking. Binding notches are present on at least two examples (Fig. 37, no. 24) and the forward edges tend to be straight or slightly ogival.

Leaf-shaped arrows. Made in the same pressure technique are one or two small pieces with a foliate outline (Fig. 37, no. 22), one with a rounded base. Not the slightest trace of anything resembling a true Egyptian type of hollow-based arrow was noted.

Awls. These form perhaps the one truly novel feature of the present assemblage, both by their size and technique. Essentially they consist of remarkably fine and delicately made points which are liable to be made on a great variety of flakes, some very small indeed. The margin in the neighbourhood of the point is equally carefully trimmed and there seems little doubt that these tools were designed for some one particular purpose of a specialized nature, tattooing perhaps, or delicate ornamentation of some kind (Fig. 37, nos. 20 and 21).

TABLE V. *Percentage of different classes of tools at Tripolitanian sites of Neolithic typology, compared with a group of Algerian sites*

	Sahel sites of Tripolitania; figures for four individual sites					Jebel sites of Tripolitania; average of seven sites	Neolithic-of-Capsian-Tradition; average of twenty-five sites from Algeria*
	I (254)	III (120)	IV (70)	V (152)	XI† (72)‡		
Blunted-backed blades (small)	13	28	26	18	39	} 37	} 37
Blunted-backed blades (microlithic)	3	18	10	14	14		
Trapezes and triangles	1½	1	1	2	1	½	3
Lunates	0	0	3	0	1	0	6
Rectangles	11	10	9	5	0	½	3
Scrapers (medium)	44	22	14	34	30	} 44	} 17
Scrapers (small)	3	4	16	14	5		
Awls	1	1	1	2	0	3	10§
Notched blades	½	0	1	4	4	1	16
Burins	0	0	4	0	0	5	1
Microburins	0	0	0	0	0	0	5
Bifacial pressure-flaked arrow heads	5	6	6	2	1½	1	1
Other pressure-flaked pieces	9	8	9	3	3	1	½
Other bifacial pieces	9	3	0	2	1½	7	½

* Calculated from data given in Vaufrey (1939).

† Roman numbers refer to reference numbers of sites given in manuscript notes preserved in Museum of Archaeology and Ethnology, Cambridge.

‡ Numbers in brackets refer to number of observations at each site.

This figure includes double-backed blades.

In concluding this description, a word may be said on the relative frequency of occurrence of the above types of artifacts, as analysed in Table V. The broad homogeneity of the material from the different find-localities is striking and reinforces the impression created by the actual scatter of specimens, which suggests that most sites represent the actual camping or occupation sites of a single culture. The figures are slightly modified from those given by the collectors (where their classification seemed to be at fault), and can in general be accepted as representing the true relative importance of the different tool groups in a tradition practised throughout the Gefara and the northern part of the Gebel plateau. It is possible that the scraper and burin classes are slightly exaggerated in the figures given for the plateau sites owing to the inclusion of a few specimens properly belonging to other groups. This is, I think, no sufficient reason for detaching the plateau sites from the remainder as far as the strictly Neolithic material is concerned.

Next arises the question of the nature of the facies represented. It is not without interest in this connexion to note, even allowing for a certain difference in raw material, the profound difference in the assemblage as a whole from that of Siwa and the Egyptian oases. In size, method of collection, and area covered, the two collections can be regarded as in every way comparable, and as representing fair samples of the surface scatter in the two regions. Apart from the obvious distinctions in general appearance, divergences are most apparent in the differential occurrence of a number of key types such as the 'plane' adze and side-blow flakes of the eastern region, and trapeze and true rectangle of the western. On the strength of the two last-mentioned tools, coupled with the evolved appearance of the miniature awls and the extreme rarity of burins, the writer would be inclined to favour a very late stage of the typical Neolithic-of-Capsian-Tradition as the culture responsible for the greater part of the Tripolitanian material.

This extension of a typically Maghrebian complex across the Gefara, coupled with the evidence of sharp cultural cleavage with eastern Libya, raises the further query of where exactly the boundary between the two provinces may have lain. At the present day the ecological contrast between the more fertile portions of the Gefara and practically the whole of the desolate country south and south-east of the Gulf of Sirte is such as to suggest inevitably that even under conditions of slightly greater humidity here was the main barrier to active coastwise intercourse along the northern edge of the continent. That *some* intercourse took place even at this early stage is of course to be expected on general grounds, and is borne out by the widespread occurrence of such forms as the tanged arrow, the trihedral rod, and the idea of impressed ornamentation of pottery, all practically ubiquitous throughout the Sahara. But that such intercourse ever reached the proportions sometimes attributed to it, or could account to any important degree for the presence of most of the Eastern Mediterranean traits observed in the west at or about this time is, I think, rendered appreciably less probable by the evidence just outlined. As a matter of general interest, it may be noted that the effect of this conclusion is to enhance the importance of the sea route along the North Mediterranean (as distinct from the African land route) as the main channel of diffusion between the early foci of cultural development in the Near East and western and southern Europe.

As far as Africa is concerned, the effect of the suggested subdivision of Libya into two fairly well-defined cultural provinces during the Neolithic, must surely be to stimulate further inquiry into the possibility of diverse origins for a number of important features of the epoch, including, perhaps, some of those of direct economic significance. After all, however obvious may be the advantages of the Nile Valley for the development of the culture traits of higher civilization, its advantages for the *initiation* of these traits in their earliest form is by no means so clear. And although it is, of course, well known that the greater prestige of an advanced culture tends to favour the diffusion of some of its elements among less developed neighbours nevertheless, as already suggested, the too ready assumption of such a process in a particular case, especially over vast areas and widely differing ecological territories, may well lead to seriously distorted conclusions.[1]

[1] No further reference is here deemed necessary to the surface finds of Neolithic character from Cyrenaica itself (described in McBurney (1947) pp. 79 and 82), since they add little to the present discussion. Some points of interest do, however, arise from the most recent finds in the Haua Fteah (see p. 218n.) which only became available after the present text was in press. Here an industry characterized by black burnished pottery in globular shapes (comparable to one of the Neolithic 'A' wares of the Fayum described in Caton-Thompson and Gardner (1934, pp. 35–6) is associated with 'plane' adzes, bifacial lance-heads, tanged arrows, large flake-scrapers, angle-burins, trihedral rods, and microlithic backed blades. Altogether this assemblage suggests an early coastwise diffusion of late 'A' or early 'B' traits to an indigenous hunting culture. Carbon readings suggest a date shortly after 4,800 ± 350 B.C. (Suess (1954)).

CHAPTER XVI

FINAL SUMMARY OF THE
ARCHAEOLOGICAL RESULTS

A list of conclusions has been provided at the end of each chapter in the archaeological section of this work, and a general review of Cyrenaican and Tripolitanian hunting cultures as seen against the background of research in other parts of North Africa, appears at the end of Chapter XIV. In addition it is thought that a final summary of the new observations, and the conclusions to which they give rise may be of use for reference purposes.

I. THE LOWER PALAEOLITHIC

There are some grounds for attributing a hand-axe industry from Northern Tripolitania to a period of considerable rainfall increase (p. 223); this would accord with the implications of the hand-axe distribution in the Western Sahara, and recently published discoveries in the Kharga Depression.

In Cyrenaica scattered finds suggest a late Acheulean occupation of unknown geological date (p. 172); the earliest datable discoveries are slight traces of a working floor of unknown culture in the beach conglomerates of a eustatic high-sea-level between 15 and 25 m. above the present (p. 160).

2. THE MIDDLE PALAEOLITHIC

Industries belonging technologically and typologically to this class, can be dated to three distinct geological phases in Cyrenaica.

The earliest comprise rare but adequately characterized traces of a flake industry with faceted striking-platforms, discoid cores, and trimmed flake-tools, found at three localities in the beach conglomerates of a 6 m. eustatic high sea-level (p. 162).

The second phase offers unusually complete data on a Levalloiso-Mousterian of Palestinian affinities (Chapter X), practised, according to the latest information, by a Neanderthaloid population, resembling in some respects at least, that found at Mount Carmel. This phase is dated geologically to the period of regression in the sea-level immediately following the 6 m. stage. Climatic evidence points to damp and relatively mild conditions, and correlation is proposed with the corresponding stage in Southern Italy, usually regarded as of early Last Glacial (Würm I) date (p. 134).

The third phase contains abundant traces of an industry of Levalloisian character (Chapter XI), all evidently representing quarry and workshop debris near the source of raw material. For this reason exact comparison is difficult with the preceding phase, but it is suggested that the basic difference in the tradition is not as great as might appear at first sight. The deposits containing this material are the Younger Gravels (Chapters V–VI) representing a relatively drier climate with marked winter frost action, accompanying a further important fall in the sea-level. It is correlated on climatic and stratigraphical grounds with similar formations in Southern Italy, generally assigned to the second advance of the Last Glacial (Würm II) (p. 134).

Correlation is proposed between the second and third phases of the Middle Palaeolithic in Cyrenaica and the later Lower Levalloiso-Mousterian, and Upper Levalloiso-Mousterian of Palestine and Syria; the former on archaeo-logical and palaeontological grounds (see for instance Miss Bate's discussion of the Skhul horizon at Mount Carmel in Garrod and Bate, 1937, p. 149), and the latter mainly on climatic and palaeontological grounds (see particularly H. E. Wright, 1951).

3. THE UPPER PALAEOLITHIC

If the above scheme of correlation is correct it would follow that the earliest blade industries in both Cyrenaica and the Levant do not occur before the second interstadial of the Last Glacial (Würm II–III), whereas a fully evolved blade industry was established in Southern Italy at least by the previous interstadial (Würm I–II).

Two distinct variants of blade industry have so far been recognized and studied in some detail in Cyrenaica; evidence regarding their absolute age and stratigraphical sequence only became available after the present work had gone to press, but has been included as far as possible. The earliest was first identified at Hagfet ed Dabba (Chapter XIII) and shows points of resemblance to the Lower Aurignacian (Emiran) of the Levant on the one hand, and to the Typical Capsian of Tunisia on the other. It is apparently separated from the latest local Levalloiso-Mousterian horizon by an interval of about 15,000 years during which nothing is yet known of the human occupation of the area. Radio-carbon readings suggest a date close to 14,000 B.C. for the first known occurrence.[1]

Some evolutionary development can be traced in this variant and it is noted that the later levels show less resemblance to the Levantine industries than the

---

[1] Suess, H. E. (1954), and McBurney, Trevor and Wells (1954).

earlier levels. Both the absolute dating and recent climatic evidence[1] from the character of the containing deposits—suggesting dry conditions with appreciable winter frosts—point to a final Last Glacial (perhaps Würm III) date.

The second variant was first recognized at Hagfet et Tera (Chapter XII); according to the latest evidence this variant overlies that of ed Dabba, and can be dated according to carbon readings from the ninth to the sixth millennia B.C. It resembles the Oranian (Ibero-Maurusian) of the Maghreb littoral in the overwhelming predominance of semi-microlithic blunted-backed blades over all other tool forms. It is noticeable however that both Cyrenaican variants differ from their nearest Maghrebian equivalents in the entire absence of geometric microliths (other than the lunate); they also differ on the other hand from the earlier stages of the Levantine Upper Palaeolithic in their much greater development of the blunted-backed blade element.

Two late microlithic cultures are the only blade industries (other than Neolithic) so far known from Sirtica or Tripolitania.

In Sirtica two widely separated sites reveal the presence of a hitherto unknown microlithic culture of which the leading type is a curious and highly character-istic arrow-head, the round-based point, which recurs in slightly modified form in the Neolithic of Siwa and Cyrenaica. Other elements are reminiscent of the Upper Capsian, although there are no true trapezes. In Tripolitania a stratified site (p. 229) has yielded a late facies of the Upper Capsian (of the type formerly known as the Intergetulo-Neolithic) superimposed on a deposit containing Aterian. Nothing is known of the time interval separating the two.

4. THE NEOLITHIC

Two large unpublished collections from the Siwa Depression and from Northern Tripolitania help to reconstruct the distribution of prehistoric cultures at the end of the stone-using period in Northern Libya, and supplement the evidence of the smaller collections from Cyrenaica. Taken together with the most recent work in the Maghreb and Egypt they suggest the subdivision of the North African littoral into two sharply defined culture provinces lying east and west of the Sirtican desert respectively. The western province—of which Tripo-litania is obviously part—is predominantly African in character and closely linked by numerous intermediate finds in the Hoggar-Tibesti massif, to the Neolithic cultures of Nigeria, the Sudan, and the Upper Nile.

The eastern province comprises the oases and habitable patches of a triangular area lying north and north east of the Libyan Sand Sea; it may be said to

[1] Briefly mentioned in McBurney, Trevor and Wells (1953).

extend from the Gebel Akhdar eastwards to the Fayum and southwards to Kharga. Within this area, relatively restricted by comparison with the vast western province, a remarkably constant culture pattern of incipient food-producing type seems to have been established starting as early as the late fifth millennium B.C. It can be ascribed to culture contact between the indigenous microlith using hunters, and intrusive food-producing groups ultimately deriving from South West Asia. The special character of this province, already noticeable in the later hunting cultures, seems to have been maintained locally until well into historical times, to judge from the uniformity and consistency of the archaeological material.

The contrast between the two North African provinces is so strong indeed that some considerable readjustment of current notions regarding the exchange of cultural elements along the North African littoral would seem to be called for. On the one hand the effect is to enhance the importance of Nubia and the central Saharan highlands as centres of secondary diffusion and even formation of not a few basic elements formally thought of as passing directly along the Mediterranean littoral. On the other hand the possibility that some significant traits were invented or considerably developed in the Maghreb itself, is also worthy of reconsideration.

5. THE PROSPECT FOR FURTHER RESEARCH

Brief allusions to work in progress, inserted while the present report was already in the press, give some idea of the promise of further interesting results which the region still has in store.

In particular it is hoped that the recently discovered site of Haua Fteah may go far towards the ideal of a complete sequence extending from the Last Inter-glacial up to the end of the prehistoric epoch. Whether this promise is fulfilled or not depends of course on a variety of practical considerations; its fulfilment will in any case necessitate research over a prolonged period. In the meantime the material in the present report is offered in the belief that it provides a reliable framework of some of the more important natural and cultural events in the area, to which later and more detailed discoveries in this and neighbouring regions can usefully be related.

VERTEBRATE FAUNAS OF QUATERNARY DEPOSITS IN CYRENAICA

By DOROTHEA M. A. BATE

British Museum (Natural History)

I. INTRODUCTION

On looking at a vegetation map of Africa it is immediately seen that the fertile areas north of the Sahara and west of the Nile Delta are two in number, and that they differ greatly in extent. The larger and westernmost includes Morocco from its Atlantic coast, northern Algeria, and Tunisia extending along the coast line to Tripoli. This section is generally referred to by French authors as Barbary. Much further east, and separated by a long stretch of inhospitable country, lies the promontory of Cyrenaica with its high land, forests and water; this fertile country soon loses itself in the desert to the south. East of Cyrenaica lies another long stretch of arid country before the Nile Delta is reached. This brief glimpse of present-day conditions of the North African littoral is important to remember in dealing with the earlier mammal faunas of this long northern border of the continent stretching between the Mediterranean Sea and the Sahara desert region. It is becoming increasingly evident that this region, together with the country bordering the southern edge of what is now a desert area, occupies a key position for the elucidation of the succession of the Pleistocene climatic changes, or fluctuations, which took place in this part of Africa, and which were without doubt correlated with the changes which occurred over a great part of the whole continent.

Mammal faunas are among the best guides towards an understanding of the earlier climatic sequences of a country. Much is already known of these faunas in Morocco and Algeria thanks to a number of splendid French workers, including P. Thomas, Pomel, Joleaud and many others, while today Professor C. Arambourg is doing invaluable work, to which the writer is greatly indebted, in studying and re-investigating the early faunas of these countries in a modern manner. He has already published the broad outline of faunal changes in North Africa (Arambourg, 1934), and has suggested (1932) that the immigration of the larger Eurasiatic species, bears, deer, oxen, pig, etc., took place in early Pleistocene (post-Villafranchian), or at latest during early Palaeolithic times. This author has also written of the uniformity of the mammalian fauna of the Pleistocene of North Africa from the Atlantic coast of Morocco to its eastern extremity (Arambourg, 1929).

It is unlikely that the immigration of Eurasiatic species all took place at exactly the same time, and the small size and isolated geographical position of Cyrenaica, today bordered by arid conditions on its three landward margins, place it in a position to have registered the times both of these immigrations and of the accompanying climatic conditions. Besides the possible opportunities for north-to-south and south-to-north interchanges the migration

274

route between North Africa and Palestine was no doubt available at intervals throughout the Pleistocene, and on to Neolithic times. In so far as the mammalian fauna is concerned there is certainly no need for raising land bridges across the Mediterranean unless these are very fully vouched for geologically. One intriguing question for which so far no satisfactory answer seems to have been suggested is: why is it that while the remains of many forms now only found living south of the Sahara have been found in North Africa, with few exceptions the Eurasiatic forms such as deer, bear, etc., have not penetrated south of the desert area?

Can it be that the latter are a residue, and that many of the forms now typically African are really comparative newcomers to this continent? It is well known that in the Pliocene, and probably many times later, Africa received immigrations, for instance of antelopes (Lönnberg, 1929 and Bate, 1940) when there was free intercontinental intercourse. This raises interesting possibilities. How long, for instance, has the now typically African lion been an inhabitant of this Continent? An attempt to answer this question is made in the section dealing with this species.

The present collection was most carefully made and recorded, by Dr C. B. M. McBurney and Dr R. W. Hey and it is regrettable that the animal remains are not more numerous in species, or more complete as specimens. This latter imperfect condition is of course due to the animals having been used for human food. Nevertheless, a sound basis for further work has been provided, and a number of new facts and fresh records of great interest have been brought to light. The animal remains fall naturally into two groups which will be described in different sections. First there are those from Wadi Derna, an open-air site containing a Levalloiso-Mousterian human industry. Secondly are the remains from two caves, one of which contained an early, and the other a late blade industry of Upper Palaeolithic or Mesolithic typology.

2. THE WADI DERNA FAUNA

As already mentioned, the deposit of Wadi Derna contains a Levalloiso-Mousterian human industry which has been tentatively correlated with the Lower Levalloiso-Mousterian of Palestine (see p. 155). The fauna associated with the latter industry in the Wadi el Mughara caves was of great interest including a number of extinct species of early type which did not survive the great change which overtook the fauna at the close of this section when *Rhinoceros* and *Hippopotamus* also disappeared from the country (Bate, 1937). In Algeria it is recorded that deposits yielding a Mousterian industry also contain a fauna which includes *Hippopotamus* and *Rhinoceros mercki* (Arambourg (1934), p. 32). The deposit of Wadi Derna is composed of a fine water-laid silt, and a considerable quantity of the matrix was collected and has yielded plant remains and fresh-water shells described in other sections of the report. A number of blocks of this matrix were dissolved and carefully searched for the remains of small mammals, but unfortunately without yielding a single specimen.

Remains of large mammals were not plentiful except in the form of broken and un-identifiable fragments. Some of these must have belonged to animals of great size, perhaps hippopotamus or rhinoceros, species which would be expected in a deposit of this age. That there had been some movement of the deposit is shown by the fact that various specimens were badly crushed, notably a portion of the skull of a Barbary sheep of large size, also one of the horn cores of an extinct buffalo whose presence certainly seems to confirm the early age of the fauna. It was only possible to determine four species from this deposit.

Homoioceras sp.	Extinct buffalo
Ammotragus sp.	Barbary sheep
Hippotigris sp.	Zebra
Testudo sp.	Small land tortoise

There is a good deal of difference in the size of the bovine limb bones, and it is possible that remains of a large antelope may be present. The three identified species of mammals are each represented by an equal number of specimens, about fifteen. There are a few fragments of the plastron and carapace of tortoise.

3. THE FAUNA FROM HAGFET ET TERA

The fauna obtained by Dr McBurney is very much later than that found in the Wadi Derna deposit, from which it is separated by a long interval of time during which considerable geological events took place. This was the first cave in Cyrenaica to be excavated by Signor C. Petrocchi, who published his results in 1940. This account includes lists of species of mammals from several levels, some earlier than those represented in the present collection. These lists were contributed by Baron G. A. Blanc, whose splendid work in the Grotta Romanelli, Southern Italy, is so deservedly well known. It is unfortunate that no figures or detailed descriptions are given of any of the specimens, for I cannot help thinking that familiarity with the Romanelli fauna may have influenced the identifications of imperfect specimens from Hagfet et Tera. It may be well to quote the lists given before voicing one or two of my doubts.

'*Equus asinus hidruntinus*	*Antilope* sp.
Bos primigenius	*Equus caballus*'
Capra sp.	

The above came from a breccia containing typical Mousterian implements.

Layer G also contained a Middle Palaeolithic industry and provided the following species:

'*Capra* sp. (noticeably of large size)	*Hystrix cristata*
Equus caballus	*Ovis*?'

Layer F was sterile, and the later Layer D contained remains of:

'*Bos primigenius*	*Capra* sp.
Bos sp.	*Antilope* sp.
Rhinoceros (*mercki* ?)	*Equus asinus hidruntinus*'
Equus caballus	

and a coprolite of hyena.

From Layers B and C with an Upper Palaeolithic industry came:

'*Bos primigenius* (of large size)	*Antilope* sp.
Bos (*primigenius* ?)	*Cervus* sp.'
Equus caballus	

It will be noticed that *Equus caballus*, the true horse, is included in the list of species from each level. It would be most unexpected to find remains of this animal from a Mousterian, perhaps even from any level, in this area. Professor Arambourg (1938) in his account of the fossil mammals of Morocco records *E. caballus* from Neolithic levels only, and then with a query. Romer (1928) was equally doubtful of its occurrence in Algeria. For this reason it seems necessary for such a record to be established by convincing figures. Zebras may attain to a considerable size, and where several equines are present it is often a matter of considerable difficulty to distinguish their remains satisfactorily.

Another species that figures in two of the lists in Petrocchi's paper is *Equus asinus hydruntinus*. This is the name used for small equines found in deposits of various sections of the Pleistocene in central and southern Europe (Stehlin and Graziosi, 1935). In view of the fact that wild asses are indigenous in North Africa and were probably represented by two species which survived until not so very long ago (Monod, 1933) it seems unnecessary to suggest the presence of a European form.

There are records of *Capra* from the three earlier levels, that from Level G being labelled as of very large size. There seems to be no other record of a large goat from the Pleistocene of North Africa, and one cannot but wonder if *Ammotragus* were not the real owner of these remains. The fact that no surprise is expressed, or stress laid on these unexpected and new records does not tend to allay one's doubts about their validity.

The mammal remains from Hagfet et Tera in the present collection come from two archaeological levels only, Layers A and B, each of which contained a Mesolithic or Upper Palaeolithic industry. These remains may be dealt with together, and among a quantity of fragmentary specimens only four species could be distinguished:

Homoioceras or *Bos*	Buffalo or ox (six specimens)
Ammotragus sp.	Barbary sheep (five specimens)
Gazella sp.	Gazelle (thirty-three specimens)
? *Hippotigris* sp.	? Zebra (six specimens)

The gazelle remains included seven fragments of horn cores; and two complete ones represented male and female Dorcas gazelles.

4. THE FAUNA FROM HAGFET ED DABBA

This cave was newly excavated by Dr McBurney, who has divided his seven Upper Palaeolithic levels into two groups, Layers I–III (Upper) and Layers IV–VII (Lower) in which he finds distinct cultural development. As in so many caves the animal remains are very fragmentary, and the identifiable specimens comparatively few. It does not seem necessary, therefore, to describe the remains level by level, for there is nothing noticeable among them to suggest any faunal change. There is, moreover, a suggestion of stability conveyed by the microtine remains which occur in Layers V–II, with maximum numbers in Layers IV and II, thus ignoring the division shown by the human industry.

In the Hagfet ed Dabba levels the number of remains of Barbary sheep far surpassed those of any other species, and there can be no doubt that this animal must have formed the staple food of the cave inhabitants. The numbers of specimens for each level are given in § 5.

Gazelle remains, with fourteen specimens, are most plentiful in Level I, the topmost. The bones of bovines are mostly foot bones, isolated teeth, or fragments of long bones, so that it is impossible to determine whether an ox or a buffalo is represented; both these animals probably lived in the country at this time.

It was of no little interest to find an unmistakable bone of a rhinoceros in Layer II. Remains of several small rodents are new to the North African fossil list, and outstanding among these is the first record of a fossil *Microtus*, a new species, probably an early form of the vole living in Cyrenaica at the present day. The remains of the other rodents were too fragmentary for it to be possible to determine whether or not they belonged to species identical with those of the present day, but they provided the first fossil record for North Africa of *Eliomys*, *Apodemus* and *Gerbillus* as well as *Microtus*.

5. NOTES ON THE SPECIES

INSECTIVOROUS BAT

Remains of a small insectivorous bat occurred sparingly in Layers II–V of Hagfet ed Dabba, generally in the form of isolated teeth which do not permit of even generic identification. A number of Recent species occur in Cyrenaica (Zavattari, 1934).

CROCIDURA sp.   WHITE-TOOTHED SHREW

A shrew is represented by two fragments of mandibular rami, one retaining part of the incisor, one unicuspid and the anterior molar. These specimens are not sufficiently complete to enable a specific determination to be made.

Red-toothed shrews are so far unknown from Africa. White-toothed shrews, *Crocidura* are found in Africa at the present day, and several species are recorded from North Africa (Heim de Balsac, 1936). Pomel (1892) in describing his discovery of '*Bramus*' (*Ellobius*) *barbarus* records that he found it not only in western Algeria, but also in Tunisia, where its remains were associated with those of a shrew and of a small bear.

PANTHERA LEO Linnaeus sp.   LION

Lion is represented by a single tooth, a lower carnassial, M 16618 from Layer I of Hagfet ed Dabba. This species is not recorded from Hagfet et Tera, but this is not surprising since remains of large carnivora are not commonly found in later cave deposits. Pomel (1896) described and figured limb bones and teeth from Algerian caves and other deposits which he considered as almost certainly those of the large *Felis spelaea*, or a related species of equal size. Later authors consider that these remains more probably represent a large race of *Panthera leo*.

Fossil remains are also recorded from a cave at Tangier (Howe and Movius, 1947) from Neolithic or later, and from two earlier levels, perhaps Upper Palaeolithic. Professor Arambourg (1938, p. 55) mentions lion from caves in Morocco and says it occurs in Middle and Upper Palaeolithic, and also in Neolithic levels. This, or an earlier large felid is recorded from earliest Pleistocene deposits.

The disappearance from North Africa of Recent lions has been due, directly and indirectly, to the influence of man. Sir Harry Johnston (1898) wrote that three lions were killed in Tunisia in 1880. Another author, perhaps not quite so reliable, remarks in a section on sport (Cook (1908), pp. 13, 14) that in Algeria lions were then nearly extinct, but adds that 'hundreds of lions...have been killed during the last twenty-five years'. This destruction was encouraged by the payment of rewards by the government. Heim de Balsac (1936, p. 98) writes that lions survived in Morocco into the decade 1900–10.

This fossil and Recent record is perhaps what might be expected of an animal so typically 'African' as the Recent lion is today. But this picture is not sustained when inquiry is turned to Africa south of the Sahara, where fossil records seem to be extremely rare. A few specimens from Serengati are very doubtfully referred to lion (Dietrich, 1942), this animal was recorded as scarce at Broken Hill (Hopwood, 1928) and is mentioned with a query from Mumbwa, Northern Rhodesia (Desmond Clark, 1950), while a few foot bones were found at the Neolithic settlement of Esh Shaheinab near Khartoum (Bate, 1950). The remains from Serengati may well belong to *Homotherium*, the machaerodont known from Omo (Arambourg, 1947), while the other records are from deposits of comparatively recent date. This is a very different picture to that shown from deposits in North Africa as briefly mentioned above.

This raises the interesting point whether lions may not be comparatively recent immigrants to Africa south of the Sahara. It will perhaps be some time before it is possible to prove this, but there are some facts that appear to support the theory. Within historical, and even comparatively recent times, lions were common in central and north-west India, south-west Asia, and they penetrated into eastern Europe. Perhaps even more significant is the fact that several extinct species of the genus *Panthera* have been described from India (Pilgrim, 1932).

VULPES sp. FOX

The presence of a species of fox is shown by a single tooth, a lower canine, which comes from Layer I of Hagfet ed Dabba.

According to Zavattari (1934) two races of true fox are found in Cyrenaica today.

GERBILLUS sp. GERBIL

This rodent is represented by a fragmentary right mandibular ramus which retains the three cheek teeth; these are considerably worn, and have a combined length of 3·8 mm. More than one species is found living in Cyrenaica today.

APODEMUS sp. FIELD MOUSE

This mouse is represented by a mandibular ramus with the two anterior cheek teeth, from Layer II, Hagfet ed Dabba, and by a single upper cheek tooth from Layer IV of the same cave.

These specimens seem to provide the first record of a fossil field mouse in North Africa.

At the present day two Recent forms are known, one in Algeria, and another in northern Morocco.

ELIOMYS sp. GARDEN DORMOUSE

A single characteristic upper premolar from Layer IV of Hagfet ed Dabba proves the presence of this dormouse.

Recent species occur in Cyrenaica today, but are not known in Africa outside the northern region.

MICROTUS CYRENAE Bate. CYRENAICA FOSSIL VOLE

The discovery of fossil remains of voles is one of the outstanding features of the present collection. A preliminary description of these has been published in *Annals and Magazine of Natural History* (Bate, 1950a) and it is with the kind permission of the editors that I am able to reprint most of that note with the figures. A considerable quantity of the matrix from the different floor levels of Hagfet ed Dabba was brought home and has been treated by the new method of breaking it down in a solution of acetic acid, which does not adversely affect either bones or teeth. This method has been developed and very satisfactorily practised by Mr A. E. Rixon, who has published a note on the subject (1949); to him the writer is greatly indebted for the assistance he has given in this way.

This method has made it possible to find remains of a number of small species of mammals, whose presence would otherwise have been unsuspected.

As long ago as 1892 Pomel published a description, unfortunately without figures, of a rodent from Pleistocene deposits in Algeria and Tunisia which he named *Bramus barbarus*. This was referred to different genera by different authors, until in 1926 Mr Hinton declared it to be undoubtedly an *Ellobius*, and therefore the first vole to be known from Africa.

In the same year (1926a) Mr Hinton described a species of *Microtus*, *M. musteri*, which Chaworth Musters had found living commonly in the inland basin of Merj[1] in north-west Cyrenaica. Now, more than twenty-five years later, remains of a fossil vole have been discovered, likewise in north-west Cyrenaica, and described as *Microtus cyrenae*.

Diagnosis. A *Microtus* belonging to the *M. guentheri* group, size small. Compared with those of Recent species of the *M. guentheri* group the cheek teeth are shorter, narrower, and the re-entrant angles smaller, the enamel noticeably thinner. Hinder loop of M^3 short. External borders of the prisms rounded in outline.

Holotype. A right mandibular ramus with the two anterior cheek teeth in place (Brit. Mus. M/16620; Fig. 38).

Locality and horizon. Cyrenaica, at present known from Layers V–II, Upper Palaeolithic of Hagfet et Dabba, north-west Cyrenaica.

It was thought at first that the small size of the vole teeth from Hagfet ed Dabba might be due to their belonging to immature animals, but while there is some variation in size there is, on the other hand, great general uniformity of size and pattern in over sixty cheek

[1] The Barce Plain delimited to the south by a high level escarpment.

teeth coming from several levels. While *M. cyrenae* clearly belongs to the *M. guentheri* group, which also includes the Recent North African *M. musteri*, and *M. philistinus* of Palestine, the differences in the teeth probably denote an earlier evolutionary stage slightly less specialized than that of the Recent species. Other differences would most likely be

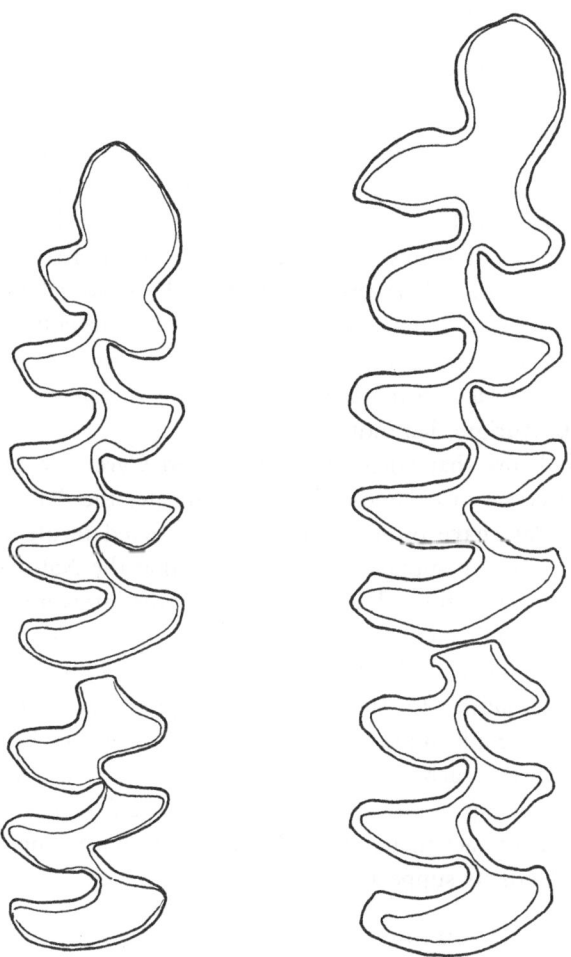

FIG. 38. *Microtus* spp.

(*Left*). *Microtus cyrenae* n.sp. Crown view of right M_{1-2}. From Hagfet ed Dabba, Level II. Holotype. B.M. M/16620. ×7. (*Right*). *Microtus musteri*. Crown view of right M_{1-2}. Recent, from Cyrenaica. B.M. 28. 8. 45. ×7.

shown in the skull, for among the vole remains from Wadi el Mughara, Palestine (Bate, 1937), it was found that the skull most clearly showed developmental changes.

In the present collection there are only two imperfect mandibular rami, each containing the two anterior cheek teeth. One of these has been chosen as the holotype (Brit. Mus. M/16620) and is shown in Fig. 38, the remainder are isolated cheek teeth. In all there

are sixty-three specimens, distributed unequally through four suc cessive levels of Hagfet
ed Dabba as follows:

Hagfet ed Dabba (Upper Palaeolithic):

Layer II	32
Layer III	1
Layer IV	28
Layer V	2

This vertical time distribution of the vole remains is of interest in several ways; it argues
for an occupation of considerable duration, and raises the question of the possible date of
immigration. Dr McBurney has divided the Hagfet ed Dabba layers into two cultural
groups, included in Layers I–III (uppermost) and Layers IV–VII below, but the incidence
of the vole remains does not follow the pattern of the cultural remains. This is not surprising, for changes in mammal faunas, which must be closely bound up with climatic
and geological changes, are not always coincident with human cultural changes. This was
brought out very noticeably in the faunas of the Wadi el Mughara caves, where the great
change in the fauna took place during Upper Levalloiso-Mousterian times, instead of with
the cultural change to Aurignacian industries.

It is important to know that voles were present in Cyrenaica in Upper Palaeolithic
times, but it is not necessary to suppose that this indicates the date of their arrival. They
may even then, as today, have been occupying a restricted fertile country bordered by
desert, and if one is correct in supposing (Bate, 1937) that the Natufian of Palestine was a
time of very dry climatic conditions this would affect all the intervening areas, and preclude
effectually any migration of voles between the two countries. Scarcely any vole remains
were recovered from the later levels of the Palestine caves. This certainly suggests a date
earlier than the Late Mesolithic for the arrival of voles in North Africa.

This discovery helps to confirm the belief, now almost universally held, that the Eurasiatic
mammals of North Africa reached that country only by way of Palestine and northern
Sinai. The close relationship of *M. cyrenae* and *M. musteri* to the *M. guentheri* group of
course shows this, and the fact that the fossil voles of Malta belong to a different genus,
Pitymys (Bate, 1935) lends its support.

HOMOIOCERAS sp. EXTINCT BUFFALO

A large block of matrix thought to contain the skull of a large animal was brought back
from Wadi Derna. When the matrix was carefully removed the block was found to contain
a fragment of skull and horn core of Barbary sheep, and two imperfect horn cores, probably
belonging to a single individual buffalo.

The left horn core, measuring 33 cm. in a straight line, is very crushed and its original
shape cannot be accurately gauged. The piece of right horn core is about 27 cm. long and
has a maximum circumference of nearly 30 cm. This last (Brit. Mus. M 16619) with its
cross-section is shown in Fig. 39. It is almost certain that these represent a species belonging
to the extinct group of long-horned buffaloes *Homoioceras* (Bate, 1949). That it was,
however, a distinct species, probably undescribed, is suggested by the differences it exhibits

from both the North African *H. antiquus*, and from *H. singae* from the Blue Nile, Anglo-Egyptian Sudan. The horn core is smaller, and the section different from the two above-mentioned species. There is evidently no roughened surface on the frontal adjoining the base of the horn core, and this of course suggests that the animal was a female.

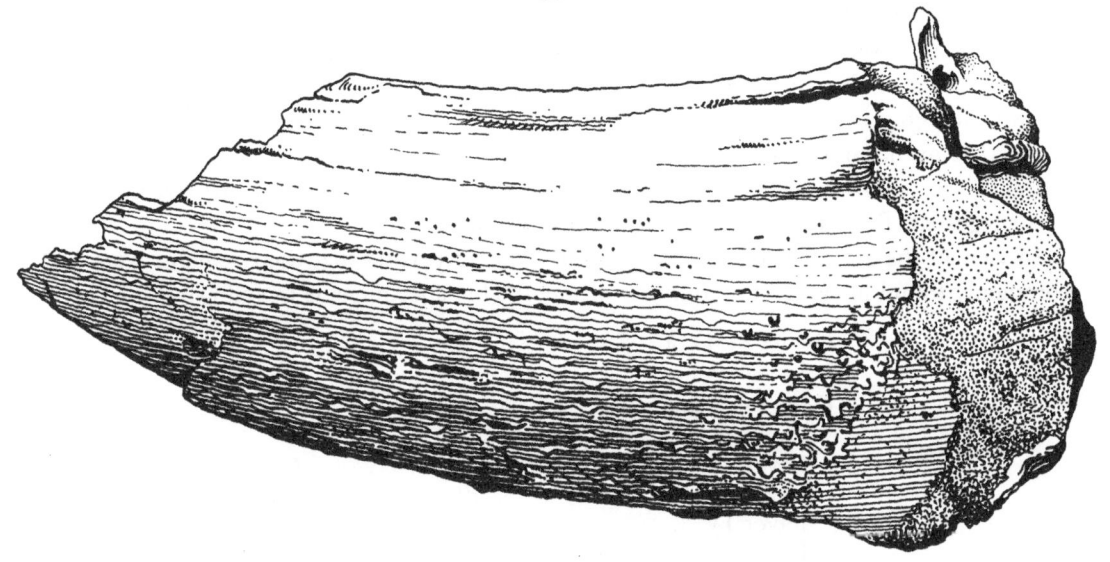

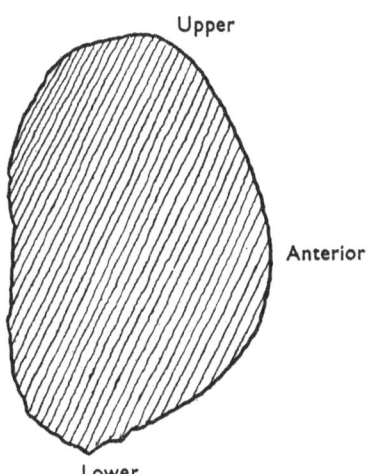

FIG. 39. Hajj Creiem: *Homoioceras* sp.
Side view and section of horn-core. Actual size.

The likelihood that this Cyrenaican buffalo represents an undescribed form seems to support the suggestion put forward in the account of the Singa fauna that the Syncerina (African buffaloes) are an entirely African group. It is only as more or less complete specimens are procured that our knowledge of the various types can be extended. The occurrence of this Cyrenaican buffalo necessarily indicates the presence of a plentiful supply of permanent water.

BOS or HOMOIOCERAS.   OX or AFRICAN BUFFALO

There are a number of specimens belonging to large bovines which it is not possible to identify more closely. The remains chiefly consist of foot bones. Both *Bos* and *Homoioceras* remains occur in North African deposits and the exact date of their extinction is not known, though it is thought that the ox survived the longest. Professor Arambourg (1932) states that oxen, with a number of other Eurasiatic forms, must have arrived in North Africa at latest during lower Palaeolithic times. The fossil long-horned African buffaloes it is suggested (Bate, 1951) are peculiar to this continent.

In view of the great size of these animals it is not surprising that their remains were not commonly found in caves. The number of identifiable specimens is as follows:

Hagfet et Tera:
Layers A 3
Hagfet ed Dabba:
Layer I 11
Layer II 6
Layer III 3
Layer IV 6
Layer V 2
Layer VI 2
Layer VII 2

AMMOTRAGUS sp.   BARBARY SHEEP

The Barbary sheep is the dominant species in the collection, its remains being more plentiful than those of other species in most of the levels, with the exception of Wadi Derna, where its numbers were equalled by those of the zebra. In the two layers of Hagfet et Tera its numbers are equalled, or surpassed, by those of gazelles, but throughout the Upper Palaeolithic layers of Hagfet ed Dabba it is by far the most numerous and must have been the chief source of food of the inhabitants.

The following records the number of identified specimens:

Wadi Derna . . . . 14 *Hippotigris* and *Homoioceras* equal
Hagfet et Tera:
Layers A, C and D . . 5 —
Hagfet ed Dabba:
Layer I . . . . . 47 *Gazella* 14
Layer II . . . . 40 Few other remains
Layer III . . . . 22 Few other remains
Layer IV . . . . 53 Few other remains
Layer V . . . . 8 Few other remains
Layer VI . . . . 58 Only three other bones identified
Layer VII . . . . 6 Few other remains

Some of the specimens are of large size compared with Recent examples, but in general are so fragmentary that it is not possible to discern if any difference exists between those of Wadi Derna and those from later levels. Pomel (1897) has described and figured a quantity of remains of this animal from Pleistocene caves and other deposits in Algeria and Tunisia. One of these specimens is a fine skull, larger than that of Recent examples, to which he gave the name *Ovis palaeotragus*, but this does not seem to have been adopted by modern authors. Great variation in size in the Recent species has long been noted; Blyth (1840, p. 75) wrote of the African 'goat-sheep', as he called it: 'This animal appears to vary considerably in size, some exceeding a Fallow Deer in stature, while others are much smaller.'

Ammotragus is one of the commonest Pleistocene fossil mammals of the whole of the North African area and was not always confined to high mountain localities. It has been recorded from throughout the Pleistocene, being common in both Middle and Upper Palaeolithic times.

Remains of *Ammotragus* have been recorded from one or two localities in Europe but the evidence seems flimsy, and is certainly insufficient in the case of the record from Hundsheim.

Joleaud (1927) remarks that the Barbary sheep was commonly represented on the monuments of the Early and Middle Empires in Egypt. Lortet and Gaillard (1903) have figured (p. 103) a complete individual mummified from Gizeh. Since only three such individuals are mentioned this suggests that they were not commonly so preserved. Barbary sheep are shown in rock paintings at Uweinat, where it is suggested that they were caught in traps (Winkler, 1939, p. 309).

The Recent Barbary sheep, *Ammotragus lervia*, is a most interesting animal. The writer has not had an opportunity of studying its morphology, but it is generally said to be related to both the sheep and the goats. Its French name is 'Mouflon à manchettes'; the adjective given in reference to the cascades of hair which decorate its throat, chest and forelegs is well earned, but the name 'mouflon' is apt to turn one's thoughts to a European origin. Asia appears to be the home of true wild sheep, but it should not be forgotten that Africa south of the Sahara has in the past had its strange sheep-like forms such as *Pelorovis* (Reck, 1928) and it is not impossible that Africa may have had its own early stock of sheep, in the same way that its buffaloes seem to be of undoubted African origin.

The Barbary sheep of today is the only species of its genus, though a number of subspecies have been named. It has a wide and interesting area of distribution: Morocco in western North Africa, parts of the Atlas Mountains, and it survives in the isolated rocky mountain areas of the Aïr District, western Sahara, the Hoggar Mountains, Tibesti, Gebel Uweinat and the Gilf Kebir. It occupies two areas in the Anglo-Egyptian Sudan, on either side of the Nile, and it may still survive, as it did quite recently, in the mountains near the Sabaluka gorge only about sixty miles north of Khartoum (Chapman, 1921, p. 33).

It is an animal amazingly tolerant of differing ecological conditions, and it is able to exist practically without water. It is found in the High Atlas close to the snowline, while on the Atlantic border of Morocco it goes down close to the sea (Joleaud, 1927). It is no doubt this tolerance that accounts for its extended area of distribution, both in Pleistocene times and at the present day.

? ANTILOPE sp. ? ANTELOPE

A few bovine remains in Layer III of Hagfet ed Dabba which cannot be attributed to *Ammotragus* or *Bos* may perhaps represent a large antelope.

GAZELLA spp. GAZELLES

Remains of gazelles do not occur in Wadi Derna, and are not common in any of the cave layers except in the uppermost one of Hagfet et Tera (twenty-eight specimens) and of Hagfet ed Dabba (fourteen specimens). There was reason to suppose that there may have been some disturbance in both these layers. The other layers of Hagfet ed Dabba provided: Layer II, six; Layer III, none; Layer IV, three; Layer V, one; Layer VI, none; Layer VII, one. None of these specimens was specifically identifiable, with the exception of two horn cores from the two layers of Hagfet et Tera, one of a male and one of a female Dorcas gazelle.

HIPPOTIGRIS sp. ZEBRA

Equine remains are not easily distinguished either specifically or generically, for few specimens are diagnostic unless fairly completely preserved. There is an incomplete mandibular ramus from Wadi Derna which from its size and tooth pattern may be considered as representing *Hippotigris* with some certainty. Fifteen other specimens from this deposit probably also belong to this form. Professor Arambourg (1938) has given figures of a number of tooth rows of African equines.

Fragmentary specimens from the two caves can only be tentatively recorded as those of zebra. These were not common in any of the cave layers, and occurred as follows:

Early Mesolithic	Hagfet et Tera:	
	A	1
	B	5
Upper Palaeolithic	Hagfet ed Dabba:	
	I	7
	II	15
	III	9
	IV	2
	V	1
	VI	1
	VII	—

According to Professor Arambourg (1934, p. 66) *Hippotigris* remains are found in North Africa throughout Middle and Upper Palaeolithic levels. Equine remains appear to be common in the Neolithic deposits of North Africa (Romer (1928), p. 121) but it seems that complete specimens are still awaited that will definitely settle the question of their identity.

APPENDIX A

RHINOCEROS sp. RHINOCEROS

Rhinoceros is represented by a single specimen from Layer II of Hagfet ed Dabba. This is the distal end of a rather small tibia, too damaged to supply measurements. From this single fragmentary specimen it is impossible to suggest to what species it belonged. It was found at the back of the cave in consolidated red earth typical of Layer II.

Remains of *Rhinoceros* have long been known from Palaeolithic deposits of North Africa (Arambourg, 1934), but it is only comparatively recently that it has been recorded from deposits of as late a time as the Neolithic (Arambourg, 1931). Some bones have even been found with pottery, but none of these specimens of late date are sufficiently complete to admit of specific determination. A specimen has been recorded from the rock shelter of Tamar Hat, one of the Beni Ségoual series, which probably dates from Mesolithic times, so that it is not surprising that remains of this animal should occur in the Upper Palaeolithic of Hagfet ed Dabba.

? GYPS sp. VULTURE

The collection includes a negligible number of bird bones, and only two of these are identifiable. They represent a large accipitrine bird, probably a vulture, and each is the distal extremity of an ulna. One of these from Layer VI of Hagfet ed Dabba has been cut off the shaft of the bone about 11 mm. from its articular end. The maximum length of the specimen is 40 mm. It is somewhat damaged, and apart from the marks of cutting for its removal from the shaft it does not appear to have been worked, or used in any way. Possibly the long tubular shaft of the bone may have been made use of. The second specimen, which is considerably damaged, is from Layer IV of Hagfet ed Dabba. It is very similar, though somewhat larger, than the example from Layer VI. Both have been compared with the corresponding bone of a griffon vulture, which they resemble except for their slightly greater size.

TESTUDO sp. SMALL LAND TORTOISE

A few fragmentary remains of a small land tortoise occurred in several of the deposits, but never in quantity, not more than about half a dozen scraps of carapace or plastron. Such a number was found at Wadi Derna, and at Hagfet ed Dabba Layers I and VI, with one or two in Layers III, IV and V. It cannot have been much esteemed as an article of food.

6. CONCLUSION

The collection of animal remains described above comes from three different sites in Cyrenaica, and is of three different ages. Most of the specimens are fragmentary, but it has been possible to distinguish seventeen species (Table IV). Previously very little was known of the Pleistocene fauna of this isolated area of North Africa. The earliest site is a fresh water deposit enclosing a Lower Levalloiso-Mousterian human industry. The most interesting specimens from here are portions of horn cores of what is believed to be an extinct and undescribed species of buffalo. This seems to add weight to the suggestion (Bate, 1950 (in the Press)) that the African buffaloes are indigenous to this continent. Of the two caves excavated, Hagfet et Tera with an Early Mesolithic Culture and Hagfet ed Dabba with a Upper Palaeolithic Culture, the latter has yielded much of interest, including remains of a field mouse, and of a garden dormouse, both new to the fossil list of North Africa. Perhaps most interesting is a quantity of isolated cheek teeth of a vole which is described as a new species, and is believed to be an early form of the vole found living in the country today. Part of a limb bone of a rhinoceros corroborates the comparatively recent occurrence of this genus in North Africa, which has already been reported for the western portion of the territory. As so often in North Africa the predominant species in

Species	Levalloiso-Mousterian Wadi Derna	Early Mesolithic Hagfet et Tera (Cave)		Upper Palaeolithic Hagfet ed Dabba (Cave)	
		B	A	Layers IV–VII	Layers I–III
Insectivorous bat	–	–	.	×	×
Crocidura sp.	–	–	–	×	–
Panthera leo	–	–	–	–	×
Vulpes sp.	–	–	–	–	×
Gerbillus sp.	–	–	–	–	×
Apodemus sp.	–	–	–	×	×
Eliomys sp.	–	–	–	×	–
Microtus cyrenae	–	–	–	×	×
Homoioceras sp.	×	×	×	×	×
Bos or *Homoioceras*	×	×	×	×	×
Ammotragus sp.	×	×	×	×	×
? *Antilope* sp.	×	–	–	–	×
Gazella spp.	–	×	×	×	×
Hippotigris sp.	×	×	×	×	×
Rhinoceros sp.	–	–	–	–	×
? *Gyps* sp.	–	–	–	×	–
Testudo sp.	×	–	–	×	×

this collection is the Barbary sheep, and for various reasons the possibility is suggested that this animal, like the African buffaloes, may be peculiar to the African continent. Reasons are given for suggesting that the lion, on the contrary, is a comparatively recent Asiatic immigrant to Africa. Except for the Wadi Derna buffalo, the collection is not sufficiently comprehensive to provide evidence of possible faunal alterations, or to give much information regarding climatic conditions which may have been much as today during the Upper Palaeolithic and Mesolithic stages. There is nothing in the collection to suggest that the domestication of animals had been begun.

Several species whose remains might have been expected to occur, such as deer and wild boar, are absent. These are commonly found in deposits further west; they may not have lived in Cyrenaica at this time, or their absence may be due to preferences of the inhabitants. It is even easy to suppose that their remains may yet be found in other similar deposits.

APPENDIX A

Grateful thanks are due to Mr D. Woodall and Mr G. Smith for the drawings accompanying this paper; also to Dr A. T. Hopwood for helpful discussions, especially regarding the equine remains, to Dr W. E. Swinton for examining the tortoise remains, and to Mr R. W. Hayman for help in comparing some of the smaller specimens, particularly those of Chiroptera. Much gratitude is felt for Dr McBurney's kindness in presenting many of the specimens to the national collection.

POSTSCRIPT
SEPTEMBER 1954

The above report, completed shortly before the death of Miss D. M. A. Bate, was based on a series of specimens, selected for the most part before preparation, as likely to offer the best chances of a specific or generic identification, for palaeontological purposes. Subsequently the entire collection amounting to several thousand specimens—the great majority unidentifiable fragments of food-bones—was prepared for archaeological examination, and some further identifications, for the most part of a more generalized character, proved possible. The figures derived from this work have been used for the statistical analyses of food-animals given in the archaeological section. More generalized classification also seemed advisable in this section; thus 'Large Bovine' is used to include remains attributed to *Homoioceros*, ? antelope, and large *Bos*. In addition three new identifications were made; a single specimen of gazelle at Hajj Creiem, a molar of hartebeeste at Hagfet ed Dabba Layer II, and an unlayered but fossilized fragment of hyena humerus at the same site.

An analysis of the individual bones by which the different species were represented at different sites and stages of culture was also undertaken, but is not quoted here in detail as no certainly significant differences were revealed. As far as the data go, they suggest that the greater part of the carcass was brought back to the living-sites at all periods.

This additional work was carried out by Mr H. P. R. Bury, with the kind help and advice of Dr A. T. Hopwood of the British Museum (Natural History).

BIBLIOGRAPHY

ARAMBOURG, C. (1929). 'Mammifères quaternaires de l'Algerie.' *Bull. Soc. Sci. Nat. de l'Afrique du Nord*, vol. XX.

—— (1931). 'Sur la longévité en Afrique du genre Rhinocéros pendant la période quaternaire.' *C.R. Acad. Sci.* vol. CLXII, p. 1044.

—— (1932). 'Les Ours Fossiles de l'Afrique du Nord.' *C.R. Sommaire Séances Soc. Biogéographie*, no. 74, pp. 29–32.

—— (1934). In Arambourg, Boule, Vallois et Verneau, 'Les grottes Paléolithiques des Beni-Segoual (Algérie).' *Arch. Inst. Paléont. Humaine*, Mem. 13 (Paris).

—— (1938). 'Mammifères Fossiles du Maroc.' *Mém. Soc. Sci. Nat. Maroc*, no. XLVI, 74 pp., 9 pls., 15 figs.

—— (1939). 'La Faune Fossile de l'Aïn Tit Mellil (Maroc).' *Bull. Soc. Préhist. Maroc*, 12° année, nos. 1–4, 1938.

APPENDIX A

ARAMBOURG, C. (1947). *Mission Scientifique de l'Omo*, tome I, *Géologie-Anthropologie*, fasc. III. (Editions du Muséum, Paris), pp. 231–562, 40 pls., 91 figs.

BATE, D. M. A. (1935). 'Two New Mammals from the Pleistocene of Malta....' *Proc. Zool. Soc. London*, pt. II, pp. 247–64, 2 figs.

—— (1937). In Garrod and Bate, *The Stone Age of Mount Carmel*, vol. I, pt. 2 (Oxford).

—— (1940). 'The Fossil Antelopes of Palestine in Natufian (Mesolithic) Times, with descriptions of New Species.' *Geol. Mag.* vol. LXXVII, no. 6, Nov.-Dec., pp. 418–43, 9 figs.

—— (1949). 'A New African Fossil Long-horned Buffalo.' *Ann. Mag. Nat. Hist.* (12), vol. II, no. 17, pp. 396–8.

—— (1950). 'The Fauna of Esh Shaheinab.' *The Archaeological News Letter*, vol. II, no. 8, Jan., pp. 128–9.

—— (1950a). 'A Fossil Vole from Cyrenaica.' *Ann. Mag. Nat. Hist.*

—— (1951). 'The Mammals from Singa and Abu Hugar.' In *Fossil Mammals of Africa*, vol. II, pp. 1–28. British Museum (Nat. Hist.), London.

BLYTH, E. (1840). 'An Amended List of the Species of the genus *Ovis*.' *Proc. Zool. Soc. London*, pt. viii, pp. 62–79.

CHAPMAN, A. (1921). *Savage Sudan* (London).

CLARK, J. D. (1950). 'The Associations and Significance of the Human Artifacts from Broken Hill, N. Rhodesia.' *J. Roy. Anthrop. Inst.* vol. LXXVII.

COOK, T. (1908). *Practical Guide to Algeria and Tunisia* (London).

DIETRICH, W. O. (1942). 'Ältestquartäre Säugetiere aus der Südlichen Serengeti, Ostafrika.' *Palaeontographica*, vol. XCIV, Abt. A, 133 pp., 23 pls.

GAILLARD, C. See Lortet and Gaillard.

GRAZIOSI, P. See Stehlin and Graziosi.

HEIM DE BALSAC, H. (1936). 'Biogéographie des Mammifères et des Oiseaux de l'Afrique du Nord.' *Suppl. Bull. Biol. de France et de Belgique*, vol. XXI; 446 pp., 15 distribution maps, 7 pls. (Paris).

HINTON, M. A. C. (1926). 'Note on the Occurrence of a Vole in Northern Africa.' *Ann. Mag. Nat. Hist.* (9), vol. XVIII, pp. 304–6.

—— (1926a). *Monograph of the Voles and Lemmings (Microtinae)*. (Brit. Mus. (Nat. Hist.)).

HOPWOOD, A. T. (1928). 'Mammalia.' In *Rhodesian Man and Associated Remains* (Brit. Mus. (Nat. Hist.)); 75 pp., 5 pls., 23 figs.

HOWE, BRUCE and MOVIUS, HALLAM L. (1947). 'A Stone Age Cave Site in Tangier.' *Papers of Peabody Museum, Am. Archaeol. Harvard Univ.* vol. XXVIII, no. 1, 32 pp., 8 pls.

JOHNSTON, H. H. (1898). 'On the Larger Mammals of Tunisia.' *Proc. Zool. Soc. London*, pp. 351–3.

JOLEAUD, L. (1927). 'Etudes de Géographie Zoologique sur la Berbérie: Le Mouflon à Manchettes.' *C.R. Sommaire Séances Soc. Biogéogr.* no. 25, pp. 43–5.

LÖNNBERG, E. (1929). 'The Development and Distribution of the African Fauna in Connection with...Climatic Changes.' *Arkiv. f. Zoologi....*, Bd. 21A, no. 4 (Stockholm), 33 pp., 5 figs.

LORTET and GAILLARD, C. (1903). 'La Faune momifée de l'Ancienne Egypte.' *Arch. Mus. Hist. Nat. Lyon*, vol. VIII, Mem. 2, pp. i–viii and 1–205, 8 pls., 82 figs.

MONOD, TH. (1933). 'Ânes Sauvages.' *La Terre et la Vie*, 3º Année, no. 8, pp. 451–62, 7 figs. (Paris).

MOVIUS, H. L. See Howe and Movius.

PETROCCHI, C. (1940). 'Ricerche prehistoriche in Cirenaica.' *Africa Italiana*, vol. VIII, nos. 1–2, pp. 1–33.

PILGRIM, G. E. (1932). 'The Fossil Carnivora of India.' *Mem. Geol. Surv. India*, n.s., vol. XVIII, 232 pp., 10 pls.

POMEL, A. (1892). 'Sur le *Bramus*, nouveau type de Rongeur fossile des phosphorites quaternaires de la Berbérie.' *C.R. Acad. Sci. Paris*, vol. CXIV, pp. 1159–63.

APPENDIX A

POMEL, A. (1896). 'Les Carnassiers.' *Carte Géol. de l'Algérie*, vol. III.

—— (1897). 'Les Ovides.' *Carte Géol. de l'Algérie*, vol. III.

RECK, H. (1928). '*Pelorovis oldowayensis* n.g., n.sp.' *Wiss. Ergebn. Oldoway Exped.* Hft. III, pp. 57–67, 1 pl., 1 text-fig.

RIXON, A. E. (1949). 'The Use of Acetic and Formic Acids in the Preparation of Fossil Vertebrates.' *Museums Journal*, vol. XLIX, no. 5, pp. 116–17.

ROMER, A. S. (1928). 'Pleistocene Mammals of Algeria.' *Bull. Logan Mus.* vol. I, no. 2, 163 pp. (Beloit, Wis., U.S.A.).

STEHLIN, H. G. and GRAZIOSI, P. (1935). 'Ricerche sugli Asinidi fossili d'Europa.' *Mem. Soc. Paléont. Suisse*, vol. LVI, 73 pp., 10 pls., 14 figs.

WINKLER, H. A. (1939). 'An Expedition to the Gilf Kebir and "Uweinat", 1938.' *The Geographical Journal*, vol. XCIII, no. 4, pp. 281–313, pls. .

ZAVATTARI, E. (1934). *Prodromo della Fauna della Libia*. Pavia.

APPENDIX B

PLEISTOCENE MARINE FAUNAS

By R.W.H.

Apart from those already mentioned, only three marine Pleistocene deposits were found from which recognizable fossils could be extracted. All three deposits were exposed in sections on the modern shore; none lay more than 3 m. above sea-level, but each one rested upon a bedrock surface which could clearly be seen to form part of the marine terrace of the 6 m. shoreline.

The localities and the forms identified are as follows:

(i) $4\frac{1}{2}$ km. W. of Apollonia (D. 5222–7):

Conus mediterraneus Brug.   *Trochus turbinatus* Born.
Cerithium vulgatum Brug.   *Patella* sp.

(ii) 18 km. W. of Apollonia (D. 5228–47):

Arca sp.   *Mitra cornicula L.
Glycimeris glycimeris (L.)   *Natica* sp.
**Amycla corniculum* (Ol.)   *Cerithium vulgatum* Brug.
Columbella rustica (L.)   *Bittium reticulatum* (Da Costa)
Conus mediterraneus Brug.   *Rissoa lineata* (Risso)

(iii) 5 km. S.W. of Tocra (D. 5248–52):

*Arca barbata L.   *Dosinia lupinus* (L.)
Glycimeris glycimeris (L.)   *Venus gallina* L.
Cardium sp.   *Loripes lacteus* (L.)

All of these forms are still living in the Mediterranean. Those marked with an asterisk have not been previously recorded from the Pleistocene of Cyrenaica. Of those which are unmarked, all except *Venus gallina* and *Loripes lacteus* are included in Blanc's fauna from

APPENDIX B

Coefia;[1] this fauna also was obtained, as has already been seen, from deposits thought to be associated with the 6 m. shoreline.[2] It may be added that the only other Cyrenaican fauna assigned to this stage, namely, that which was collected from the upper marine deposits at 'Locality B', near Benghazi,[3] consisted entirely of forms already recorded from Coefia.

APPENDIX C

MOLLUSCAN FAUNAS FROM FOSSIL DUNES

By R. W. H.

Land-snails were collected from the fossil dunes at two localities, the dunes in both cases being younger than the 6 m. shoreline. The forms identified are as follows:

(i) Derna, Ras bu Azza (D. 5136–55):
    Helicella variabilis (Drap.)   *Helix melanostoma* Drap.
    Xerophila cyrenaica Mart.

(ii) Benghazi, 500 m. S.E. of railway station, between railway and sea (D. 5156–88):
    Rumina decollata (L.)   *H. cespitum* (Drap.)
    Helicella variabilis (Drap.)   *Xerophila cyrenaica* Mart.
    H. pyramidata (Drap.)   *Helix melanostoma* Drap.

All of these species, with the exception of *Xerophila cyrenaica*, have already been identified by Gambetta in a collection from the fossil dunes at Punta Giuliana, Benghazi, together with seven other species not found by the present writer.[4] These latter dunes are also certainly younger than the 6 m. shoreline. In all these collections, *Helicella cespitum* is the only form no longer living in or near the Gebel Akhdar.

APPENDIX D

DESCRIPTION OF A NEW VARIETY OF SNAIL

By R. W. H.

XEROPHILA CHADIANA Pallary var. DARNENSIS nov.

(Plate 15 a)

1911. *Hygromia sordulenta* (Morelet): Newton, op. cit. p. 619 and Pl. XLIII, figs. 1 and 2.

Description. The original species is described in Pallary, P., *Journal de Conchyliologie*, vol. XLVI (1898), pp. 87–8. The main respect in which the new variety differs from the typical *Xerophila chadiana* is the possession of a deeper suture. In addition, most of the

---

[1] Blanc, A. C. (1936).   [2] See above, p. 43.
[3] See above, p. 52.   [4] Gambetta (1934), p. 272.

specimens examined show some or all of the following differences: umbilicus narrower, shell more elevated, striations coarser. Dimensions of largest specimen[1] (here illustrated): diameter 10 mm., height 7 mm.; number of whorls of same specimen: 5½.

Locality. The specimens were found in a bed of marl on the west bank of Wadi Gahham, 250 m. upstream from its confluence with Wadi Derna.

Remarks. The diameter of the new variety is greater than that of *Hygromia sordulenta*, which is given as 7 mm. in Tryon, G. W., *Manual of Conchology*, 2nd ser., vol. III (1887), pp. 177–8. On comparison with specimens of *H. sordulenta* in the British Museum (Nat. Hist.), the new variety was also found to show the following differences: umbilicus less open, aperture less nearly circular, striations coarser.

The identification of the specimens as *Xerophila chadiana* was carried out at the Musée de l'Histoire Naturelle, Paris, with the kind assistance of Professor Fischer. According to Pallary, this is a species which now lives in western Morocco.

The proposed varietal name is derived from 'Darnis', the classical name of Derna.

APPENDIX E

PLANT-IMPRESSIONS FROM THE WADI DERNA TUFAS

By R. W. H.

As stated on p. 109 above, a collection of plant-impressions was obtained from the tufas of the higher terrace of Wadi Derna, near its confluence with Wadi Gahham, and these impressions have been identified as belonging to four different species. In this appendix, the nature of the specimens assigned to each species will be described in some detail; in each case, some mention will also be made of the grounds upon which the identification was based.

Rubus ? ulmifolius (Pl. 15 *b* and *c*).

The specimens are in the form of impressions of individual leaflets, both upper and lower surfaces being represented. None of the impressions shows a complete leaflet, either the specimen or the original leaflet having invariably been damaged; in particular, the margins are scarcely ever preserved. Nevertheless, the margin is present in two specimens, and can be seen to be serrate. The leaflet from which the largest specimen was derived cannot have been less than 6 cm. wide.

The venation is clearly visible, especially on impressions of upper surfaces, and this was the feature upon which the generic identification was based. For this purpose, an examination was made of specimens of the majority of the dicotyledonous genera now living in North Africa, southern Europe and the Near East, in so far as these were represented in the Cambridge herbarium; only the more unlikely families were passed over. The identification

---

[1] D. 5030.

was later confirmed by the discovery that some of the impressions showed traces of prickles along the undersides of the mid-ribs.

Certain external moulds of stems were found, showing a polygonal cross-section. These, also, may belong to *Rubus*.

The only living brambles reported from Cyrenaica are *R. ulmifolius*,[1] *R. cyrenaica* Hruby, and various hybrids between these two species and *R. sanctus* Schreb.[2] The fossil leaflets, in all their visible characteristics, are identical with those of *R. ulmifolius*, the species found in Wadi Derna at the present day.

Laurus canariensis (Pl. 16).

The specimens are again in the form of leaf-impressions. Many of the specimens show almost complete impressions of individual leaves. The shapes of the leaves were highly variable; some were lanceolate, with acute apices, others were elliptic, with obtuse apices. All the larger specimens are damaged, but they show that the leaves could exceed 9 cm. in width. The margins of the leaves were entire, and not wavy. Details of the venation are again clearly visible, especially, in this case, on impressions of the undersides. Many specimens carry traces of small protuberances in the angles between the mid-rib and the secondary veins.

It was at first thought that these specimens, with their very different shapes and sizes, must belong to several different genera. They were then subjected to the same process as was used with the impressions of *Rubus*, and it was concluded that all must be assigned to the genus *Laurus*. This conclusion was based on the venation of the leaves, the nature of their margins, and the presence of the protuberances; these latter features are present in *Laurus*, and may be extra-floral nectaries.

The two living species of *Laurus* are *L. canariensis* and *L. nobilis* L. A comparison of the two showed that the leaves of the latter were always more or less lanceolate and seldom more than 4 cm. wide, whereas those of the former were often both elliptic and much wider. On these two differences alone, most of the fossil specimens could immediately be assigned to *L. canariensis*, the species already identified from this locality by Tongiorgi. The remainder of the specimens, from their shape and size, might equally well have belonged to *L. nobilis*. Generally, however, though not invariably, the leaves of this species have wavy margins. Since this feature was completely absent from the fossil specimens, it was finally decided that the whole collection could be assigned to *L. canariensis*.

A few moulds and casts of berries were found within the tufa. These have been examined by Miss M. E. J. Chandler, who has kindly reported that they might perhaps belong to one of the Lauraceae.

Arundo sp. (Pl. 5 a).

Incrustations around reeds are common within the Wadi Derna tufas, and are not confined to any one locality. In many places, the incrustations were formed while the reeds were still in their positions of growth, and have never subsequently been moved;

[1] Pampanini (1930), p. 129.   [2] Pampanini (1936), pp. 28–30.

many extensive and well-defined beds were found, 20–30 cm. thick, entirely composed of tufa which had originated in this way.

As a rule, these incrustations are thin and fragile, and have often split longitudinally, to reveal impressions of the leaves and stems. From their size alone, all the specimens collected have been assigned to *Arundo*, rather than *Phragmites*. Specific identifications cannot be attempted. The specimens, however, show no visible differences from *Arundo donax* L., the species which now grows in Wadi Derna.[1]

Pinus halepensis

Two specimens have been assigned to this species; both are external moulds of cones. One of the two is badly weathered, but shows the whole length of the original cone and nearly half of its greatest circumference; the length is 11 cm., and the greatest diameter about 7 cm. The other specimen shows only a part of the proximal end of the cone, but carries remarkably well-preserved impressions of the surface detail (Pl. 15a). The cone is closed; the exposed portions of the scales are rhomboidal or roughly hexagonal, and are often nearly as broad in the direction of the length of the cone as in the direction at right-angles. Each of these exposed areas is almost flat, except for the presence of a very low umbo, centrally placed, from which radiate a number of low ridges.

These specimens have been compared with cones of all the species of pine now living in the Mediterranean region, the Near East and temperate Europe. Of these species, *P. halepensis* was the only one whose cones showed the same characteristics.

APPENDIX F

REPORT ON THE SAMPLES OF CAVE DEPOSITS FROM CYRENAICA

By PROFESSOR F. E. ZEUNER AND DR I. W. CORNWALL
Department of Environmental Archaeology, London University Institute of Archaeology

I. HAGFET ED DABBA

Question I. Is there evidence to suggest the relative ages of samples 2 and 13, sample 2 coming from the light-reddish earth (Layer V) at the outer end of the section and sample 13 from the innermost part of the section, one and a half feet south of trench 5?

Question II. Is the firm dark-red earth (Layer VI) of different origin from the overlying light-reddish earth (Layer V)?

Question III. Is there any indication of weathering in Layer V suggested by the pairs of samples 7 and 8 and 10 and 11?

---

[1] Pampanini (1930), p. 112.

INVESTIGATION

All the samples were subjected to a number of preliminary tests and microscopic examination, and the amount of alkali-soluble humus was determined. The results are shown in Fig. 40. It was found that all the samples coming from Layer V, with the exception of the middle portion of the section (samples 7, 9 and 10) vary only from 0·6 to 0·9, irrespective of the depth at which they were taken in the deposit. This suggests an originally uniform distribution of humic matter in this layer. Samples 1, 5 and 6, however, which come from Layer VI, are consistently poorer in humus (0·25–0·5). This indicates a difference between Layer V and Layer VI which is due either to soil formation processes or, in view of the consistent values obtaining over the whole of Layer V, to a covering of Layer VI by a later deposit, Layer V. There is a disturbance in the area of samples 7, 9 and 10. Sample 7 is unusually rich in humus (1·4) whilst samples 9 and 10 are deficient in humus (0·1–0·3).

Three samples from Layer V were selected for special examination (samples 2, 9 and 13), samples 2 and 13 marking the extreme ends of the section, while sample 9 comes from the middle and differs by being cemented into a hard calcareous breccia. The carbonate content was very high in all three samples, being 76 per cent for sample 2, 90 per cent for sample 9 and 54 per cent for sample 13. The high figure for sample 9 was not unexpected, since calcium carbonate was obviously the cementing material which made it into a hard rock. It appears, therefore, that near the location of sample 9 an exceptional supply of calcium carbonate, probably from a source of water, was available. This also explains the exceptionally low humus content, which is due to the prevalence of calcium carbonate per unit weight compared with the other samples.

Despite the presence of biggish bone fragments, especially in sample 9, the brown sifted soil (less than 1 mm. grain size) is poor in phosphates. No significant difference could be found, the amount of P_2O_5 being about 0·6 per cent. This rules out bat guano as a significant constituent of the material.

The mechanical composition is very similar in the samples in question, as shown in the following table:

	Sample 2 (%)	Sample 9 (%)	Sample 13 (%)
Above 1 mm.	24·1	34·2	28·1
1 mm.–0·5 mm.	6·1	6·01	9·9
0·5 mm.–0·1 mm.	15·05	18·05	24·1
Below 0·1 mm.	54·75	41·74	37·9

There is a surprising deficiency of the grade between 0·5 and 1 mm. in all the samples, whilst a third to about a half of the weight of the sample belongs to the silt grade. The grading suggests a common origin of the materials composing the samples.

Further tests were made, for example sieving and washing, which showed that the coarse particles consisted of fine material cemented with calcium carbonate, and also confirmed that the fine particles are mainly silt, whilst there is very little clay present. The weathering of limestone would produce a clay and such clay would, in a cave, perhaps be mixed with calcareous grit produced by dripping water, as described by Lais. But it is extremely improbable that any silt deposit would be formed in a cave *in situ* as a result of decomposition of limestone.

The presence of large quantities of silt is most easily explained by assuming that wind was the transporting agency.

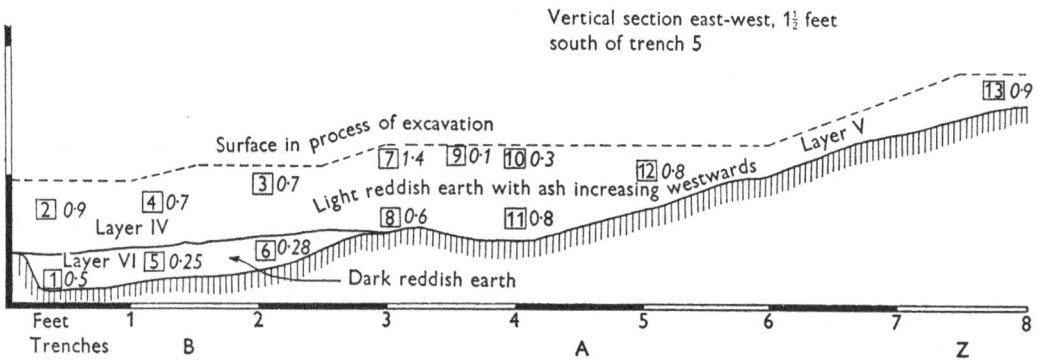

FIG. 40. Hagfet ed Dabba: positions of soil samples
Sample numbers in squares; indices of alkali-soluble humus in italic.

In order to investigate this point still further, the residue of sample 2 insoluble in hydrochloric acid was examined. The particles which would not pass the 70 B.S. sieve proved, under the binocular, to consist of quartz grains, vegetable debris and charcoal. There was very little of this grade. Over 80 per cent, however, passed even the 120 B.S. sieve, but settled in water within the hour, showing that practically all the acid-insoluble material belongs to the silt grade.

There is little difference between Layers V and VI. An examination of sample 4 (from Layer V) and sample 5 (from Layer VI) showed that the acid-insoluble residue is almost the same. Moreover this residue consists almost entirely of silt, as shown in the following table:

	Sample 4 (%)	Sample 5 (%)
Remaining on the 120 B.S. sieve	3	1·43
Coarse silt	15·72	19·6
Fine silt and clay	23·8	21·05

There is, however, more iron in sample 5 than in sample 4. Both samples consist mainly of small friable aggregates of colloidal silica of silt grade. There are also a few quartz grains and charcoal, the latter being more frequent in sample 4.

CONCLUSIONS

Question I. The great similarity of samples 2 and 13 suggests that there is no reason to regard them as formed under different conditions or belonging to different layers. In fact it appears that Layer V is a fairly homogeneous deposit of a silt consisting of fine silica aggregates and a small quantity of sand, probably blown into the cave by wind and subsequently more or less solidly cemented by calcium carbonate. The presence of human occupation is indicated throughout by charcoal and bone fragments. Near samples 9 and 10 a source of calcium carbonate, possibly from dripping water, appears to have existed. Sample 7, immediately in front of sample 9, is somewhat richer than the others in organic matter. This suggests perhaps the presence of a hearth or occupation site nearby.

Question II. No significant difference between the materials of Layers V and VI can be detected. Both contain charcoal, making the presence of man probable, but the amount is greater in Layer V. It should be pointed out, however, that in the dried samples the colour of Layer VI is distinctly lighter than Layer V and in accordance with their relative contents of humic matter. It is suggested, therefore, that the darker colour of Layer VI as seen in the field was due to a higher content of water.

Question III. The sample pair 7/8 suggests that Layer V suffered weathering after its formation, sample 7 having the lower pH, containing less calcium carbonate and having a higher humus content. In pair 10/11, however, the pH is virtually the same in both, but there is more calcium carbonate and less humus in the upper sample. Moreover the amount of charcoal is greater in sample 10, which suggests that this sample came from a spot on or near the surface of Layer V, where the occupation was at one time more intense than during the formation of the lower part of Layer V (sample 11).

2. HAGFET ET TERA

Question. Immediately north-north-east of the eastern extremity of the wall is a patch of extremely hard breccia resting on a rock shelf in the side of the cave. This breccia yielded Mousterian implements to C. Petrocchi in 1938, and was equated by him with the lower of two implementiferous layers contained in the terrace deposits south of the wall. The latter yielded only implements of characteristically Upper Palaeolithic (later) typology, and the object of the series of samples is to test the validity of Petrocchi's correlation. Are both breccia and hardened cave-earth likely to have been formed simultaneously under conditions of slightly different humidity in the two different parts of the cave, or are they two entirely distinct formations laid down at different times?

INVESTIGATION

Alkali soluble humus of all the samples submitted was determined. It is on the whole higher in those coming from outside the Roman wall and lower in the Mousterian breccia. Since the position of the samples relative to the cave wall varies, this may mean that human occupation was more intense away from the wall and possibly to the outside, than inside and near the cave wall. Since, however, vegetation has been present outside the Roman wall where the high values were encountered, the values may be due to plant remains of a comparatively recent date. These tests, therefore, cannot be regarded as conclusive.

Three samples, 20, 24 and 27, were taken as being representative of the supposedly different deposits, sample 20 coming from outside the Roman wall, sample 24 from its side near the rock wall of the cave and sample 27 from the Mousterian deposit inside the cave. Examination in the lump was not very informative, all three being more or less hardened by infiltration of calcium carbonate, sample 24 most of all.

Five-gramme samples of each were treated with boiling concentrated HCl; the residue was washed and subdivided with the 120 B.S. sieve into two fractions, which were then dried and weighed:

	Sample 20 (%)	Sample 24 (%)	Sample 27 (%)
> 120 B.S. sieve (A)	4·8	1·6	3·6
< 120 B.S. sieve (B)	27·2	22·0	25·0
Total HCl-insoluble residue	32·0	23·6	28·6
Extracted by HCl and loss	68·0	76·4	71·4

The coarser residues (A) consisted mainly of quartz and siliceous aggregates. A few grains were wind-rounded and there were small splintery fragments of a colourless chalcedony, the material, apparently, of the stone industries. In all there were a few grains of charcoal, plant fibres and, in sample 20 in particular, traces of unburnt wood. Sample 20 also contained the largest number of chalcedony chips.

The finer residues (B) consisted almost exclusively of granular aggregates of silica. Even charcoal was not detected in more than stray grains. There was very little material of the clay grade; the suspension of the finer material settled out almost completely in a matter of half a minute.

CONCLUSIONS

There does not seem to have emerged from this analysis any feature which would distinguish between these samples. Both from the colour of the residues and from the humus determination it is clear that sample 24 contains the largest amount of organic matter, sample 27 next and sample 20 least. As all come from about the same depth below the surface of the present deposits the significance of this is not clear and is, probably, that sample 24 comes from nearer a hearth than either of the others.

APPENDIX G

LATE POTTERY FROM HAGFET ED DABBA

By KATHLEEN M. KENYON

The sherds are all small, and the majority are the wall-sherds of jugs. There are very few sherds which appear to come from storage or other large jars or jugs. With two exceptions, all are wheel-made.

I. CLASSES OF WARE

A. The largest number are from thin-walled vessels of a clay which when well-fired is buff to reddish, soft in texture with large numbers of small white grits.

The following are subdivisions of this main class:

Ai. Rather dirty buff. Many of the grits have calcined out, leaving a pitted surface.
Disturbed Zone, 1 sherd; Layer II, 19 sherds; Layer II/III, 2 sherds; Layer III, 5 sherds.

Aii. Dirty grey ware, mostly firing grey-buff at surface, which is pitted as in Ai.
Layer IIA, 2 sherds; Layer II, 3 sherds; Layer II/III, 2 sherds; Layer III, 3 sherds; Layer IV, 1 sherd.

Aiii. Buff ware, similar to Ai, but surface not pitted.
Layer I, 1 sherd; Layer II/III, 1 sherd.

Aiv. Similar to Aiii, ware rather pinker.
Layer IIA, 1 sherd; Layer II, 1 sherd; Layer II/III, 2 sherds, Layer III, 2 sherds.

B. Rather similar ware, but harder and redder. Many small white grits.
Layer IIA, 2 sherds; Layer II, 8 sherds; Layer III, 2 sherds; Layer IV, 1 sherd.

C. Buff ware, fine and well-levigated, rather soft. This may be similar to Ai, but better levigated.
Layer IIA, 1 sherd; Layer II, 7 sherds; Layer II/III, 2 sherds; Layer III, 4 sherds; Layer III/IV, 1 sherd.

D. Hard reddish-buff ware, well levigated, slightly coarse texture, surface smoothed. This ware is found in the Late Hellenistic and Early Roman periods at Sabratha.
Layer IIA, 1 sherd; Layer II, 16 sherds; Layer II/III, 1 sherd; Layer III, 7 sherds.

E. Hard ware, grey at core, well levigated, firing reddish, surface smoothed from fairly large vessels.

Ei. Smooth texture.
Layer II, 3 sherds.

Eii. Slightly coarse texture.
Layer II, 2 sherds; Layer III, 6 sherds.

300

F. Hard dark grey ware, well levigated, rather coarse texture, surface smoothed.
   Layer II, 3 sherds.

G. Fine thin hard grey ware, rather coarse texture, but no visible grits. This type of ware is probably early Roman in period.
   Layer IIA, 1 sherd; Layer II, 4 sherds; Layer III, 4 sherds; Layer III/VI, 1 sherd.

H. Ware with many small, shining grits.
   Hi. Hard buff ware, well levigated, slightly coarse texture.
   Layer II, 1 sherd.
   Hii. Hard grey ware, firing reddish at surface.
   Layer II, 1 sherd.

J. Hard buff ware, texture close but gritty, surface rough.
   Layer II, 1 sherd; Layer II ?, 1 sherd; Layer III, 1 sherd.

K. Reddish-brown ware, with many small reddish and white grits, some of which have calcined out. Soft rough texture. Sherds from large vessels, one of which has flat base.
   Layer II, 2 sherds.

None of the sherds can be very closely dated. One is Hellenistic, possibly as early as the second century B.C. But though exact parallels would be difficult to find, my feeling is that the majority are late first century B.C. to first century A.D., roughly of the Augustan period. This is based on general resemblances in ware and type to thin fine vessels which appear in layers at Sabratha, and which preliminary examination suggests belong to this period. The almost complete absence of published details of coarse pottery from North Africa makes any closer dating impossible. It is quite certain that with the exception of one specimen, all the sherds are either imported ware, Hellenistic or Roman, or else local products under the influence of such imports.

2. ANALYSIS OF THE OCCURRENCE OF DIFFERENT WARES

Disturbed Zone

Neck and rim of jug, fine hard pinkish ware, some small white grits, rather coarse texture. Such fine jugs appear in the late first century B.C. to first century A.D. groups at Sabratha.

Layer IIA (Upper unconsolidated portion of Layer II in interior of cave)

Shoulder fragment of pot, with place of attachment of handle. Fairly hard, purplish ware, firing pinkish on surface. The surface has small rounded ribs. This type of ribbing, and also the rather purplish tinge of the ware, is typical of cooking-pots in Palestine, also I think found in Egypt, of the first century B.C. to first century A.D. (Samaria, Roman I–II groups). The position of the handle also agrees. Whether there is a connexion or not there is not sufficient evidence at present to say.

Small fragment of lid, of plain straight form, thin section. Dirty grey ware, some small white grits, surface pitted. Such lids are common in first century B.C. to first century A.D. deposits at Sabratha, but the ware is different. Ware Aii.

Layer II

Handle probably of small jug or bottle. Hard grey ware, grey matt surface. Ware G. Handle of similar type and ware, but smaller.

Sherds of similar ware. Such fine jugs, of similar hard ware, are found late first century B.C. to first century A.D. at Sabratha, with Arretine and other sigillata.

Base of deep bowl. Good hard well-levigated pinkish ware, good high brownish-black glaze, with bright red zone in interior base. Such zoned glaze is found on Hellenistic pottery in the second and first century B.C.

Base of small bowl, with wide, concave disc base. Dirty grey ware, with white grits, many of which have calcined out. Ware Aii. Such bowl bases appear at Sabratha, late first century B.C. to first century A.D.

Two sherds of neck of jug. Signs of attachment of handle, but it is not clear that handle 16, which is of similar ware, belongs to this jug. Hard ware, reddish core, firing grey on both surfaces.

Rim of bowl. Ware with dark grey core, firing pinkish buff on both surfaces, with many shining white grits. This bowl may be hand-made. The ware, of which there was one other sherd, is coarser than any of the other examples, and may represent the native tradition.

Base of small bowl, with small, high, slightly concave, disc base. Softish dirty buff ware with small white grits, traces of pinkish buff slip. Ware Ai.

Base of coarse bowl. Softish red-brown ware, many small reddish grits. Wet smoothed in and out.

Rim of small, fine jug. Hard, well-levigated buff ware, smoothed externally.

Rim of small bowl. Red ware with small white grits. Medium hard, rather rough texture. Ware B. Such out-curved rimmed bowls are typical of the late first century B.C. to first century A.D. at Sabratha.

Base of jug, base sharp-cut and slightly concave. Dark grey ware, well-levigated with coarse texture, smoothed externally. Ware F. Such sharp-cut bases, usually less concave, occur at Sabratha in the late first century B.C. to first century A.D. levels.

Base of bowl with base and ring of narrow section. Ware similar to last.

Layer II/III

Stump base of small pot, buff ware firing slightly pinkish on surface, rather soft. Well-levigated, slightly coarse texture. Probably a Roman form.

Layer III

Light grey ware, firing grey-buff on surface, thin buff slip out. The form could be Hellenistic or Roman.

APPENDIX H

CHARCOAL FRAGMENTS FROM HAGFET ED DABBA

By Dr C. R. METCALFE

Samples of charcoal from various layers were submitted to Dr C. R. Metcalfe, Royal Botanic Gardens, Kew, who kindly reports as follows:

'The charcoals have now been examined and identified as far as possible. Some of the smaller fragments could not be identified because of their size, whilst others were difficult because the material was apt to crumble, thereby making the structure obscure. Most of the samples consisted of one or other of two distinct coniferous woods. One, in which resin canals could be seen, is probably a pine, and might quite well be *Pinus halepensis* Mill., the Aleppo pine, which is common in Cyrenaica. This was found in the following layers (the number in brackets indicates the number of specimens):

Layer II (4)   Layer III (3)   Layer IV (3)   Layer V (2)   Layer VI (1)

The second coniferous wood, without resin canals, and which frequently showed very narrow growth rings, might quite well be *Cupressus sempervirens* L., the Mediterranean cypress. This is also common in Cyrenaica. The only other species from which this material is likely to have been derived is *Juniperus phoenicea* L., the Phoenician juniper. As the nature of the pitting and other diagnostic characters were no longer visible in the charcoal, it was impossible to determine if either of these species is the one concerned, but, on the whole, *Cupressus sempervirens* appears to be the more likely. The coniferous wood without resin canals was seen in samples from the following layers:

Layer II (6)   Layer III (2)   Layer IV (2)   Layer VI (1)

Charcoal, probably derived from coniferous wood, but which was either too fragmentary or decomposed for more precise identification, was seen in samples from the following layers:

Layer II (2)   Layer III (2)   Layer IV (3)   Layer V (4)   Layer VI (1)

Charcoal derived from hardwoods was uncommon, but was detected in samples from the following layers:

Layer I (1). Probably *Pistacia lentiscus* L.

Layer IV (1). Probably *Phillyrea* sp., similar to *P. latifolia* L. var. *media* Schneid.

Layer IV (1). Probably *Olea europea* L. (Olive wood).

Layer VI (1). Appears to be derived from a palm, but it could not be identified more precisely. One would imagine that it is likely to be a date palm (*Phoenix dactylifera* L.).

The remaining samples could not be identified.'

BIBLIOGRAPHY

[N.B. (i) The following list includes only those publications to which reference has been made in the text; for a more comprehensive bibliography of the prehistory of Libya and adjacent territories, see McBurney (1947) and Bates (1914).

(ii) Appendix A has been provided with its own bibliography, the majority of whose items do not appear below.]

AHLMANN, H. W. (1928). 'La Libye septentrionale.' *Geografiska Annaler*, H. 1–2. Stockholm.

ALMAGRO BASCH, M. (1946). *Prehistoria del Norte de Africa y del Sahara Español.* Barcelona. Consejo Superior de Investigaciones Científicas.

ARAMBOURG, C. (1934). 'Les Grottes Paléolithiques des Béni-Segoual.' *Archives de l'Institut de Paléontologie Humaine*, vol. XIII.

ARKELL, A. J. (1949). *Early Khartoum.* Oxford University Press.

BARNES, A. S. (1947). 'The technique of blade production in Mesolithic and Neolithic times.' *Proceedings of the Prehistoric Society*, vol. XIII.

BATES, O. (1914). *The Eastern Libyans.* London.

BAULIG, H. (1928). *Le Plateau Central de la France et sa bordure Méditerranéenne.* Paris.

—— (1935). *The Changing sea-level.* Inst. Brit. Geogr., London, Publ. no. 3.

BAUMGARTEL, E. J. (1947). *The cultures of prehistoric Egypt.* Oxford University Press.

BEAN, W. J. (1919). *Trees and shrubs hardy in the British Isles*, vol. II (2nd ed.). London.

—— (1933). *Trees and shrubs hardy in the British Isles*, vol. III (1st ed.). London.

BLACKWELDER, E. (1928). 'Mudflow as a Geologic Agent in Semi-arid Mountains.' *Bull. Geol. Soc. America*, vol. XXXIX, pp. 465–84.

—— (1933). 'The Insolation Hypothesis of rock weathering.' *Amer. Journ. Sci.*, 5th ser., vol. XXVI, pp. 97–113.

—— (1935). 'Talus slopes in the Basin Range province' (Abstr.). *Geol. Soc. America, Proc.*, 1934, p. 317.

BLANC, A. C. (1936). 'Contributo alla conoscenza del Quaternario marino della Cirenaica occidentale.' *Boll. Soc. Geol. Italiana*, vol. LV, fasc. 2, pp. 279–82.

—— (1937). 'Low levels of the Mediterranean Sea during the Pleistocene Glaciation.' *Q.J.G.S.* vol. XCIII, pp. 621–51. London.

BLANC, G. A. (1921). 'Grotta Romanelli, I. Stratigrafia dei depositi e natura e origine di essi.' *Arch. Antrop. Etnol.* vol. L. Florence.

—— (1928). 'Grotta Romanelli, II. Dati ecologici e paleontologici.' *Arch. Antrop. Etnol.* vol. LVIII. Florence.

BLANCKENHORN, M. (1921). *Handbuch der Regionalen Geologie*, VII Band, 9 Abteilung, 'Aegypten'. Heidelberg.

BORDES, F. (1950). 'L'évolution buissonnante des industries en Europe occidentale.' *L'Anthropologie*, vol. LIV, pp. 393–420.

BOURCART, J. (1943). *Rev. Sci.*, Paris, no. 3224, pp. 311–36.

BRANNER, J. C. (1911). 'Aggraded Limestone Plains of the Interior of Bahia and the Climatic Changes suggested by them.' *Bull. Geol. Soc. America*, vol. XXII, pp. 187–206.

BRUNTON, G. and CATON-THOMPSON, G. (1929). *The Badarian Civilisation.* London.

CATON-THOMPSON, G. (1946a). 'The Levalloisian in Egypt.' *Proc. Prehist. Soc.* vol. XII, p. 57.

—— (1946b). 'The Aterian industry, it's place and significance.' *Huxley Memorial Lecture, 1946.* The Royal Anthropological Institute. London.

CATON-THOMPSON, G. (1952). *Kharga Oasis.* London.

CATON-THOMPSON, G. and GARDNER, E. W. (1932). 'The prehistoric geography of Kharga Oasis.' *Geogr. J.* vol. LXXX, no. 5.

——— ——— (1934). *The Desert Fayoum.* The Royal Anthropological Institute, London.

CHILDE, V. G. (1947). *The Dawn of European Civilisation* (4th ed.). London.

——— (1952). *New Light on the most ancient East* (2nd ed.). London.

COON, C. S. (1951). *Cave explorations in Iran 1949.* Philadelphia.

CROWFOOT, J. 'Notes on the flint implements at Jericho' in Garstang (1936).

DE LAMOTHE, L. (1911). 'Les anciennes lignes de rivage du Sahel d'Alger et d'une partie de la côte algérienne.' *Mém. Soc. Géol. France,* 4th ser., vol. I, mem. no. 6. Paris.

DENIZOT, G. (1935). 'Observations sur le Quaternaire moyen de la Méditerranée occidentale et sur la signification du terme de Monastirien.' *Bull. Soc. Géol. France,* 5th ser., vol. V, pp. 559–71.

DEPÉRET, C. (1918). 'Essai de coordination chronologique générale des temps quaternaires.' *C.R. Acad. Sci., Paris,* vol. CLXVII, pp. 418–22.

DESIO, A. (1935). 'Studi geol. sulla Cirenaica, sul deserto Libico, sulla Tripolitania e sulla Fezzan orientali.' *Miss. sci. della R. Acc. d'Italia a Cufra,* vol. I. Rome.

——— (1939). 'Studi morfologici sulla Libia orientale.' *Miss. sci. della R. Acc. d'Italia a Cufra,* vol. II. Rome.

DE STEFANI, T. (1946). 'L'evoluzione biologica e geografica della Sicilia dall'inizio del Quaternario ad oggi.' *Il Naturalista Siciliano,* 3rd ser., ann. 3. Palermo.

DINES, H. G., EDMUNDS, F. H. and CHATWIN, C. P. (1929). 'The geology of the country around Aldershot and Guildford.' *Mem. Geol. Survey, England and Wales,* Expl. of Sheet 285.

EMIG, W. H. (1917). 'Travertine deposits of the Arbuckle Mountains.' *Bull. Oklahoma Geol. Survey,* no. 29.

EVANS-PRITCHARD, E. (1949). *The Sanusiya Sect.* Oxford University Press.

FLINT, R. F. (1947). *Glacial Geology and the Pleistocene Epoch.* New York.

GAMBETTA, L. (1934). 'Su alcuni molluschi continentali subfossili e viventi della Cirenaica e della Tripolitania orientale.' *Miss. sci. R. Acc. d'Italia a Cufra,* vol. III. Rome.

GARROD, D. A. E. and BATE, D. M. A. (1937). *The Stone Age of Mount Carmel,* vol. I. Oxford University Press.

GARSTANG, J. (1936). 'Jericho: City and Necropolis'. *Liverpool Annals of Art and Archaeology,* vol. XXIII.

GAUTIER, E. F. (1946). *Le Sahara.* Paris.

GIGNOUX, M. (1913). 'Les formations marines pliocènes et quaternaires de l'Italie du Sud et de la Sicile.' *Ann. Univ. Lyon,* n.s., fasc. 36.

——— (1950). *Géologie stratigraphique* (4th ed.). Paris.

GOBERT, E. C. (1912). 'L'abri de Redeyef.' *L'Anthropologie,* vol. XXIII, pp. 151 *et seq.* Paris.

GOBERT, E. C. and VAUFREY, R. (1932). 'Deux gisements extrêmes d'Ibéromaurusiens'. *L'Anthropologie,* vol. XLII, pp. 449–90. Paris.

GREEN, J. F. N. (1943). 'The age of the raised beaches of South Britain.' *Proc. Geol. Assoc.* vol. LIV, pp. 129–40. London.

GREGORY, J. W. (1911). 'The Geology of Cyrenaica.' *Q.J.G.S.,* vol. LXVII, pp. 572–615. London.

HALLER, J. (1946). 'Notes de préhistoire phénicienne—l'abri de Abou-Halka (Tripoli).' *Bull. Musée de Beyrouth,* vol. VI, pp. 1–19.

HENRI MARTIN. See Martin, Henri.

HOWE, B. and MOVIUS, H. L. (1947). 'A stone age cave site in Tangier.' *Papers of the Peabody Museum,* vol. XXVIII, no. 1. Harvard University Press.

HUME, W. F. (1925). *Geology of Egypt,* vol. I. Cairo.

HUZAYYIN, S. A. (1941). 'The place of Egypt in prehistory.' *Mem. Inst. Egypte,* no. 43. Cairo.

BIBLIOGRAPHY

JOHNSON, D. W. (1931). 'The correlation of ancient marine levels.' *Union Géogr. Internat. 3me Rapp. de la Comm. pour l'Etude des Terrasses Pliocènes et Pleistocènes*, pp. 42–54.

—— (1932). 'Rock fans of arid regions.' *Amer. J. Sci.*, 5th ser., vol. XXIII, pp. 389–416.

—— (1944). 'Problems of terrace correlation.' *Bull. Geol. Soc. America*, vol. LV, pp. 793–818.

JUNKER, H. (1929, 1930, 1932, 1940). 'Vorläufige Berichte über die Gräbung der Akademie der Wissenschaften in Wien auf der neolithischen Siedelung von Merimde-Benisalame.' *Anzeiger d. Akad. d. Wiss. Wien, phil.-hist Kl.*

KELLEY, H. (1934). 'Harpons, etc., de Tafergit et Tamaya Mellet.' *Journal de la Société des Africanistes*, vol. IV, pp. 135–43.

KNOWLES, F. (1947). 'The manufacture of a flint arrow-head by a quartzite hammerstone.' *Occasional papers of the Pitt-Rivers Museum*, no. 1. Oxford.

KRIGE, A. V. (1927). 'An examination of the Tertiary and Quaternary changes of sea-level in South Africa, with special stress on the evidence in favour of a recent world-wide sinking of ocean level.' *Ann. Univ. Stellenbosch*, vol. V, pp. 1–81.

KULP, J. L., TYRON, L. E., ECKELMAN, W. R., and SNELL, W. A. (1952). 'Lamont natural radio-carbon measurements'. *Science*, vol. CXVI, pp. 409–14.

LAGAAIJ, R. (1952). *The Pliocene Bryozoa of the Low Countries*. Maastricht.

MARCHETTI, M. (1934). 'Note illustrative per un abbozzo di carta geologica della Cirenaica.' *Boll. Soc. Geol. Italiana*, vol. LIII, fasc. 2, pp. 309–25.

—— (1935). 'La necropoli sommersa di Appollonia e gli spostamenti del livello marino in Cirenaica.' *Boll. Soc. Geol. Italiana*, vol. LIV, fasc. 1, pp. xciv–xcviii.

—— (1938). *Idrologia Cirenaica*. Florence.

MARINELLI, O. (1920). 'Sulla morfologia della Cirenaica.' *Riv. Geogr. Ital.* annata XXVII, fasc. IV–VII, pp. 69–86.

MARTIN, HENRI (1909). *L'Evolution du Mousterian*, vol. II. Paris.

McBURNEY, C. B. M. (1947). 'The Stone Age of the Libyan Littoral.' *Proc. Prehist. Soc.* pp. 56–84.

McBURNEY, C. B. M., HEY, R. W. and WATSON, W. (1948). 'First report of the Cambridge archaeological expedition to Cyrenaica (1947).' *Proc. Prehist. Soc.*, n.s., vol. XIV, pp. 33–45. London.

McBURNEY, C. B. M., TREVOR, J. C. and WELLS, L. H. (1953). 'The Haua Fteah fossil jaw.' *J. R. Anthrop. Inst.* vol. LXXXIII, pp. 71–85.

McCOWAN and KEITH (1939). *The Stone Age of Mount Carmel*, vol. II.

Mediterranean Pilot, vol. V (3rd ed.), 1937.

MOND, SIR R. and MYERS, O. H. (1937). *Cemeteries of Armant I*. Egypt Exploration Society, London.

MOUNTFORD, C. P. (1951). *Brown men and red sand* (2nd ed.). London.

MURRAY, G. W. (1951). 'The Egyptian climate: an historical outline.' *Geogr. J.* vol. CXVII, pp. 421–34.

NEUVILLE, R. and RUHLMANN, A. (1941). *La place du Paléolithique ancien dans le Quaternaire Marocain*. Inst. Hautes-Etudes Maroc, Casablanca.

NEWTON, R. B. (1911). 'Kainozoic mollusca from Cyrenaica.' *Q.J.G.S.* vol. LXVII, pp. 616–53. London.

OAKLEY, K. P. (1939). 'The nature and origin of flint.' *Science Progress*, vol. XXXIV, p. 279.

—— (1943). 'A note on the late post-glacial submergence of the Solent margins.' *Proc. Prehist. Soc.*, n.s., vol. IX, pp. 56–9. London.

PAMPANINI, R. (1930). *Prodromo della flora cirenaica*. Forli.

—— (1936). 'Aggiunte e correzioni al "Prodromo della flora cirenaica".' *Arch. Bot.* vol. XII, pp. 17–53. Forli.

PERICOT, L. (1942). *La Cueva del Parpallo*. Madrid.

306

PETROCCHI, C. T. (1940). 'Ricerche preistoriche in Cirenaica.' *Africa Italiana*, vol. VII.

PFALZ, R. (1931). 'Note geologiche sui terreni di El Gubba e Derna.' *Boll. Geogr. del Serv. Studi del Governo della Cirenaica*, no. 13, pp. 7–27. Benghazi.

REISNER, G. A. (1911). *Survey of Nubia.*

RICHARDS, H. G. (1936). 'Fauna of the Pleistocene Pamlico formation of the Southern Atlantic Coastal Plain.' *Bull. Geol. Soc. America*, vol. XLVII, pp. 1611–56.

ROMANO, M. (1933). *Cirenaica Nuova.* Benghazi.

RUHLMANN, A. (1945). 'Le Paléolithique marocain.' *Publications du Service des Antiquités du Maroc.* Fasc. 7.

SANDFORD, K. S. (1936). Problems of the Nile Valley. *Geogr. Review*, vol. XXVI, pp. 67–76. New York.

SANDFORD, K. S. and ARKELL, W. J. (1933). 'Palaeolithic man and the Nile Valley in Nubia and Upper Egypt.' *Univ. Chicago Orient. Inst. Publ.* vol. XVII.

SILVESTRI, A. (1929). 'Sulla *Heterostegina costata* d'Orb.' *Riv. Ital. di Paleont.* vol. XXXV, pp. 69–78. Pavia.

SOERGEL, W. (1922) *Jagd der Vorzeit.* Jena.

SPRATT, T. A. B. (1865). *Travels and researches in Crete*, vol. II. London.

STEFANINI, G. (1930). 'I terrazzi fluviali e marini dell'Africa Italiana.' *Int. Geogr. Union, 2nd Rep. of the Commission on Pliocene and Pleistocene Terraces*, pp. 23–39.

—— (1935). 'Breve guida alla escursioni geologiche in Cirenaica.' *Boll. Soc. Geol. Italiana*, vol. LIV, pp. xxxviii–cxviii.

SUESS, HANS E. (1954). 'U.S. Geological Survey Radiocarbon Dates I.' *Science*, vol. CXXI, (in the Press).

THOMSON, D. F. (1949). *Economic structure and the ceremonial exchange cycle in Arnhem Land.* Melbourne.

TONGIORGI, G. (1935). 'Una formazione travertinosa con impronte di *Laurus canariensis* Webb in Cirenaica (presso la cascata dell'U. Derna).' *N. Giorn. Bot. Ital.*, n.s., vol. XLII, pp. 665–6. Florence.

VALORI, B. (1927). *Atti della prima riunione del Istituto di Palaeontologia Umana.* Florence. 'Osservazioni sui rapporti preistorici fra l'Egitto e la Libia.'

VAUFREY, R. (1929). 'Les éléphants nains des îles méditerranéennes.' *Arch. Inst. Pal. Hum.*, Mem. 6. Paris.

—— (1933). 'Notes sur le Capsien.' *L'Anthropologie*, vol. XLIII.

—— (1934). 'Stratigraphie Capsienne.' *Swiatowit*, vol. XVI, pp. 15–34.

—— (1936). 'L'âge des hommes fossiles de Mechta el-Arbi.' *Bulletin de la Société Géographique de Sétif*, 1935, pp. 1–25.

—— (1938). 'Le Capsien des Environs de Tebessa.' *Recueil de la Société de préhistoire de Tebessa*, vol. I, pp. 41–82.

—— (1939). 'L'Art rupestre Nord-Africain.' *Archives de l'Institut de Paléontologie Humaine*, no. 20.

—— (1950). 'La question du Capsien ancien.' *Proceedings of the International Congress of Prehistoric and Protohistoric Studies.* Zürich.

VIGNARD, A. E. (1920). 'Station Aurignacienne à Nag Hamadi (Haute Egypte).' *Bull. Inst. Fran. d'Arch. Orientale*, tome XVIII, pp. 1–20.

WATSON, W. (1949). 'The surface flint implements of Cyrenaica.' *Man*, vol. XLIX, pp. 100–4.

WOOLDRIDGE, S. W. and LINTON, D. L. (1939). 'Structure, surface and drainage in South-east England.' *Inst. Brit. Geogr.*, Publ. no. 10. London.

WRIGHT, H. E. (1951). 'Geologic setting of Ksar Akil, a palaeolithic site in Lebanon—preliminary report.' *Journal of Near Eastern Studies*, vol. X, no. 2, pp. 115–19.

WRIGHT, W. B. (1937). *The Quaternary Ice Age* (2nd ed.). London.

BIBLIOGRAPHY

WULSIN, F. R. (1941). 'The prehistoric Archaeology of North-West Africa.' *Papers of the Peabody Museum of American Archaeology and Ethnology*, vol. XIX, no. 1. Harvard University Press.

ZAVATTARI, E. (1930). 'La schistosomiasi in Cirenaica.' *Boll. Soc. Med. Chirurg. Pavia*, vol. XLIV, pp. 356–83.

—— (1934). *Prodromo della fauna della Libia.* Pavia.

ZEUNER, F. (1945). *The Pleistocene Period.* London.

PLATE I

(a) Coastal escarpment 4½ km. west of Apollonia. The gorge is Wadi Haula; the alluvial fan of the same wadi is seen in the foreground.

(b) Typical 'staircase topography' 18 km. west of Apollonia. The landward edge of the terrace in the foreground (here covered with a thin layer of alluvium) lies about 35 m. above sea-level. A higher terrace can also be seen, whose landward edge lies at 55 m.

PLATE 2

Section on west side of mouth of Wadi Haula, 4½ km. west of Apollonia. The upper
half of the section is composed of normal alluvial gravel. Below this lies a roughly
equal thickness of unstratified material believed to represent a mud-flow. The
mud-flow rests upon a thin layer of beach-deposits (on which lie the hammer
and the rolled-up towel), and this in turn rests upon a surface of Lower Eocene
limestone which forms part of the terrace of the 6 m. shore-line. Total height of
section: 4 m.

PLATE 3

(a) Small wadi one kilometre south-east of Derna bridge, showing a terrace composed of Younger Gravels.

(b) Modern marine cliff composed of Younger Gravels, 5 km. west of Derna. The gravels at this locality yielded the artifacts described on p. 163.

PLATE 4

(*a*) East bank of Wadi Derna, just outside Derna town walls, showing deposits of higher terrace. The deposits here are well bedded, and consist mainly of broken and rolled fragments of tufa. A remnant of the lower terrace is just visible on the left.

(*b*) View up Wadi Derna, 8 km. from the sea. Photograph taken from the surface of the higher terrace. The deposits consist largely of tufa, with bedding scarcely visible.

PLATE 5

(a) Close view of deposits of higher terrace of Wadi Derna, near mouth of Wadi Gahham. The photograph shows a bed composed of tufa formed around reeds in their original positions of growth, and subsequently only slightly disturbed. Actual size.

(b) West bank of Wadi Gahham, just above its confluence with Wadi Derna, showing sections of curved sheets of petrified moss.

PLATE 6

(*a*) Hajj Creiem: bone *compresseur* (slightly crushed in deposit)
showing two zones of utilization.

(*b*) Hajj Creiem: cortical surface of large flake with intact
adhering tubules of marine worms.

PLATE 7

(a) Hajj Creiem: view of section when first discovered. The overlying pebbles and boulders belong to a very local deposit which may be an isolated remnant of the Younger Gravels.

(b) Hajj Creiem: excavation in progress (1947). The cultural layer, partially excavated, is seen to the right of the seated figure.

PLATE 8

(*a*) Hagfet ed Dabba: general view of work looking north towards entrance of inner cave. Figure in left background is kneeling on partially excavated pier of C4; central figure is beginning work on layer VI in square B7; right-hand figure is completing removal of layer IV in square C7.

(*b*) Hagfet ed Dabba: closer view of partially excavated pier at C4, showing layer III exposed after removal of layers I and II. Box rests on large limestone block, and tins on surface of layer I with surface dust removed.

PLATE 9

(b) Hagfet ed Dabba: general view of trial trench at entrance after removal of large blocks.

(a) Hagfet ed Dabba: close-up of section, square C2.

PLATE 10

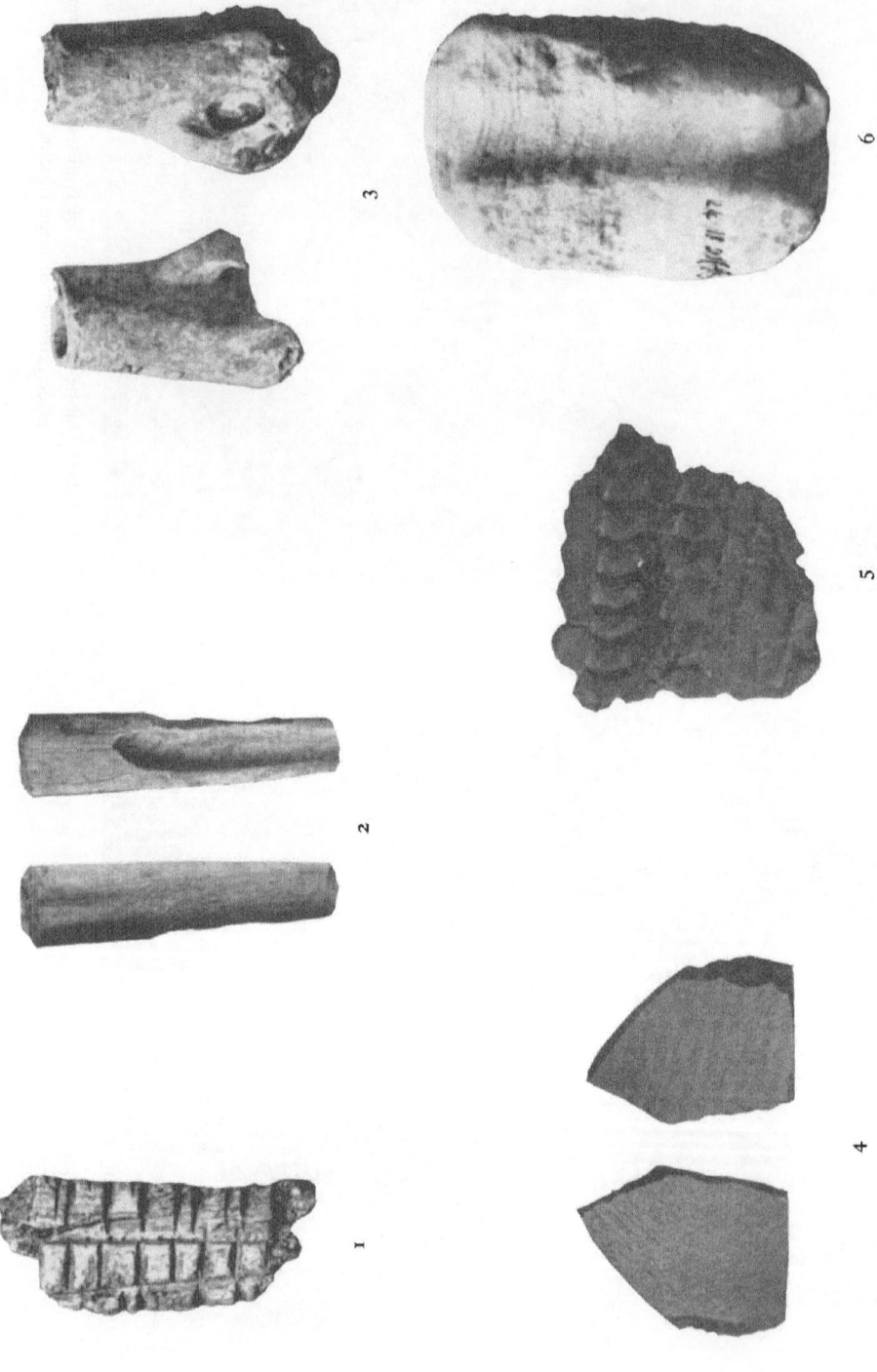

Hagfet ed Dabba: bone industry. Siwa Oasis: flint knife, decorated pottery and fragment of mace-head. No. 1, Hagfet ed Dabba: fragment of bone with grooved decoration. No. 2, Hagfet ed Dabba: awl of bird-bone ground with grooved-cut base. No. 3, Hagfet ed Dabba: extremities of two bird limb-bones with grooved-cuts. No. 4, Siwa Oasis: reverse and obverse of fragment of flint knife, polished before being trimmed by chipping. No. 5, Siwa Oasis: fragment of pottery with impressed decoration made with articular end of bird bone. No. 6, Siwa Oasis: fragment of mace-head. All actual size except No. 1 enlarged × 1½.

PLATE II

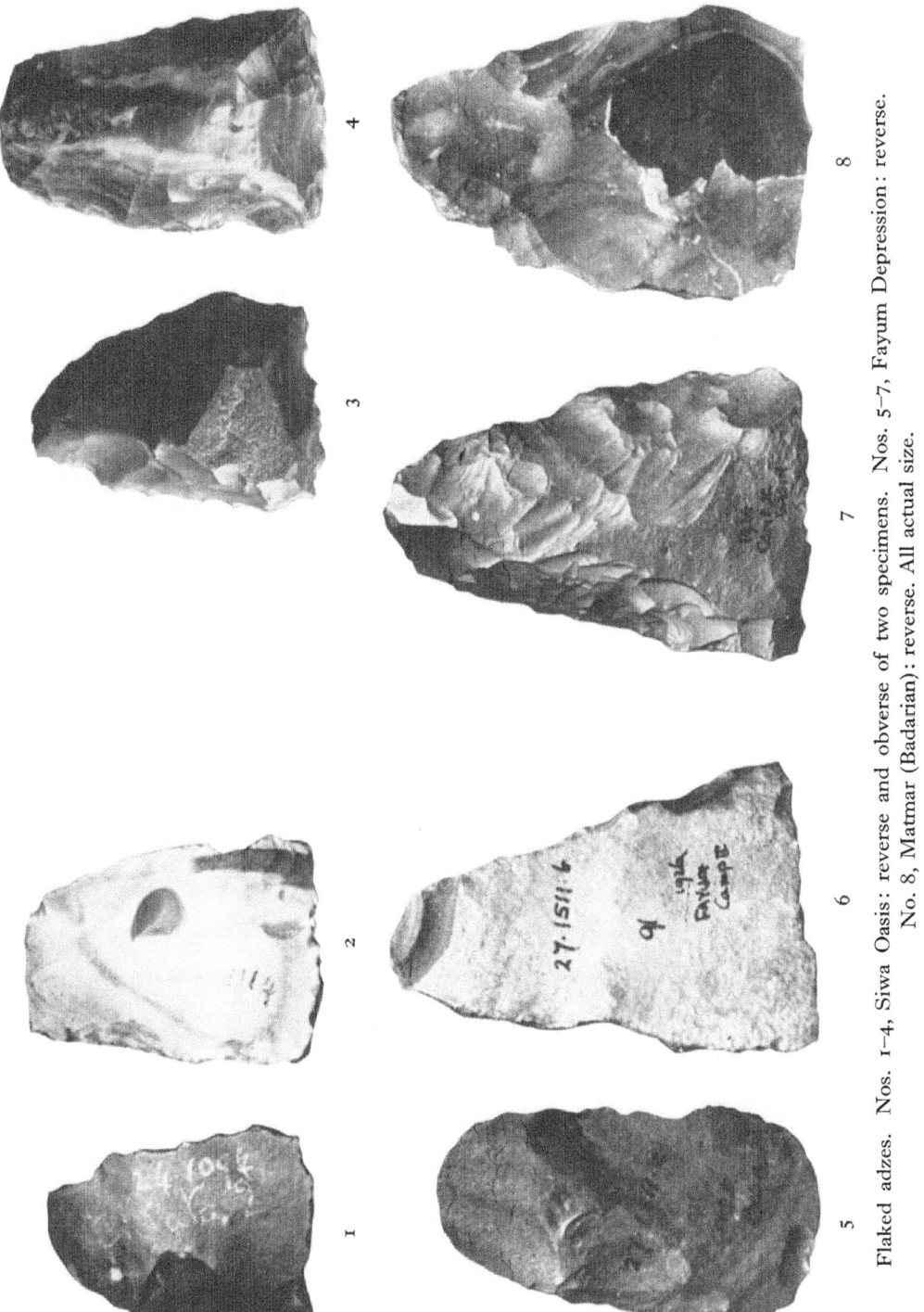

Flaked adzes. Nos. 1–4, Siwa Oasis: reverse and obverse of two specimens. Nos. 5–7, Fayum Depression: reverse. No. 8, Matmar (Badarian): reverse. All actual size.

PLATE 12

Siwa Oasis: obverse of flaked adzes and chisel. All actual size.

PLATE 13

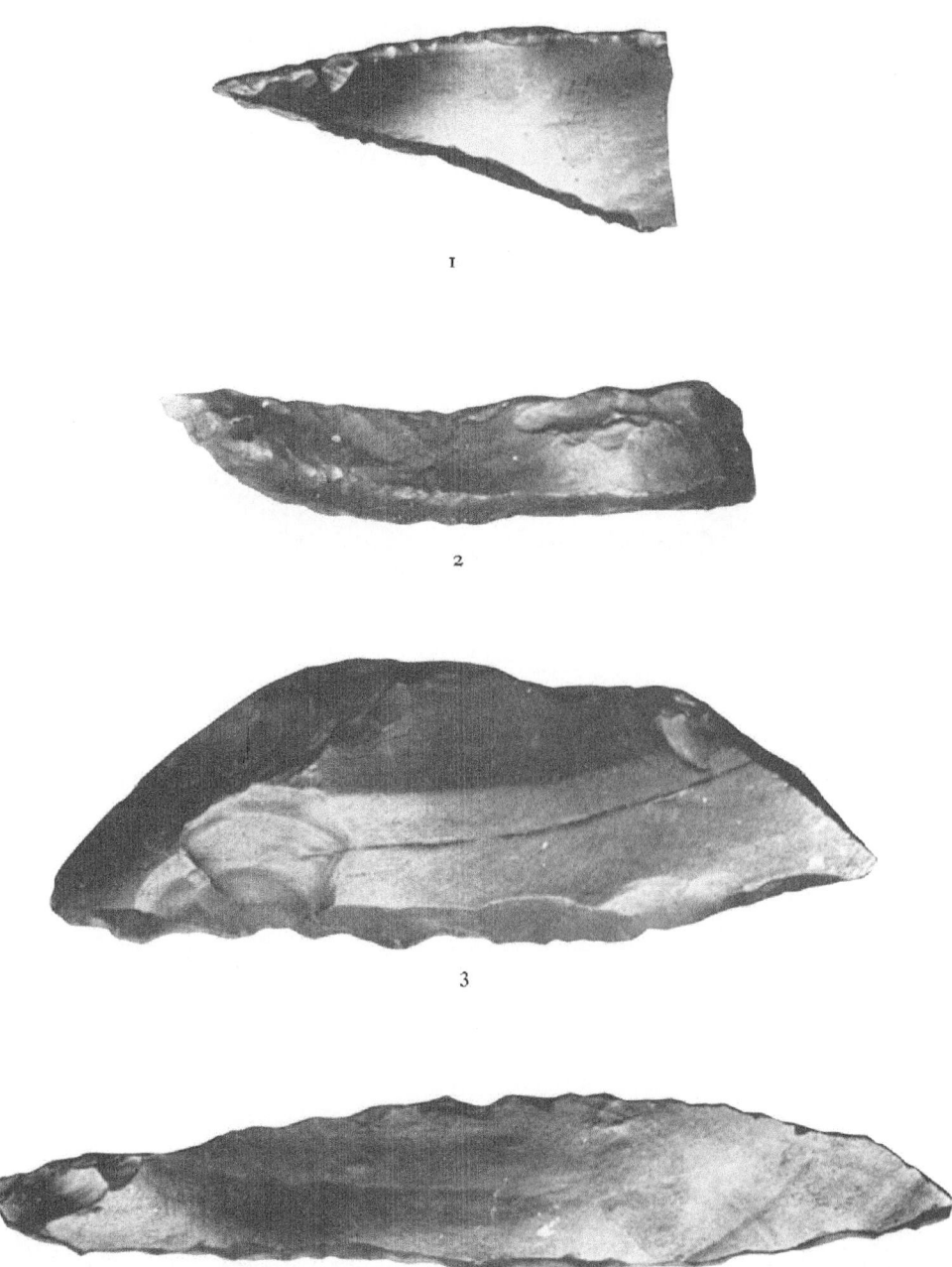

1

2

3

4

Obverse of concavo-convex scrapers from Egyptian oases. Nos. 1 and 2, Siwa Oasis. Nos. 3 and 4, Fayum Depression. All actual size.

PLATE 14

1

2

3

4

Reverse (bulbar face) of specimens shown on Pl. 13. All actual size.

PLATE 15

(a) *Pinus halepensis*: external mould of cone. The details of the scales are faintly visible. The impression of the stem can also be seen, near the lower edge of the photograph. Reg. no. K. 4007. × ¾.

(b) *Rubus*: impressions of leaflets. Both upper and lower surfaces are represented. Reg. no. K. 4008. × ½.

(c) *Rubus*: impression of upper surface of leaflet, showing details of venation. Reg. no. K. 4009. × 1¼.

(d) *Xerophila chadiana* var. *darnensis*. As with all the fully grown specimens that were collected, the aperture is slightly damaged. Reg. no. D. 5030. × 2.

PLATE 16

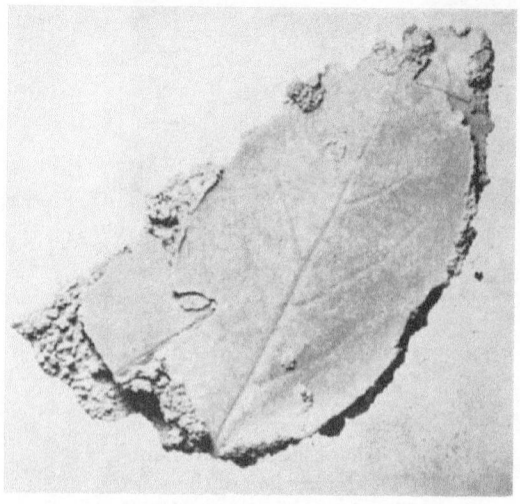

(b) *L. canariensis*: impression of upper surface of small leaf. Reg. no. K. 4016. Actual size.

(a) *Laurius canariensis*: impression of lower surface of leaf, showing the typical non-wavy margins. The so-called 'extra-floral nectaries' appear as projections, having presumably become perforated on the original leaf. Reg. no. K. 4015. Actual size.

(c) *L. canariensis*: part of impression of lower surface of leaf, showing details of venation. Reg. no. K. 4017. × 3.

INDEX

NOTE. For names beginning with El (Ed, En, etc.) or Wadi, see under next name. Thus for El Hania, Wadi Haula, Wadi en Naga see Hania, El; Haula, Wadi; Naga, Wadi en.

Abou Halka, 216–17

Abri Clariond, 181

Acheulean: hand-axes referable to, 172–4; dates at Kharga and Mount Carmel and in Nile Valley, 173; date of final stage in Libya, 233; summary of Libyan occurrences, 270; date in Morocco, *see* Neuville and Ruhlmann

Adzes: flaked, 245, 256, Pls. 11, 12; pressure-flaked, 258; *see also* Planes

Agedabia, 43, 85, 86

Agheila, El, 89, 90, 95, 96

Ahlmann, H. W., 17, 304

Ain bu Mansur, 101–5 *passim*, 114, 119, Fig. 8

Ain Derna, 101, 119, Fig. 8

Ain el Bled, *see* Ain Derna

Ain Mara: deposits, 120–3, 128, 129; cores, 157; primary flaking, 157, Fig. 15; retouched specimens, 159

Ain Metherchem, 215

Ain Zeiana, 43

Air photographs, 27–8, 37, 39, 45

Akhdar, Gebel, *see* Gebel Akhdar

Algiers, 66–7

Alluvial deposits, 21, 44, 135; nature, 74–5; mode of occurrence, 75–6; fauna, 77; chronology, 77–8; conditions of deposition, 78–83; *see also* Younger Gravels

Alluvial fans, 44, 75–6, 77, 78, 79, 130, 135, Pl. 1a

Almagro Basch, M., 243 n., 304

Alps, correlations with, 63, 134, 136

Amato, A., 48 n.

Ammotragus, 108, 239, 284; *see also* Hagfet ed Dabba; Hagfet et Tera; Hajj Creiem

Amratian, 216

Angil, Wadi el, 87

Antelat, 9, 41, 46, 57, 60

Antilope, 276, 286

Apodemus, 279

Apollonia, 19, 20, 95; marine terraces and shore-lines near, 23, 25, 30–1, 35, 37, 60, 291; fossil dunes, 86, 88–9, 97, 99; *see also* Haula, Wadi

Arambourg, C., 174, 275, 278, 279, 290, 304; *see also* Beni Seguoal

Arkell, A. J., 210, 304

Arkell, W. J., 66, 307

Armant, pre-dynastic industry at, 216

Arouia, Cave of, 239

Aterian: Cyrenaican, 170, 175, Fig. 18; bifacial tools, 174, 180; Siwan, 220; geographical distribution, 233; final stage, 240; Tripolitanian, *see* Gan, Wadi

Atrun, El, 19, 20, 31; tufaceous deposits, 123–8, Figs. 11a and 11b; industry of tufaceous deposits, 159; industry of 6 m. shoreline deposits, 162

Atrun, Wadi el, 123, 124, 127, 128, Fig. 11a

Augila, 7, 9, 262

Badarian culture, 246, 247, 249; origin of, 250

Barber, F., 143 n.

Bardia, 95

Barnes, A. S., 230, 234, 304

Bat, insectivorous, 278

Bate, D. M. A., xi, 108, 115, 176, 274–91, 305

Bates, O., 10, 304

Baulig, H., 65–6, 70 n., 304

Baumgartel, E., 241, 246, 304

Beduin Microlithic, *see* Kharga Oasis

Bean, W. J., 115 n., 116 n., 304

Bei el Kebir, Wadi, 224–5

Ben Gebara, Wadi, 87

Benghazi, 42, 46, 57–60 *passim*, 71, 92, 98, 292; geology, 47–56, 68, Fig. 5; fossil dunes near, 85, 92, 94, 95, 292

Benina, 41, 46; 'step', 40, 45–6, 47, 51, 57

Beni Segoual, 181, 187, 304

Bent, Wadi, 100

Berka, 52, Fig. 5

Bersis, 42

Bifaces, 188, 229, 258, 266; *see also* Acheulean; Aterian; Neolithic-of-Capsian-tradition

Bir Dufan, 223–4

Bird remains; Hagfet ed Dabba, 210–11; absence at Hajj Creiem, 155

Blackwelder, E., 75, 80, 81, 82, 304

Blade manufacture, as criterion of date in Egypt, 234, 250–1

Blanc, A. C., 42, 43, 70 n., 131–2, 136, 291–2, 304

Blanc, G. A., 132, 276, 304

INDEX

Blanckenhorn, M., 66, 304
Blyth, E., 285, 290
Bomba, Gulf of, 9
Bordes, F., 233, 304
Bos, 239–40, 276, 284
Bourcart, J., 173, 304
Bouyssonie, J., 183
Branner, J. C., 117, 304
Brunton, G., 246, 304
Bu Msafer, Wadi, 32, 87, 92, 163, Fig. 7; industries in Younger Gravels near, *see* Derna
  West
Bu Rueis, Wadi, 107
Bury, H. P. R., 289
Bu Sahela, Wadi, 86
Bu Sheefa Rock, 89, 90
Bwadi, 11

Calabrian stage, 66–7, 69
Capra, 276–7
Capsian: *Pointes scalènes*, 184; distribution, 187,
  214; type of backed blades, 200, 206; relations
  to Hagfet ed Dabba, 214, 219; classification,
  214; composition, 215, 230; radiocarbon date,
  235
Carbon 14, *see* Radiocarbon
Carmel, Mount, 131 n., 147; *see also* Skhul;
  Tabun
Caton-Thompson, G., 172, 228, 229, 242, 243,
  261, 304
Cervus, 276
Chandler, M. E. J., 294
Chapman, A., 285, 290
Châtelperron, backed blades of Hagfet ed Dabba
  compared to, 200
Chatwin, C. P., *see* Dines *et al.*
Chersa, 32, 35, 37, 87, 93, 110
Childe, V. G., 241 n., 242 n., 305
Clactonian, 223
Clark, J. D., 279, 290
Coastal plain, western, of Cyrenaica, 9; topography, 40–1, 44–5; solid geology, 41; origin,
  45–7, 56–7, 60–1, 68
Coefia, 42, 57, 59, 292
Compresseurs, bone, 154, Pl. 6
Concavo-convex flakes at Siwa, 256
Continental deposits, Pleistocene, 72, 130–7; *see
  also* Alluvial deposits; Dune-deposits, consolidated; Tufaceous deposits
Cook, T., 279, 290
Coon, C. S., 241, 305
Cornwall, I. W., 295
Crocidura, 278
Crowfoot, J., 248 n., 305
Cuf, Wadi, *see* Hagfet ed Dabba

De Lamothe, L., 66, 69–71, 305
Denizot, G., 63, 305
Depéret, C., 69–71, 305
Derna, 19, 20, 68, 78, 82–3, 100, 107, 110, 111,
  Figs. 6 and 8; marine terraces and shorelines
  near, 32–3, 35, 37–9; fossil dunes near, 86–7,
  89, 91, 92, 97, 110, 292, Fig. 6
Derna brickworks, industry, 170
Derna East, industry in Younger Gravels, 166–7,
  169
Derna, Wadi, 32–3, 35, 76, 77, 87, 122, 127,
  Figs. 6 and 8; description, 100–1; lower
  terrace, 102–3, 110; higher terrace, morphology, 103–4, Fig. 9; higher terrace, deposits,
  102–19, Fig. 10, Pl. 4; Pleistocene faunas, 108–9,
  275–6, 282–6, 287, 292–3; Pleistocene flora,
  109–10, 129, 136, 293–5, Pls. 15, 16
Derna West, industry in Younger Gravels, 163–6,
  169, Fig. 17, Pl. 3b
Desio, A., 10, 17, 30, 41, 43, 45, 46, 47, 53, 81, 84,
  88, 89, 90, 95, 305
De Stefani, T., 137 n., 305
Dietrich, W. O., 279, 290
Dines, H. G., Edmunds, F. H., and Chatwin,
  C. P., 67, 305
Domestic animals and plants, origin in Egypt,
  251
Double-backed blades, 254
Driana, 42, 46, 85, 89, 92
Driana, Secche di, 89, 95–6
Dune-deposits, consolidated, 30, 31, 33, 42, 43,
  44, 127, 130, 135, Figs. 6 and 7; lithology, 23;
  mode of occurrence, 84–90; faunas, 84, 292;
  chronology, 91–4; 'Younger', definition of, 91;
  conditions of deposition, 94–8; *see also* Dunes,
  fossil
Dunes, fossil; mode of occurrence, 84–90; submerged, 87–90, 95–6, 98–9, 135, 136; conditions of deposition, 94–8; *see also* Dune-
  deposits, consolidated
Dunes, modern coastal, 96

Egypt, 66–7, 81–82, 137; earlier food-producing
  societies in, 241–51; *see also* Acheulean;
  Kharga Oasis; Natufian; Qara; Siwa
Elfie Rock, 89
Eliomys, 280
Emig, W. H., 117, 118, 305
Equus asinus hydruntinus, 276–7
Equus caballus, 276–7
Escarpment, lower, 9; solid geology, 19, 41;
  topography, 19, 21, 22, 37, 39, 40–1, 44–5;
  origin, 21–5, 46–7, 57, 60–1, 68; partial submergence, 37–9
Escarpment, upper, 10

310